The Concept of Matter

EDITED BY

ERNAN MCMULLIN

CONTRIBUTORS

RICHARD J. BLACKWELL

V. C. CHAPPELL

LEONARD J. ESLICK

MARIE BOAS HALL

P. M. HEIMANN

MARY B. HESSE

ROBERT O. JOHANN

N. LOBKOWICZ

J. E. MC GUIRE

ERNAN MC MULLIN

RICHARD RORTY

KENNETH SAYRE

JOHN E. SMITH

JAMES A. WEISHEIPL, O.P.

The Concept of Matter

IN MODERN PHILOSOPHY

UNIVERSITY OF NOTRE DAME PRESS

NOTRE DAME ~ LONDON

Revised Edition

Copyright © 1963, 1978 by
University of Notre Dame Press
Notre Dame, Indiana 46556

Library of Congress Cataloging in Publication Data
Main entry under title:

The Concept of matter in modern philosophy.

  Much of the material found herein originally
appeared in *The concept of matter,* edited by
E. McMullin, which consisted of rev. papers from
a conference held at the University of Notre Dame,
Sept. 5–9, 1961.
  Includes bibliographical references.
  1. Matter—Congresses. 2. Philosophy, Modern—
Congresses. I. Blackwell, Richard J., 1929–
II. McMullin, Ernan, 1924–   III. McMullin,
Ernan, 1924–   ed. The concept of matter.
BD648.C66   117   74–27891
ISBN 0–268–00706–3
ISBN 0–268–00707–1 pbk.

Manufactured in the United States of America

CONTENTS

PREFACE

This book contains a selection of articles from the second half of *The Concept of Matter* (ed. E. McMullin, Notre Dame, 1963), which has been out of print for some years. A good deal of new material has also been added. The first half of *The Concept of Matter* has already appeared as *The Concept of Matter in Greek and Medieval Philosophy* (Notre Dame, 1965), a reprint, except for a shortening of the Introduction, of pages 1–343 of *The Concept of Matter*. The essays in the new volume by Richard Blackwell and by J. E. McGuire and Peter Heimann were specially written for this new collection, as were the Introduction and Epilogue. Almost half of the material in *The Concept of Matter in Modern Philosophy* is thus new.

The Concept of Matter was originally the record of a conference held at the University of Notre Dame in 1961. The papers were reworked in the light of the discussions at that conference; these discussions were transcribed, and a few edited segments have been included here to give a feel for the kind of wide-ranging debate that went on. Participants at the conference were: Joseph Bobik, A. R. Caponigri, V. C. Chappell, Leonard Eslick, Herbert Feigl, Milton Fisk, John J. FitzGerald, M. B. Hall, N. R. Hanson, Mary Hesse, Robert Johann, Czeslaw Lejewski, Nikolaus Lobkowicz, Norbert Luyten, Richard McKeon, Ernan McMullin, Edward Manier, Cecil Mast, Charles Misner, Joseph Owens, Harry Nielsen, Richard Rorty, Kenneth Sayre, Wilfrid Sellars, John E. Smith, James Weisheipl, Allan Wolter and A. E. Woodruff.

One point of editorial usage may be noted here. Typography has been pressed into the service of clarity in a way which is happily becoming standard in philosophical works. Single quotes are used for *mention* only, i.e. in order to name the expression they enclose. Double quotes are used not only for quoting material but also (in the case of words or short phrases) to indicate that the expression that they enclose is being used in some special sense. Italics are used for emphasis, or to indicate that the expression italicized is borrowed from a language other than the main one of the text, or to warn that the term, instead of pointing to its ordinary concrete reference, denotes a concept instead.

Acknowledgment is made to the University of Notre Dame, which funded the original conference, and to the National Science Foundation, whose grant allowed me the needed time to write the Introduction and Epilogue for this new collection. My thanks are also due to the staff at the University of Notre Dame Press for their patience at the delay in preparing the manuscript for them. This was, in part, due to the original inclusion in the collection of a lengthy study of Newton, *Newton on Matter and Activity*, now to appear as a separate monograph.

ERNAN McMULLIN

INTRODUCTION:
THE CONCEPT OF MATTER
IN TRANSITION

Ernan McMullin

§1 *Foreword*

In an earlier volume, *The Concept of Matter in Greek and Medieval Philosophy*, we have followed the career of the concept of matter from its first introduction as a tool of speculative thought by the Ionian physicists of the sixth century B.C. up to the beginnings of its decline as a technical cosmological category in the seventeenth century. In the present volume, the story will be carried from the seventeenth century up to the present day.* The seventeenth century itself will appear as the decisive period in the history of the concept. It was during this century that the empirical sciences began to dissociate themselves from their parent natural philosophy. The effect of this was to call all the older cosmological concepts into question. If they could not find a place in the new sciences, they tended to lapse into vague generality in everyday usage, or else to be confined to quite technical metaphysical contexts whose very meaningfulness came gradually to be called into question as new criteria of significance were adopted for terms purporting to describe or explain the physical universe. If they did find a place in the new sciences, it was as a rule only after a considerable sharpening of their definition, and a shift in their relation with empirical evidence. Or as also happened, a new term was introduced to mark off some precision in a much older concept; the older, more complex concept was then dropped from the daily practice of science, and (to all appearances at least) ceased to play any further role in it.

Thus, *element* and *atom*, both of them crucial concepts in Greek natural philosophy, were carried over into the new science in something close to their original sense (though later developments in that science would force

---

*This Introduction was written during a period of research made possible by a grant from the National Science Foundation. The author would like to thank the History and Philosophy of Science division of the Foundation for their continuing encouragement and aid.

them to be rather radically modified when "elements" turned out not to be really elements, and "atoms" not to be atoms). Whereas *nature, principle* and *substance* had been perhaps the most important categories of Greek cosmology, they were dropped from active service in empirical science, and left to do vague duty in the metalanguage only. *Nature,* for instance, became in one of its uses a convenient label for the world that science is about, carrying no specific connotation of its own. *Substance* remained as a loose general concept to denote the things that chemists work on, but it likewise carried no technical connotation. It was also retained by many philosophers, though its utility as a strictly *cosmological* category ("which *are* the substances in Nature?") was more and more questioned. *Principle* gradually became a purely logical and methodological notion, eventually losing the cosmological sense it had earlier possessed. *Species* and *genus* were transformed for technical use in biology. Discussions of *place* gave way to theories of *space. Essence* was largely abandoned by the scientists, as was *form,* though the new notion, *structure,* did some of the work of the two older concepts. The broad concept, *motion,* was limited to the precise and operationally manageable idea, *velocity.* The central Aristotelian concept, *dynamis* (potency), was relegated to adjectival status, and limited to one type of active potency only. *Energeia* and *vis* (force), on the other hand, were promoted to center stage, and transformed to meet highly specific new theoretical needs. The terminology of the new science was thus in a continuity of sorts with that of the natural philosophy that had preceded it. But very complex changes took place, changes that repay close study because they give us a clue to the profound shift that occurred at this time in man's approach to the knowledge of Nature.

One of the most revealing of these changes was that which transformed *matter* into *mass.* The new physics (like the old) was concerned with motion, but its practitioners were less interested in definitions of motion than in exactly describing the motions of a variety of bodies and reducing these under a few quantitative formulae of extraordinary predictive power. The optimistic belief that such a reduction could be brought about depended on the existence of an intrinsic "motion-factor" peculiar to each body, one that could be operationally defined as velocity had already been. In the long search for this factor—a search which is not yet satisfactorily concluded—the concept of matter played an indispensable role. Mass (as this motion-factor came to be called) was first grasped as "the quantity of matter".

The study of the concept of matter in the modern period may thus serve two important functions. It should help us to understand how and why science and philosophy separated, and what their present relations are.

And since *matter* has continued to play a central part in most (though not all) post-Cartesian systems of philosophy, to trace its philosophical fortunes is to follow the career of philosophy itself through some of its liveliest centuries.

But before we begin, a preliminary question has to be faced. What does it *mean* to study "the concept of matter" in the seventeenth century? There are two things it does *not* mean, one too narrow, the other too broad. It does not mean that we search out the various uses of the term 'matter' in philosophy and science during that period, and try to correlate them. This is part of what we shall do, but it is not sufficient. There are other analogous terms (like 'space', 'body', 'substance', 'aether') which have been interlocked so closely with the term 'matter' that any adequate consideration of the history of the concept of matter would have to take them into account also. At the other extreme, it is not necessary to study all the "theories of matter" (i.e. the entire range of physics and chemistry) in this period either.[1] One must distinguish between a "theory of matter" (where the word 'matter' is simply used as a vague label for physical objects, and one is concentrating on the theories, perhaps quite complex ones, in terms of which these objects may be understood), and the "concept of matter" (where the emphasis is on what the word 'matter' itself conveys, on the characteristics things have to possess in order to qualify as "matter"). There is obviously a great difference between, say, Descartes' *concept* of matter (he made matter equivalent to extension), and his *theories* of matter (his theories of percussion, vortex action and the like). It would be possible

---

[1] The sort of inquiry carried through by S. Toulmin and J. Goodfield in their useful book, *The Architecture of Matter*, London, 1962, which studies "theories of matter" (views on "what things are made of") from the Greeks to the present day. This incorporates the history of physics (excluding mechanics and optics), chemistry and physiology. For the period up to 1600, the concept of matter is explicitly discussed; the authors ask what role the notion of matter itself played in the natural philosophies of the day. But for the period from 1700 onwards, the term 'matter' is equivalently assumed by them to be a convenient name to denote the objects studied by the physicist, and their attention switches to the various structural theories of the physical world that thereafer succeeded one another. This curious and significant switch from an account of the concept of matter to a review of theories of matter occurs even more clearly in S. Toulmin's article "Matter" in *The Encyclopedia of Philosophy*. It is almost as though the history of the concept of matter ended around 1650; no philosopher after Descartes is discussed. The author clearly feels that a historical elucidation of the concept of matter would cover the philosophers up to Descartes, and the scientists from Descartes onwards, even though the term 'matter' is rarely a technical one for the scientists, usually no more than a vague expression of the metalanguage.

to understand his concept of matter and the reasons that led him to adopt it without knowing much of his theories of matter, even though these (in his view, at least) derived from his concept of matter.

This distinction becomes even more obvious in later thought as the term 'matter' came to be used in a progressively vaguer way, especially by scientists. A study of the "theories of matter" (i.e. the physics) of such nineteenth-century physicists as Faraday or Maxwell does not necessarily tell one what their concept of matter was, i.e. what they meant when they spoke of something as "material". If 'matter' comes to mean anything a physicist can study, the theories of physics may be as complex as one wishes, but the concept of matter has become almost entirely vacuous. 'Matter' in recent usage is often no more than a referring phrase like 'thing' or 'object' or 'physical world', one which serves to specify a possible object of scientific inquiry while saying as little about it as possible.

What, then, *does* it mean to trace the history of the concept of matter? It is a question in the first instance of locating the set of problems to which the notion of matter first responded. We have to trace the history of this cluster of interrelated philosophical problems, and discover what modifications of the original notions of matter became necessary. In addition, we can watch an evolution of a rather different sort in the uses of the terms 'matter' and 'material' in the context of the new science; the terms did not in this case form an explicit part of a functional explanatory system. Yet their senses changed slightly from time to time as problems of the "materiality" of various physical realities (aether, radiant heat, caloric, energy ...) forced a clarification in the criteria of what it was to be "material". The oft-repeated question one hears from eighteenth- and nineteenth-century scientists: "is so-and-so to be accounted as *material*?" was a methodologically complicated (and indeed often a confused) one, because it was by no means obvious *how* one would decide what the criteria of "materiality" ought to be. To what could one appeal when asserting, say, "no, the aether is *not* material"? Was it simply a question of checking the common usage of the term 'material', of seeing what physical characteristics it usually connoted, and then deciding whether aether possessed these or not? Or was it not, sometimes at least, a question of radically *altering* the lexical usage of the term, or of stipulating a sharper sense for it, perhaps, than it had previously possessed? But in this latter case, what sort of criteria could validly guide the scientist in making such meaning-changes? The criteria could not be first-order scientific ones, since the term 'matter' did not (as we shall see) occur in the technical language of any physical theories, so that it could not be directly affected by changes in these theories. Nor could the criteria be of the sort that philosophers like

Kant called upon in deciding what to make of matter in their ontology, since there was no longer any question in empirical science of a precise technical explanatory usage of this sort for *matter*.

The essays in this collection will cover some of the main developments in the concept of matter over the past four centuries. There are many lacunae inevitably; in particular, eighteenth and nineteenth century science (on which so much new light has been cast by the efforts of historians in the past decade) would need a much fuller treatment than it has been given here. In this Introduction, I wish to focus mainly on the decisive, but immensely complicated, period from Descartes to Newton, in an attempt to untangle some of the most important threads linking old to new. In order to do this, it will be necessary to summarize the wide variety of ways in which the concept of matter figured in Greek and medieval philosophy.

§2 *The Greek and Medieval Inheritance*

When searching for traces of the concept of matter in modern philosophy, it may be useful to recall for just which problems a concept of matter was invoked in Greek philosophy. Eight of these can be listed.

Substratum of change: The Ionians sought to "explain" physical changes by postulating one or more types of universal underlying "stuff"; if the chaotic multiplicity of change could be reduced to the unity which a single underlying stuff suggested, then the physical world had been "explained" in a very strong sense of that term. Aristotle objected to the "stuff" mode of explanation, because it led, in his view, to a denial of the reality of unqualified ("substantial") change. If the death of a man is no more than a transformation of an underlying stuff which itself remains unchanged in its nature, then death is not a fundamental change. It is no more than a redistribution of materials, and the only true "substance" is the underlying stuff itself. Aristotle thought this to be counter-intuitive, and sketched a theory of a primary "matter" which serves as substratum of change but is not itself qualified by any properties of any sort (unlike the Ionian "stuff").[2] It can be regarded as a "constituent", in some analogous sense, of bodies subject to substantial change, i.e. physical ("corruptible") bodies, and it serves as the principle of continuity in elemental changes, at least. Matter-substratum was the response to two different problems: the Ionian

---

[2] Aristotle himself devotes only a few lines to this topic; it clearly did not have for him the centrality that it came to have for his followers. See *The Concept of Matter in Greek and Medieval Philosophy*, ed. E. McMullin, Notre Dame, 1965.

one, how do we reduce changes to the unity the human mind desires? and the Aristotelian one, how are substances composed, if on the one hand they are genuinely diverse from one another, and yet on the other hand they can change into one another?

Principle of individuation and multiplicity: Plato emphasized Form as the basis of intelligibility and therefore of reality in everything. But Form is one. How can it be replicated an indefinitely great number of times? How can there be many triangles, many horses? Whatever it is that makes each thing the *kind* of thing it is, is something that it somehow shares with others. What, then, is the basis for the thing's *uniqueness* as an individual, distinct here and now from all other individuals of the same kind? Plato answered in terms of a Receptacle, a matrix of becoming; Aristotle relied on the material element in every sensible (replicable) thing. Since matter has no characteristics of its own, it must individuate not by means of distinguishable characteristics but by being somehow a ground for distinguishable beings. In medieval philosophy, this problem became quite central in discussions of various theological issues, especially those related to the Eucharist. It was generally conceded that primary matter alone would not suffice to individuate; the quantitative aspects of the sensible thing also played a role.

Space as a plenum: The ontological status of space became an important issue early in the history of Greek speculation. Is the space in which bodies move real? Is it simply created by those bodies and their interrelations? Or could it exist independently of them? The atomists added to the One-Being of Parmenides the "Non-Being" of space, and were thus able to defend local motion against the Eleatic critique: space allowed a multiplicity of Beings (atoms) and of changes, as atoms moved about.[3] But the paradox generated by treating space both as real and as (relative) Non-Being proved a troublesome one; both Plato and Aristotle had much to say about it. Aristotle excluded the possibility of a vacuum in Nature, but the *plenum* for him was constituted by the spatial relations of sensible bodies in contact with one another. Plato, on the other hand, and after him the neo-Platonists, tended to make of space a sort of featureless arena, yet a real ground in its own right for the replication of Form.

---

[3] Harold Johnson's article, "Three ancient meanings of matter: Democritus, Plato, and Aristotle", *Journ. History of Ideas, 28,* 1967, 3–16, contains some good insights on this point. The account given above of the relations between the atomists and the Eleatic school is challenged by T. G. Sinnige in his *Matter and Infinity in the Presocratic Schools and Plato,* (Assen, Netherlands, 1968).

The locus of potentiality: one of the most important clarifications brought about by Greek philosophy was in the status of *potentiality* or *capacity*. To say of something that it is potentially an X (i.e. that it can become an X), means more than simply saying that it is not (now) X. Thus there is an intermediate state between the possession of some property and its simple absence, namely, the absence of the property combined with the intrinsic capacity to develop this property, given favorable extrinsic conditions. To say of a particular seed that it has the capacity to develop into an elm immediately marks it off from another seed which does not have this capacity. But it does not mark it off precisely in terms of present actuality, as an elm would be distinguished from an oak. Rather the two are distinguished in terms of what they can become, and only secondarily in terms of what they now are. But what they now are does, of course, indirectly tell us what they can become.

Plato and Aristotle analyzed this "potential" aspect of physical things in characteristically different ways. Plato saw in Form the principle of actuality, and in matter the principle of potentiality. It was the material aspect of sensible things that made them liable to change, "corruptible" and unable to maintain the integrity of their present Form. The Platonic Form was, of itself, unchanging; its "insertion" by way of image in the matrix of the sensible world, however, made it subject to the hazards of change. Aristotle, the biologist, could scarcely accept this account of form, though he was clearly tempted by its simplicity. For him, the forms themselves of natural things were principles of change, of growth, of natural motion. The reason why an acorn changes into an oak is not just because it is a material being, but because the acorn-form naturally leads into the oak-form. So the locus of potentiality for Aristotle was the matter-form composite, strictly speaking, not just matter alone. Yet matter and form bore different responsibilities for the capacity that things have to become specifically different. The matter-aspect of a thing was that which made it capable of taking on different forms, and capable of being acted upon by an unlimited number of outside agencies. The form-aspect was that which made it what it was, thus specifying just what it could and could not become, as well as "pointing" it towards one particular line of "natural" becoming.

The source of defect: One of the most obvious things about the world is the presence in it of features that we would wish to be otherwise: of catastrophe and suffering, of things which seem to fall away from their "natural" goals. To a philosopher who sees the world as suffused by purpose, as Plato and Aristotle did, these will appear as *defects*, as something which require explanation, since presumably they might have been otherwise. One must ask why such things should occur in a universe where natural agency is directed to the good. They must have a source; there must be some feature of the world which tends, occasionally at least, to defeat the generally dominant intelligibility of natural process. Both Plato and Aristotle identified this source with a "material" factor, but attributed quite different roles to it. Plato saw in matter a source of opposition to the imposition of form on sensible things. Such things never realize form perfectly; in the myth of the Demiurge, the matrix is not wholly receptive of the replication of Form, so that individual sensible instances are not perfect images of the original. Further, there is question not just of a passive inability to reproduce the exact intelligibility of Form but also of some "shaking motion" in the matrix which actively interferes with the normal development of natures. Aristotle, however, denied that sensible things are imperfect copies; in his view, each natural being fully instantiates a substantial form. Thus, the natural world was not the defective thing for him that it was for Plato. But there still were chance events, unforeseeable developments such as biological monstrosities where nature clearly seemed to have gone wrong. Aristotle attributed these to the capacity things have to be acted upon by entities unrelated to their own proper finality; unavoidably, therefore, the normal development of a nature will be interfered with in occasional cases. The reason for this "breakdown" lies in the potential or material factor common to all physical beings. The idea of a matter-principle as source of defect (an idea with many roots in Eastern thought) is fully explored by Plotinus, who makes matter evil in itself. This is later rejected by Augustine and the Christian tradition because of its implicit attribution of evil to the direct work of the Creator.[4]

Over against life, mind and Divinity: Certain sorts of being exhibit life, and the distinction between them and things which lack the characteristic activities of the living being is one of the most obvious features of Nature. Long before the beginnings of Greek physics, men had seen in *pneuma,* breath, the source and sign of life. When breath ceased, life itself seemed to depart, and what was left was an inert heap of material. There must, therefore, be a distinction in the living thing between two principles, one the source of life, and the other a material factor which can persist after

---

[4] Plotinus hints (as Plato had done) that the soul also must bear some responsibility for defect. What Augustine and Proclus do is to treat matter simply as negation, as wholly potential being in which the Creator can achieve his designs, and to shift responsibility for evil entirely to the weakness that lies in the human soul itself. See Denis O'Brien, "Plotinus on evil", *Downside Review,* 87, 1969, 68–110.

the vital principle has gone. Furthermore, certain living things exhibit a quality of rationality and insight that seems to transcend the mutable order of life and sense entirely, leading one to suppose that the source of this power must be very different in kind from any mere component of the corruptible limited "material" order. Indeed, this power may seem so radically other in kind as to appear like a spark of Divinity, a participation in the proper activity of God. Dichotomies of this sort, between living and non-living, reason and sense, soul and body, played an important role in sharpening the conceptual tools used by the Greeks in their analyses of man. And on one side of them was always a material factor which was set over against life, reason, soul, spirit, a factor which was known by what it was not rather than by what it was.

Some philosophers made this set of dichotomies much sharper than others did. Plato (and even more emphatically, the later neo-Platonists) saw in the human spirit the traces of a mode of being so far transcending the shadow-world of sense that its relationship with that world was of a tenuous and temporary sort, at best.[5] Aristotle, on the other hand, made the human soul no more than the formal principle of a particularly complex sort of natural being, not something existing in its own right. Yet he insisted that since intellect, the highest power of soul, did not depend on bodily organs for its expression, it at least was "immaterial" and had "something of the Divine" about it. The Stoics stressed the agency of *pneuma* wherever purpose and organized action were to be found, but they refused to draw the sharp distinction that Aristotle had between soul and the matter-factor. Instead *pneuma* was represented by them (after the analogy of breath) as an active, tenuous medium differing in degree rather than in kind from other sorts of material agency, and responsible for the organized character of vital activity. Being a substance, it could exist on its own after bodily death, but it did not thereby cease to be a material being. Finally, ranged against this entire way of understanding man were the atomists and the Epicureans who denied the reality of soul and reduced the activities of life and mind to the motion of atoms.

What was at stake in all of this discussion was the irreducibility to the

---

[5] However, in Plato's own work there is a crucial ambiguity concerning matter. Can the "matter" which is the Receptacle-space in which forms are replicated also be a contrast-factor, set over against life and mind? The answer is: no, and herein lies one of the principal tensions in Plato's ontology (one resolved by later neo-Platonists). Life, *pneuma*, only occurs *in* the Receptacle; the Forms themselves are non-living, changeless. There are thus no Forms of living things as such, and soul has to be regarded as "material", in that it belongs to the domain that lies over against Form. Only the Forms are strictly "immaterial"; soul, as the origin of movement, is a "material" being.

level of ordinary physical changes of certain activities characteristically associated with man. Did purposive and intelligent behavior (and the design so evident in the products of such behavior) require an ontologically special source or not? Those who said it did made matter a contrast-factor, something which of itself lacked the directive drive of spirit, but which was indispensable as a partner of spirit in its spatio-temporal career. Thus matter was defined precisely by the lack of the characteristics asserted to be peculiar to soul or *pneuma* or spirit, and these varied a good deal from one philosophical school to another.

A factor in explanation: Aristotle listed four types of explanation of physical changes. One of these was in terms of the "materials" of the change, that to which the change occurred. This was to him the least interesting type of explanation, and quite separate from the other three which were interdependent. "Material" explanation merely sets the limiting conditions, so to speak, within which the causalities of form, agency and end have to work. The material factor is the "given" of the explanation, that which is taken for granted, like the bronze of the statue or the stuff of the seed. It is thus relative to the formal level of explanation decided upon; it is specified but not inquired into. If for example, one invokes the peculiarities of bronze in explaining the sort of statue one has, the material factor of such an explanation is no longer bronze (since the formal features of bronze are now being called upon to explain the resultant). There is thus something peculiar about calling the material factor an "explanation"; it is, rather, a postulated precondition for explanation, itself not explanatory in the strict sense.

The ultimate subject of predication: In his analysis of the subject-predicate relation in the *Metaphysics*, Aristotle asserts that though accidents are predicated of substance, substance itself is predicated of "matter", so that "matter" becomes the ultimate subject of a predication, itself therefore lacking in any predicable properties. It is apparently the "subject" in some sense of all predicable properties, specifically the essential properties constituting substance. To be a subject here is to be that in which the properties "inhere", after the analogy of an accident inhering in a substance. Aristotle seems to identify this matter-subject with the substratum-matter of substantial change, although it is not immediately evident that the arguments by means of which he reaches the two, one of them cosmological and the other purely logical, lead to the same "matter".

There are, however, two even more serious difficulties about the Aristotelian analysis of predication here. One is that it seems to understand predication as "restoring" a property to the subject, so to speak, so that the subject-term is understood to refer to the concrete individual considered in abstraction from the particular property here predicated of it. But this model of predication leads to obvious difficulties and seems to conflict, indeed, with Aristotle's own insistence on the concrete particular thing as the referent of all predication. Furthermore, it does not seem possible to find any linguistic examples of substance being predicated of a featureless matter. What subject-term would one call upon in such a case? If the claim that matter rather than substance is the ultimate subject of predication is to be validated, one would expect that there would be some instances of such a predication in ordinary linguistic usage. The analysis of predication, after all, is the analysis of how the subject-predicate forms of ordinary language actually function. Can predication really be said to involve a mental "stripping-away" and "restoration" of the property to be predicated? Yet if not, how can a logical analysis of predication arrive at a featureless "matter" as the ultimate referent?[6]

Aristotle drew a further distinction between "first", or primary matter, and "second" matter. Primary matter is ultimate, ontologically simple, with no admixture of form, and is reached conceptually by pushing to its limits the particular problematic within which matter is being postulated. Second matter is proximate; it is matter relative only to a particular form. It is substance, not just a principle of substance. It already possesses form, and is thus not ontologically simple, but plays the role of matter with regard to the specific form under consideration. This primary-secondary distinction arose especially in the context of the "Aristotelian" problems of change, predication, explanation and potentiality. It did not arise in the more "Platonic" problems of spirit, space, defect, individuation, where only an ontologically primary matter would suffice. It is worth emphasizing that the controversial aspects of the matter-tradition were all centered on primary matter, because of the ontological claims made for it. "Second" matter offered no real problem; to say that something is "matter-relative-to-a-specific-change" (or predication or explanation) makes no particular ontological claim. "Second" matter can be regarded as the term of a *logical* analysis of change-description, predication or explanation. What was controversial about the Greek matter-doctrine was the claim that there must be a "material component" of the physical universe which is in some basic ontological fashion distinguishable from the "formal" (intelligible, spiritual, actual, predicable . . .) component.

---

[6] These difficulties are the main topic of the essays by Fisk, Owens and Sellars in *The Concept of Matter in Greek and Medieval Philosophy*.

These are the contexts in which the concept of matter played an explicit role in Greek and medieval thought. Several points ought to be noted. It was *assumed* without any extensive discussion that the "primary matter" which was postulated in each of these contexts was, in fact, the same principle (though what could even be *meant* by 'same' in this context was not at all clear). The hylemorphic distinction was applied across the entire range of cosmological and logical problems; it was expected to show up, in one way or another, in every solution given to such problems. But on closer scrutiny, it is not easy to see how the same matter-factor *could* play all of these roles. How could it, for example, be entirely featureless and yet serve as substratum (i.e. the principle of continuity) of a particular substantial change, or as that which individuates? How could it be the contrast-correlate of spirit and the ultimate subject of predication as well? It sounds as though some untangling was needed, and that instead of being a single explanatory principle, the "matter" of Greek philosophers was really a cluster of analogous "principles" postulated to fill a whole series of not quite identical cosmological and logical roles.

Another problematic feature of the earlier matter-analysis was that matter always seemed to come out as indeterminate, featureless, impossible even to name or describe, strictly speaking. In what sense then, did it "explain"? If it is present in the same way in all physical beings, to say that it is there, is to say no more than that they *are* physical beings. But even more seriously, what do we mean by speaking of ontological "principles" here? When we distinguish between the "matter" and the "form" of a thing, are we thinking of separate components? different aspects? How are claims about such "principles" to be verified? Are they not purely conceptual and postulatory? The entire enterprise of pre-Galilean natural philosophy was built around the assumption that the defining of such "principles" constituted *explanation* in the realm of nature.[7] The direction taken by physics in the seventeenth century called this assumption into radical question.

§3 *Transition: Matter and Mechanics*

Though Galileo and those who followed him described their mechanics as a "new science", this was not wholly true. The Aristotelians had for centuries been endeavoring to give some account of the regularities observable in physical motion. Some had asserted that the speed of fall is proportional to a body's weight; others made it proportional to its density,

---

[7] This is the topic of McMullin, "Matter as a principle", *ibid*.

and others to the excess in density of the body over the density of the surrounding medium.[8] There had been no attempt to verify these suggestions observationally, however, and motions other than "natural" ones like *fall* were not thought to be analyzable even in approximate terms. This rather modest quantitative "mechanics" was conceived as being quite peripheral to physics, to the proper *understanding* of change. The Aristotelian was not much concerned with specific changes, except insofar as they indicated differences of "nature". Nor was he really interested in quantitative analysis of motion, because quantity was not the road to essence.[9] The natural philosopher, in his view, had to abstract a general notion of change and derive the necessary and sufficient ontological and logical conditions for its occurrence. The concept of matter was central to this latter inquiry, but it played no part in mechanical speculations, partly because of its generality, partly because of its indeterminate character, but especially because it had been postulated as an explanatory concept in a quite different context from that of mechanics.

The "new scientists" wanted a mechanics which would unite the different types of motion, celestial and terrestrial, either in a single set of descriptive quantitative "laws" or, even more ambitiously, in a single explanatory theory. It would be "explanatory" in a new and much more powerful way than the physics of Aristotle, because it would be able to specify the exact motion that should ensue, once the initial mechanical conditions be given. This was a new kind of question, in one respect especially: operationally definable parameters were needed to answer it. Could the concept of matter, or any variant of it, be pressed into service in this new context? One clue was that bodies of different sizes or different densities respond differently to the same impressed force; they also produce different effects on other bodies. Was it possible that there should be some intrinsic measure of this differential response? Something like that would be needed if a unified mechanics were to be constructed. It was tempting to think of the needed parameter as having something to do with matter, since the more "stuff" (in some intuitive sense) there is in a body, the harder it is to move it. And it was coming to be realized that it was *change*

---

[8] For a fuller discussion of this and the remaining points made in this paragraph, see "The medieval background", in *Galileo, Man of Science*, ed. Ernan McMullin, New York, 1967, pp. 7–9.

[9] There was a gradual change in this regard during the later medieval period, principally due to Platonic influence. But the mathematical *calculationes* of fourteenth-century science must be regarded much more as a mathematical exploration of the consequences of different geometrical definitions of local motion than as a genuine physics of moving bodies.

of motion (acceleration) that needed explanation in the neo-Platonic matter-tradition. The matter-spirit dualism was much emphasized, and the
idea of matter as that which is inert, resistant to change of movement,
seemed an appropriate metaphor.

Two successive efforts were made to utilize the concept of matter in the
new mechanics. The first, that of Descartes, was ultimately unsuccessful,
but it had far-reaching effects, as we shall see, on the everyday usage of
the term 'matter' in modern Western languages. The second, that of Newton, succeeded, but only at the cost of shifting the focus from the concept
of matter to that of mass. Mass was in the first instance a measure of the
effect that the inertial stuff-aspect of a body had upon the body's motion.
Being an operational measure-concept, it could (unlike the older matter-
concept) play a *direct* role in the new mechanics, and was rapidly recognized as the third of the three basic parameters of mechanics (the other
two being space and time). While all of this ferment went on in the
new mechanics, philosophers like Spinoza, Locke, Leibniz and Berkeley
still made use of the more traditional matter-concepts (or near variants
of them) when responding to the "philosophical" problems (now gradually dissociating from "scientific" problems) of change, individuation and
knowledge. Thus, the seventeenth-century has three main points of interest
for us: Descartes' attempt to construct a mechanics and metaphysics, starting from a Platonic notion of matter; the transformation of the traditional
cluster of matter-concepts at the hands of the philosophers; Newton's
doctrines of matter and mass.

§4 *Descartes: A New Philosophy of Matter*

After Aristotle himself, there is probably no more crucial figure in the
entire story of the concept of matter than Descartes. His system was a
bridge between old and new, an attempt to do a new mathematically defined physics while staying as close as possible to the older concepts and
methods. His aim was to geometrize mechanics in order that there could
be a science of motion of the same *a priori* form as geometry itself. Galileo
had already gone a long way in this direction by formulating the two basic
laws of *in vacuo* motion on the earth's surface, the law of free fall (making
the space fallen proportional to the square of the time taken), and the law
of projectile motion (postulating a parabolic relation between the spatial
co-ordinates of the trajectory). In both laws, only geometrically represent-
able factors (space and time) were mentioned; the older belief that the
speed of fall depended on the weight of the body (a non-geometrical factor)
seemed to have been refuted. It was this that led Galileo to be so sanguine

about the possibilities of his "new science", since if it were similar to geometry in structure, an unlimited array of theorems could be derived and the science itself had a satisfyingly self-evident basis. He was aware that he had left aside various factors that did not seem to be geometrically reducible, notably *gravità* and the resistance of the medium (the two factors Aristotle had emphasized in his analysis of fall), but it seemed plausible to him to regard these rather as "disturbing" features which would not have to be taken into account in describing the idealized motions of frictionless bodies in zero-density media.[10]

Descartes was repelled by what seemed to him a lack of conceptual elegance in the Galilean system. Galileo did not ask *why* bodies fell or *why* planets move around the sun.[11] And his laws of motion seemed empirical and thus contingent in their warrant. Descartes decided upon a far more radically geometrical approach, one which would provide a conceptually necessary structure, and would unite the cosmic system in one explanatory mechanical framework. He asserted that the only basic feature of matter is its extension and that all its other characteristics are reducible to (logically derivable from) that one single feature. This went far beyond Galileo, who had merely suggested that the "secondary" qualities of color, temperature, etc., are reducible to the "primary" features of quantity. Descartes, however, reduced even these latter to a single property, extension, the correlate of Euclidean geometry, thus leaving himself with the task of deriving impenetrability, weight and the other apparently "primary" properties of matter from extension alone.

The difficulties in the way of this wholesale reduction were plain. His universe had to be a *plenum*, filled with a "matter" that is perceptible in some regions but not in others. Variations in density of this "matter" (between a stone and a corresponding volume of apparently empty space, for instance) had to be somehow explained away. Motion could take place only by percussion, so that the basic laws of dynamics would have to be the laws of percussion. Yet when he attempted to formulate these laws, using only the *volume* of a body as a measure of its response to percussive action, his

---

[10] It later turned out that he had in this respect carried his idealization too far; one can validly abstract from considerations of weight only if the motions one is describing occur at roughly the same distance from the center of the earth. Indeed, it was the very *dependence* of motion, in a more general context, upon the non-geometrical factor of weight (or more exactly, mass) that gave Newton the clue he needed for his successful physics of gravitation.

[11] He had, indeed, speculated in passing that the answer to either one of these questions might well provide the answer to the other, but he resolutely refused to spend time seeking such answers, contenting himself with the idealized *description* of motion only.

results were clearly inadequate, even at the commonsense level of observation. Descartes knew quite well that geometrically similar bodies do *not*, in general, respond in the same way to impressed forces. To get around this, he invoked "hidden percussions" of the medium on the body, percussions that were not observable, nor amenable to mathematical treatment.[12] Furthermore, the phenomenon of weight entirely defied analysis in these categories.

In his earlier work, Descartes suggested that gravity is an extra "dimension" of bodies, but with his later exclusion of all factors save volume and motion, this idea was abandoned. The vortices that he so readily postulated in trying to account for planetary motion or free fall were, indeed just as "occult" as the substantial forms of the scholastics that he rejected. But this could be overlooked as long as the *possibility* of a mathematical description of such vortices could be held out. Eventually, however, the utter inadequacy of Cartesian physics in the face of even the simplest natural phenomenon began to be realized. But so great was its attractive coherence and potential explanatory power that it dominated physics until Newton managed to unite the terrestrial and celestial dynamics in a very different conceptual system where "matter", which somehow had an active gravitational power associated with it, and a non-geometric measure-index (mass) replaced the inert geometric "matter" of Descartes.

In one respect, Descartes' system constituted a sharp break with all previous philosophical traditions. In it, there was nothing opposed to *idea*. Only thought and extension existed, and extension could be fully grasped by thought. To achieve this confident rationalism, matter in both the Platonic and the Aristotelian senses had to be eliminated, since for each of these (though in different ways, as we have seen) there is a basic obstacle to a complete grasp of the universe, and this obstacle is partly what is meant by "matter". Descartes could just as easily have dropped the matter-category entirely, and claimed that his reduction of physics to mathematics constituted a physics *without* "matter" (i.e. with no intrinsic barrier to

---

[12] As R. Blackwell points out in his essay elsewhere in this volume, Descartes was forced to rely upon all sorts of ingenious elaborations of motion to make up for the deficiency in his idea of matter. In a sense, motion played a larger explanatory role in his system than did extension. Of itself, matter for him was inert, entirely lacking in motion. This idea, incorrect as it was, still led him in the direction of two fundamentally important mechanical principles, that of inertia ("matter of itself tends to continue in whatever state of motion it possesses"), and conservation of momentum ("motion is communicated to matter by God at the creation, and its amount thereafter remains constant").

total mathematical intelligibility), as indeed some modern exponents of the Cartesian ideal have recently put it.[13]

Why, then, did he preserve the name 'matter' at all here? Because of a certain conceptual continuity with three of the traditional matter-problems, it would seem. First, he was giving a very definite answer to the old question about the reality of space by affirming not just the being but even the "materiality" of all that is extended spatially. Second, he thought of his matter-space as a sort of stuff, in some broad sense, a substratum of change of a very different sort, admittedly, from Aristotle's featureless substratum. The Scholastics had made quantity the "first accident" of material substance; what Descartes did was to push this one stage further and make it equivalent to the essence of material substance. Like his Scholastic mentors, he identified quantity with extension. Had they treated density, weight and the rest as quantitative factors instead of as irreducible qualities, Descartes might have been warned to take them more seriously in his "quantification" of essence.

But it was on neither of these two strands of the matter-tradition that Descartes mainly relied. To see this, let us ask what he *meant* by the term 'matter'. When he said that "matter is extension", he surely did *not* mean that the two terms are synonymous. "Matter" for him was in the first instance "that which is opposed to thought"—not the substratum or principle of individuation of the Aristotelian tradition, therefore, but the substantial over-against-spirit of the neo-Platonists. His starting-point, the *cogito* analysis, led him to a separation between thinking and non-thinking beings even more radical than theirs. Beings incapable of thought were for him just as substantial as thinking beings, so that the distinction between them was not just that between shadow and substance but that between two kinds of being, each existing (subject to God's conservation) in its own right. The absolute distinction between the sciences of spirit and of matter was thus a cardinal point in Descartes' proposal. The term 'matter' in this way for the first time became a convenient cover name for all those entities to which the science of physical motions, i.e. mechanics, applies. The concept of matter functioned, therefore, for Descartes as a means of dividing the universe into two sorts of substance, and consequently of dividing knowledge into two sorts of science.

This use of the concept had always been implicit in the neo-Platonic tradition, but it was now for the first time clearly developed. Plato had

---

[13] See C. Misner, "Mass as a Form of Vacuum", in *The Concept of Matter,* ed. E. McMullin, Notre Dame, 1963.

not believed matter to be substantial, nor had he believed a science of matter (or even of sensible things) to be possible. Descartes, however, made matter a substance, and constructed an attractively complete account of its motions. The shift here is a crucial one, because it means that (1) matter is no longer either a Receptacle or a co-principle with form, incomplete in itself and mysterious in its ontological indeterminacy; and (2) it is that which physical science may claim to describe and explain. There had been no single term or concept for this in pre-Cartesian natural philosophy. Aristotle had made physics a study of "mobile being", but there were two quite different sorts of mobile being, terrestrial (i.e. corruptible or "material") and celestial (incorruptible and thus in this respect, at least, "immaterial"; lacking in primary matter, yet still movable spatially). Physics was not, therefore, for Aristotle a study of "material" being only, much less a study of "matter" (which, of itself, was not for him a being, nor something of which there could be a science). The bond holding the object of Aristotle's physics together was the concept of *physis*, nature, the source of motion characteristic of a particular being. Physics was a study of everything that had a "nature". But what of the First Mover? and of soul? The First Mover was not itself mobile, yet Aristotle found it necessary to devote one whole book of his *Physics* to showing how the First Mover is ultimately required to explain *any* physical motion. And soul, though mobile in one sense, is a special sort of source of motion in another, so that there can be a separate science of the soul, yet physics applies to beings that have a soul just as much as to those which do not.

Descartes adopted a very different solution here, removing the First Mover and soul entirely from the ambit of physics, and eliminating the sharp distinction between celestial and terrestrial entities which had been so central to Greek thought. The domain of physics was now *matter*, nothing more and nothing less, the *plenum* of entities that can move one another by mechanical contact. It was convenient to have a single term to denote the class of objects to which physical science applies, especially since it did so in a quite unspecific way, not on the basis of some particular property (such as density or impenetrability or the like). To ask whether a particular entity was "material" was precisely to ask whether it could be treated in terms of mechanics, not whether it was extended or of finite density. This Cartesian usage of the term 'matter' proved so useful in an age when mechanics was revealing its possibilities that soon the older, more technical senses were forgotten, save by philosophers. In this way, 'matter' passed into general use in Western languages, no longer as a clear-cut technical term, but rather as a vague practical label for the varied array of things the physicist speaks about, one that does not commit the speaker to any particular theory about the nature of these things. This is the sense it retains in ordinary usage today.

It was a decisive turn in the history of the concept of matter because this vague unproblematic usage conveyed no sharp concept at all. The original Cartesian sense of "that-which-is-opposed-to-spirit" gradually became lost, though an overtone of it remained in certain contexts. There was some analogy, too, between the new usage and the older Ionian one which made matter simply the stuff underlying all physical changes. But the post-Cartesian everyday usage of the term 'matter' no longer responded to a question; it was no longer the case that matter was postulated to explain some general feature of the world. In a sense, then, this concept of matter had no subsequent history; it could not develop in any significant way. The many new theories of matter that have arisen since the seventeenth century tell us much about matter, but do not modify to any serious extent the concept associated with this ordinary-language use of the term 'matter', one which today (as it did three hundred years ago) refers simply to that which is governed by the laws of mechanics.

The changes that have taken place in these laws do not alter the concept of matter, although problems arose in the eighteenth and nineteenth centuries as to whether or not certain entities postulated by the physicist (aether, caloric, etc.) were "material" or not. These problems do indirectly affect the sense of the term. But it is important to note that to the extent that a term like 'matter' ceases to respond to a particular problematic, there can be no theoretical reasons which compel its redefinition. Further modifications of its sense are likely to come only if practical issues about its unambiguous application to concrete contexts force one to stipulate the boundaries of its reference more sharply. But modifications of this kind are not of great theoretical interest, and do not usually enlighten one very much about the "dynamics" of the concept one is studying.

§5 The Problems Change

Descartes was by no means the only philosopher who contributed to the rapid evolution of the concept of matter in the seventeenth century. One reason for this evolution has already been discussed: the search for a parameter that could be of service in the new science of mechanics. But there were two even more fundamental reasons for the shift: the conviction, first of all, on the part of all the major philosophers of the early part of the century that the substantial forms of the Aristotelian tradition really did not "explain" in any useful way, and secondly, their rejection of the whole idea of incomplete co-principles, like matter and form, essence and existence, "principles" that were not things, nor properties, nor concepts. Skepticism about the status of these alleged ontological constituents had been a notable feature of the nominalist philosophies of three centuries earlier. To empirically-minded thinkers, like Bacon, Galileo and Hobbes, they seemed no more than illegitimate and unfruitful reifications of the theoretical concepts of an empirically-vacuous metaphysics. Instead of explaining sensible beings by hylemorphic composition, these men sought to characterize such beings by metrical properties which would allow their specific motions to be logically derived. There was an overtone here of the old Aristotelian ideal of a physics in which properties would be logically derivable from essence, except that now certain properties would be postulated as common to all "material" beings and the logical (or mathematical) relationships would hold, rather, between specific values of the properties. Questions about the necessary and sufficient conditions for change-in-general, involving ontological distinctions within essence itself, would be laid aside as idle.

In this climate, primary matter, indeterminate co-principle of substantial form, could hardly survive. And it did not. But this was not the only role that *matter* played in Greek thought, so that a review of the eight problem-areas discussed in §2 is needed. Did they cease to be problems? or were different analyses of them given, analyses not employing the notion of matter? The answer, in brief, seems to be that four of them ceased to be of interest, for one reason or another; in two of them, the concept of matter still played something of its traditional role; in two, it was radically transformed.

Defect: In the Christian world view, matter was God's direct responsibility. Augustine was thus led to question the neo-Platonic identification of it as the source of defect. He argued (and Christian theologians almost unanimously followed him in this) that the responsibility for evil and deficiency must lie rather in the human will. Another quite different motive for eliminating the link between matter and defect was that with the growth of the "mechanical philosophy", teleological modes of understanding the operation of the cosmos were rejected, and thus defect did not appear to call for any sort of special ontological source in physical beings themselves.

Potency: The potency-actuality distinction, so much stressed by Aristotle, also dropped out of sight. It had originally been introduced to solve the Eleatic paradoxes, and to give a schema for the general understanding of change, notably biological change. Now, with the emphasis on local motion as the primary sort of change, and on actuality (properties presently possessed) as the sole appropriate mode for its understanding, the notion of potency came to seem a merely verbal one, of no explanatory force. It was not that the potential aspect of physical beings was denied; it would have made no sense to deny that things have the capacity to change into other things and that this capacity is conditioned by what they actually *are*, here and now. Rather, it did not seem worth saying this unless one could go on to specify what it actually is in the present thing that allows such changes to occur. To a hard-headed empiricist, potency seemed no answer to someone who asks: "What is it that allows this acorn to become an oak?" Besides, most mechanical philosophers accepted ultimately only one sort of change, i.e. local motion, the one to which the whole notion of potency seemed least relevant. There was no need, then to postulate a special ontological source of potency other than corpuscules and their motions.

Predication: After a golden period in the fourteenth and fifteenth centuries, in the seventeenth century, logic reached one of its lowest ebbs. Both scientists and philosophers were distrustful of logic, impatient of semantic analysis, convinced that the syllogism had been the death of ancient science. There was little inducement to reopen the old issue of predication. The new science had substituted mathematics for logic, and even among philosophers the most rudimentary sort of subject-predicate logic was thought to suffice. As we have already seen, Aristotle's identification of primary matter with the ultimate subject of predication had raised grave difficulties even for his own followers. In the new philosophy, where all things Aristotelian were thought to have been superseded, it was not likely that a claim so bound up both with Scholastic logic and with the repudiated ontology of indeterminate primary matter would evoke interest. Locke and Leibniz (as we shall see below) were among the very few who devoted serious attention to it.

Individuation: One final problem that the seventeenth century tended to set aside was that of individuation. Greek philosophers had seen in Form the knowable element. But what is it that makes it possible for the Form to be replicated over and over in the physical order? How can there be *many* equilateral triangles, *many* horses, since what is known is strictly speaking only one? The easy answer was "matter": the "realization of form in matter" could differentiate the instances. But now with form being replaced by property, it was not so obvious that a special principle of individuation had to be found. The nominalists had earlier claimed that since knowledge is of individuals, there is no need to seek such a principle. The mechanical philosophers took a different line. All physical beings have

certain properties (impenetrability, inertia, etc.) in common. The properties that are most useful from a scientific standpoint are those that lend
themselves to metrical specification. Any physical body must be capable
of *unique* specification in terms of such measures, if the whole notion of
state-description on which the new mechanics rested were to work at all.
If the physicist is to predict the path of a body, it is essential that he can
somehow mark off this body from all others. And this, it seemed, could
easily be done by means of geometrically defined space and time parameters. What makes one instance of impenetrability (i.e. one body) differ
from another, then, is simply their separation in space and time. The
principle of individuation for the mechanical philosophers was not an
indeterminate matter, but rather a mathematically determinate network,
analogous to what the older philosophers would have called "quantity",
but differing in that it was no longer regarded as an accident of some underlying substance. So obvious did it seem that any individual could be uniquely located in this way that discussion of individuation almost ceased.

The only metaphysician of the century for whom it remained a major
issue, almost *the* major issue, was Leibniz.[14] He rejected Newton's idea of
an infinite real space, and the supposition that position in such a space
sufficed to individuate the bodies that moved in it. The points in space
have no ontological reality in their own right, he argued. They cannot,
therefore, provide the ontological basis for individuality. When we distinguish between two leaves, say, it is ordinarily because of some properties
that one has and the other lacks. But even where this is not the case, and
we still say there are two not one, the grounds for this judgment would
be that the perceived spatial relations between us and the two leaves are
not the same. Thus the distinction here is in terms of relation, which still
seems to leave open the question whether another set of bodies sharing
the *same* relations could also exist. Or to put this in another way, is it
possible that a thing can be *completely* described in terms of properties
and relations (i.e. predicates) so that there is no logical possibility of there
being another instance that would satisfy the same set of predicates?

Leibniz was torn in two opposite directions in his many attempts to
answer this question, and in consequence never did achieve a consistent
position regarding it. On the one hand, the starting-point of his entire
logic and metaphysics was an analysis of subject-predicate propositions
which made the predicates part of the subject-concept, adequately understood, thus suggesting that every true affirmative proposition is in some
sense an analytic one. This led him to the curious doctrine of monads, the

---

[14] See W. von Leyden, *Seventeenth Century Metaphysics*, London, 1968, esp. chap. 9.

simple unextended spiritual substances of which the world is made up. Since "every predicate, past, present, or future is contained in the concept of the subject", every subject (soul or monad) must be a "world apart", independent of everything else except the Creator. These monads do not come to be nor pass away, thus they are not material; they are not externally related with one another, nor can they interact upon one another. Every change in a particular monad must come from within the monad itself, otherwise there would be true propositions about it which are not derivable from the idea of this monad (i.e. predicates not derivable from the subject). Thus all apparent relationships between monads (such as spatial or temporal ones) must be reduced to non-relational properties of the individual monads (such as perceptions thought to be of spatial or temporal relation), truly predicable of them quite independently of any empirical knowledge of other monads. The monads, being spiritual, must all have some degree of consciousness and appetition; though they are simple, they can have a plurality of perceptions and ideas which serve to differentiate them (though, of course, the perceptions have to be interpreted as somehow self-induced).

Among the many difficulties facing this metaphysics, surely one of the most counter-intuitive philosophical positions ever adopted by a major thinker, one of the most serious is that of individuation. There is no room in Leibniz' system for a matter-principle; he has eliminated space, and the monads do not of themselves begin nor cease to be. What makes a monad an individual, then? How would two monads having the same predicates be differentiated from one another? Leibniz' usual answer to this was his Principle of Identity of Indiscernibles, which would simply claim that for two monads to be different, they have to differ in at least one predicate. This suggests that individuation is achieved wholly in predicable or formal terms, or to put this in an ontological way, that the individual is nothing more than the sum of its attributes. Leibniz did not believe the Principle to be "logical" (i.e. derivable from the Principle of Contradiction), but only "contingent" (derivable from the Principle of Sufficient Reason), yet he held it to be "most certain". At first sight, the Principle seemed sufficient to define what would count as a single individual monad, thus solving the problem of individuation, always a grievous problem for thinkers in the rationalist tradition.

But it appeared to raise more problems than it solved. Leibniz did *not* wish to admit (as some later thinkers, like his admirer Bertrand Russell, would) that substance can be reduced to a bundle of qualities. For one thing, this would be to eliminate altogether the subject-predicate distinction (as well as the substance-attribute distinction which he took to be its ontological counterpart). If a "subject" is *entirely* reducible to predicates, if reference is entirely reducible to sense, then there *is* no "subject" as such in which predicates inhere, no substance to serve as support for the attributes. Leibniz could not accept this, since one of his primary assertions was that the monad-substance was something existing in its own right as an indivisible perceiving subject, and not just a set of states of awareness. Yet what *is* this over-and-above that constitutes the individual substance as individual, and would its existence not contradict the Principle of Identity of Indiscernibles? Aristotle could answer in terms of primary matter, but Leibniz had refused this answer by making the monad spiritual.[15] It would seem that if he had consistently pursued *this* line of thought, his only choice would have been to individuate the monad somehow in terms of spirit. But would this not require an unacceptably "indeterminate" or "un-predicable" dimension of spirit, something analogous to a hylemorphic composition of spirit itself, with a "quasi-matter" still serving as subject of predication? There were parallels for something like this in earlier Augustinian and Scotist thought, but in the general context of Leibniz' doctrine of monads it made little sense. One can only conclude, then, that Leibniz' elimination of individuating matter created grave problems of consistency for his entire system of thought.

Space: One of the problems inherent in Greek atomism was that of assigning ontological status to the space in which the atoms were said to move. On the one hand, if some sort of being were attributed to it, there would be two sorts of "matter" in the system (space and atoms), and the analogy with Eleatic metaphysics would break down. On the other hand, if space were to be strictly taken as non-being, in what sense could it be a "receptacle", a place in which atoms could move? For the mechanical philosophers of the seventeenth century, this problem once again became a pressing one.[16] Many of them defended a "corpuscularian" model, preferring it to the *plenum* model of Descartes because of its greater explanatory

---

[15] He does, of course, use the term 'matter' quite frequently, sometimes speaking, for instance, of a "primary matter" which is the source of passive inertial resistance in the monad. Yet such verbal usages cannot undo the fact that he had made the monad an unextended spiritual substance. Russell and more recently Rescher have argued that Leibniz allows two sorts of space, one perceptual and internal to each monad separately, and the other public and objective, though relational. This would give him a "second matter" which is the aggregate of monads making up the macroscopic bodies of ordinary experience, related through their "primary matters".

[16] See M. B. Hall's paper in this volume, and R. Harré, *Matter and Method*, London, 1964, pp. 63–72.

power. (Indeed, Descartes himself was forced to fall back on corpuscular ideas constantly in his endeavor to explain different sorts of change). Some, like Gassendi, even defended a strictly atomistic view, though the majority agreed with Boyle that there was no real evidence for the indivisibility of the elementary particles.

Gassendi adopted a forthright Epicurean atomism, and immediately faced the classical Aristotelian arguments against the notion of strict void entailed by it. He argued with considerable ingenuity that space could be real, yet not a substance, not an accident, not "matter", not "corporeal". So well did he succeed in persuading many of his contemporaries that the notion of a void is not logically contradictory that from then onwards it became more and more widely accepted that it was an empirical (rather than a logical) question whether some of the space of the universe is, in fact, completely empty, or whether there is a "subtle matter" present in even the emptiest-seeming natural *vacua*. At the end of the century, there were three views of space still in contention: (1) space is absolute but of itself void and "non-material", in the sense of possessing zero density and zero inertia (Newton); (2) space must be a *plenum*, since the notion of a void is self-contradictory (the Cartesians); (3) space is constituted by relations (either internal or external), and thus need not necessarily be a *plenum* in the substantial sense (Leibniz argues at times for this position). Those who defend the first view are likely to identify matter and body, and will define matter in terms of mechanical properties like density. Those who defend the second view will tend to make use of the concept of matter to describe what it is that makes space a *plenum*, and will thus be unable to link it with specific mechanical properties (since the space in which the planets move, for instance, offers no apparent mechanical resistance to the motion). In such a view, matter would be no more than "space-filler", and there will have to be all sorts of "subtle" and "aethereal" varieties of matter to account for the facts of observation. In the third (relational) view of space, the concept of matter may be dropped altogether, depending upon the type of space-relation that is postulated.

Spirit: Though the old matter-form distinction lost much of its force in the seventeenth century, another of the traditional matter dichotomies exerted a great deal of influence. This was the spirit-matter dichotomy of the neo-Platonic tradition. During the Renaissance, this took on a new importance not only because of the growing emphasis on human freedom and human uniqueness, but also because of a new enthusiasm for the Hermetic tradition of panpsychism which held for a direct "sympathetic" contact between man and Nature through a sharing of soul, as well as for alchemy, with its belief in the analogy between the physiological development of an organism and the chemical formation of a mineral or a salt in the "womb of the earth". Bruno in his cosmology, Paracelsus in his medicine, Gilbert in his theory of magnetism, all stressed the primacy in Nature of the active agencies of soul. Soul was regarded as the origin of all energy, movement, life; matter was correspondingly taken to be inert, resistant to the forces of life and purpose. This kind of sharp dualism, which we have already noted in Descartes, was gradually eroded by the new mechanical philosophy. As confidence in the explanatory possibilities of mechanically moved corpuscles increased, the necessity diminished for postulating life-forces of one sort or another operating in every natural body, including the apparently inanimate ones. Yet, the neo-Platonic ascription of a kind of "inertial" character to matter in the writings of thinkers like Kepler and Henry More undoubtedly helped in the formulation of the new concept of inertial mass, as well as of the correlative principle that it is not motion but change of motion that requires an explanatory cause.

Though the mechanical philosophers had little use for vital forces, their colleagues in physiology and chemistry could not do without them. It seemed obvious to most of those who (like Harvey) actually worked with physiological problems, that the explanatory resources of atomism and mechanism were too meager to account for development and function, "as if, forsooth, generation were nothing in the world but a mere separation or collection or order of things. I do not indeed deny that to the production of one thing out of another, these forementioned things are requisite, but generation itself is a thing quite distinct from them all".[17] And Harvey was as good as his word, because he relied upon mechanical principles where possible, but insisted upon the obvious presence in the living body (by contrast with the dead one) of various "vital spirits" such as the one we breathe and upon which our life depends. He attacked the indiscriminate use of the notion of "spirit" as a "common subterfuge of ignorance", noting that some "make the spirits corporeal, others incorporeal, and those who want them corporeal sometimes make the blood . . . the link with the psyche. . . . Those who declare the spirits incorporeal have no ground to stand on. . .". He himself concludes that somehow "blood and spirit mean one and the same thing", as "a hand that is dead is no longer a hand, so blood without the spirit of life is no longer blood". But he is puzzled to know whether one should think of the relation of spirit to blood as bouquet

---

[17] Quoted in Toulmin and Goodfield, *The Architecture of Matter*, p. 161.

to wine, or as wine-fed flame to the wine, a rather significant difference of status.[18] On the other hand, his contemporary, Van Helmont, was convinced that there had to be an "immaterial" agent working upon the passive material of the seeds from which everything, even mineral, comes to be formed; he collected a wide range of experimental evidence (from organic growth, fermentation, trapped vapors in mines, burning of organic materials) to support the claim that "vital spirit" is responsible for all substantial comings-to-be. He ends on a prophetic note: "I call this spirit, unknown hitherto, by the new name of 'gas'". One can foresee how the notion of "spirit" among these chemist-alchemists (much more than among the mechanical philosophers) is going to bear fruit in the "pneumatic chemistry" of the following century, which will find its clues in an enormously wide knowledge of the kinds of chemical change, rather than in the narrow range of the physics of gases.

The interest of this long and complex development for us lies in the debate about whether, and in what sense, the "vital spirits" could be said to be "material" and thus subject to mechanical laws (as the word 'material' was coming to signify). The fact that this question could be raised indicates how far from the ancient matter-spirit dichotomy these thinkers had moved and how the notion of matter was gradually shifting to allow life and activity to be included in it directly. Nevertheless, there was a clear realization on the part of most chemists and physiologists that the inert qualities attributed to "matter" by seventeenth-century mechanics were simply not going to be enough to explain the wide variety of changes beyond the narrow limits of purely spatial motion. It was all very well for the corpuscular philosopher-physicists to *talk* that way, but then they had never succeeded in actually *explaining* any chemical changes.

Materialism: The concept of matter played another role in the context of matter-spirit distinctions which was destined to have a more enduring significance. Already in Epicureanism, it had been utilized to make a central metaphysical affirmation: non-matter (spirit) does not exist; there is nothing *but* matter. 'Matter' here meant atoms, bodies whose only modes of interaction were collision or interlocking, and whose only properties were density, shape, motion. The obvious unities of the organism had to be either explained by the atomic motions, or else attributed to a special mind-stuff "formed of particles exceedingly minute" (Lucretius). But how are the mind-particles themselves held together in a unity? It required a

---

[18] See *The Architecture of Matter*, pp. 164–65.

substance "altogether without name", a "force of minute particles"—but the regress could scarcely be ignored. It is not to be wondered at that relatively few took the Epicurean-atomist cosmology seriously: its inadequacies as an explanatory structure for *all* change were too obvious. The Stoic *pneuma* was a much more likely story, since it took account of a much wider range of facts. Why, then, did Epicureanism have even the vogue it did? Given its implausibility as a general theory of change, it must have had some strong motives underlying it.

There appear to have been two major motives, and they will accompany every later form of materialism too. First, there was an undeniable intellectual attraction in the explanatory *reduction* worked by atomism. It eliminated all casual agencies whose operation was not fully describable in terms of interlocking and collision. Of course, by postulating that complex wholes are completely explainable in terms of the properties of their hypothetical parts, it went against the commonsense holistic approach which until then had dominated natural science. All the allegedly irreducible features of natural wholes—form, soul, *pneuma,* substance and the rest—were rejected and a quite different form of understanding was substituted. It was audacious, but in an age when clarity and simplicity of postulate were so much prized, the counter-empirical aspects of atomism were forgiven by many.

The other factor that told in its favor was the widespread skepticism about the religious forms of the day, and a clear realization that any religious claim of transcendence, whether for God or for man, must ultimately rest on holistic modes of thought. Thus by denying these, religion itself was cut at the root. There is no more emphatic way of rejecting transcendence claims of every sort than by beginning with the "lowest" (simplest, most inert) elements in the universe and then claiming that *everything* that exists, no matter how complex its activity, is of exactly the same nature as these elements, since it is nothing more than a spatial configuration of them. Atomism was the first half-way viable cosmology that the religious skeptic could call upon in his support. It may not have been much in those days from the empirical standpoint, but if one had lost faith in man or in the gods, it was the intellectual "leveler" that one needed in self-defense.

The revival of Epicureanism during the Renaissance was closely associated with the growing religious skepticism of the day, but it could not really prosper: it reduced *too* much, not only the gods but also man. Its implicit rejection of the uniqueness and dignity of man ran counter to the central article of faith of Renaissance humanism. But with the waning of the Renaissance, the impressive success of the new mechanics once again made materialism possible. It did not *have* to be atomistic or corpuscularian (after all, a *plenum* model of mechanical explanation would also have sufficed), but in practice it usually was, because this was the simplest sort of reductionist hypothesis, and reductionism is one of the two constant strands of materialism. The new mechanics posed an awkward dilemma for the philosophy of man, for the understanding of the nature of man himself.[19] Only two clear-cut options appeared to emerge, those of Hobbes and of Descartes.

Hobbes asserted that man is nothing more than a complex of particles governed by mechanical laws; mind is no more than a secondary reducible property, and the most complex goal-directed human behavior can in principle be explained by the simple laws of interaction of man's bodily elements. Hobbes was concerned to reduce *all* changes, even political ones involving countless human beings, to the simple motions of corpuscles obeying the laws of Galilean mechanics. His primary category is for this reason *body* rather than *matter*; he defines body as "that which having no dependence on our thought is coincident with some part of space". For him, 'incorporeal' is an incoherent term, because the primary intelligible features of our world are mechanical and spatial; thought is derivative and transspatial, being unknowable and thus unreal. There has been much debate in modern times about the extent to which his "materialism" was irreligious in origin; it seems safe to say, in any event, that reductionism was the dominant strand in it. In answer to a question about the nature of God, a topic which he tended to avoid, he once responded that God is a "corporeal spirit", thus apparently subject to the mechanical constraints in terms of which corporeality had been defined. Though Hobbes had no more empirical warrant than had Epicurus earlier for the claim that all change can be understood in terms of the spatial motions of imperceptible constituents, he did at least have the advantage of being able to specify some of the mechanical laws that the material constituents would presumably obey, which Epicurus had been unable to do. His doctrine is primarily a form of "scientism", an ontological counterpart to the positivism of a later day. For Hobbes, all that is can be understood, not just in terms of empirical science, but in terms of a particular empirical science as it then existed. His "materialism" is thus the denial of the reality of any alleged being whose activity is not entirely explicable in terms of mechanics of the Galilean style. Thus, although he defines "matter" in terms of extension, figure, and motion, the concept of matter most appropriate to his

---

[19] See E. A. Burtt, *The Metaphysical Foundations of Modern Physical Science*, London, 1924, chap. 8.

"materialism" would make matter "that which can be understood in terms of a reductionist mechanics". His world view did not gain wide acceptance until a century after his death when the triumph of the new "mechanical" ways of thought was assured.

The other option for a philosophy of man was that of Descartes, which was to cut man into two parts, abandoning one part to the mechanical philosophers, and placing the other (spirit) in a realm entirely independent of mechanical science. This radical solution did save the uniqueness of man and the transcendence of God by postulating a domain of being where the ambitious reductionism of the new mechanics, because of its own methodological restrictions, could never penetrate. But the cost was to sunder the two realms and to force the most improbable sorts of causalities at the boundaries between them. Hobbes and Descartes agreed in defining "matter" as the exclusive domain of mechanics; they disagreed as to whether something "non-material" could exist. Both solutions were reductionist in tendency; one denied the uniqueness of man, and the other undermined his unity. It seemed that the only possible way of maintaining both human uniqueness and human unity was to reject the reductionist approach entirely. But such holism smacked so much of the derided substantial forms of the Aristotelian tradition that no one made a serious attempt to reconstruct a philosophy of man which would take the substantial unity of man as its starting point, rather than as something to be rejected or explained away. It is clear that where such a philosophy would disagree most with that of Hobbes or of Descartes would be in its concept of matter.

Immaterialism: There is, however, one seventeenth-century philosophy that does lie at the other end of the spectrum from the Hobbes-Descartes position, if the "spectrum" be defined in terms of the concept of matter. It begins not from the unity of man, but from the much more striking unity of *all* that is. Like Leibniz, Spinoza begins with an ideal of rational understanding modeled on Aristotle's notion of *epistēmē*: the logical derivation of *all* properties from substance. But if this is to be possible, then there can be only one substance—if there were more, their mutual relations would not be derivable. He is thus much more thoroughgoing than Leibniz, who is prepared to allow a multiplicity of monads (though he has to pack the entire universe into each). Spinoza permits only a single monad, thus avoiding Leibniz' difficulties regarding individuation, but creating some awesome new difficulties of his own, since everything that is (including each individual man) becomes a mode of God. All true propositions constitute a single deductive system; there are no external relations and no statements at once contingent and true. This means that thought and extension must be identical, as must God and Nature. He criticizes Descartes for dividing them and thereby making it impossible to claim that every truth is a necessary truth. Spinoza is particularly uneasy about an epistemological starting-point such as the *cogito*: once one gives a higher status to spirit than to matter in the scale of certitude, there is no way of bringing them together again in a single science.

Spinoza is thus rejecting the dichotomies of classical metaphysics in his attempt to carry the classical ideal of "science" to its limits, as well as to deny the implicit limitation placed upon God by Christian metaphysicians who claim that created Nature somehow has a being distinct from that of God. Though 'immaterialism' is a label not usually attached to Spinoza's system (if for no other reason than that, unlike Berkeley, he very rarely uses the term 'matter'), it would seem that it is peculiarly appropriate since he rejects the possibility of a "matter", in most of the traditional senses we have been examining. There are no individuals other than God, so no principle of individuation is required, or even conceivable. There are no contingencies, no "dark spots" or indeterminacies, in his cosmos which is wholly interlinked in a logical structure of necessary propositions. There is nothing over against spirit or thought. There is no inert or resistant element. Every body has an "idea" corresponding to it which is its "soul"; these are, however, nothing more than two indissoluble aspects of the same reality. To many of his contemporaries, Spinoza seemed as much a "materialist" as did Hobbes; did he not equate God with extended mechanically-governed Nature? But in fact, the two thinkers were worlds apart, even though their systems of thought were structurally so similar. It is because both Spinoza and Hobbes reject the matter-dichotomies in order to affirm that only one sort of being exists that we might tend to collapse their systems into a single monistic framework, neutral as between matter and spirit. But in actual fact, this would be quite incorrect. If the universe were best described in terms of Spinoza's categories, then 'immaterialism' would be the correct label, since the emphasis is upon those features (logical necessity, unity, creativity . . .) that have traditionally characterized the "spiritual" counterparts of matter; whereas if Hobbes' conceptual framework is more appropriate, then one could properly use the term 'materialism', since his emphasis was always on those features (impact as the source of motion, reduction of higher-order unities) which have been taken to characterize the "material" counterparts of spirit. For Spinoza, unity was something given, whereas plurality constituted a problem. For Hobbes just the opposite was the case, and the difference between the "immaterialism" of the one and the "materialism" of the other developed directly from this difference in starting points.

§6 Matter, Perception and Reduction

Before leaving this review of the career of the concept of matter in seventeenth-century philosophy, it is worth bringing together some hints in the
previous section in order to show that the role accruing to *matter* at that
time in connection with the problem of sense-knowledge was in many
respects a new one. In Greek philosophy, the object of intellectual knowledge was form, and the object of perception was the concrete individual,
the matter-form composite taken as a whole. The notion of matter did not
play a prominent part in the incessant discussions of knowledge-problems
except in that as the source of individuation and potency, the matter-aspect
of things was regarded as the barrier to their total comprehension by the
mind. It is also true that matter, in effect, was a necessary condition for
perception; it is by being instantiated in space and time that a form enters
into the sensible world. But primary matter, being the indeterminate aspect or constituent of sensible substance, does not of itself constitute an
object of perception, nor does it figure in reports of what we perceive, except indirectly insofar as the space-time context becomes relevant.

The emphasis given to the distinction between "primary" and "secondary" qualities right from the beginnings of the new science, by Bacon,
Galileo, Hobbes, Boyle, Locke and a host of others, had troubling implications both for epistemology and for ontology.[20] There were three rather
different sorts of consideration involved, as a rule, in drawing the primary-
secondary distinction; it is important to note this (though it was not explicitly noted at the time) because they gave rise to three different forms
of distinction, with importantly different consequences. First, there was
the observation, dating back to ancient times, that many perceptual qualities of things are dependent on the perceiver. Perception of color, for example, not only depends on conditions of lighting, but on the state of the
organism itself. The same is true of temperature, taste and so forth. This
suggests that these sense-qualities are not absolute properties of the object,
but are somehow rooted in a combination of organism and object. Some
were led to conclude that such qualities can be said to "exist", properly
speaking, only in the organism; they are not "in" the object at all. 'Primary'
in this context means: an intrinsic property of the object itself independently of an observer; 'Secondary' means: dependent, at least partly, for its
existence or nature on an observer, thus variable independently of the object and not ascribable, without qualification, to the object.

---

[20] Harré devotes Part 2 of his *Matter and Method* to a discussion of their implications for his notion of a "general conceptual system".

This is an ontological distinction between two sorts of property, absolute and relational. An ontologically-primary (O-primary) quality may itself be perceptible, though in a way which is observer-invariant (solidity and shape were the examples usually given), or may not be perceptible though capable of instrumental verification (the attraction exerted by a magnet, for example), or may not even be instrumentally verifiable but postulated on the basis of indirect evidence (e.g. the property of containing a specified number of elementary particles). The only requisite for an O-primary property is that it should be intrinsic to the object, and in no way observer-dependent.[21] For a property to be O-primary, it does not have to be possessed by all physical objects. There will be an advantage in basing any physical science upon O-primary properties, although reliance on O-secondary properties need not necessarily be excluded, provided that their observer-dependence can be allowed for.

The second way of drawing a primary-secondary distinction is to ask whether there is a subset of the properties being considered in terms of which the remaining properties can be explained. The latter properties, because they can be "reduced" are said to be "secondary", and it is assumed that any theory which purports to explain the domain of the two sets of properties will make use of the "primary" properties only. Just as the earlier distinction arose in the context of perception, this one arises in the context of *explanation* (specifically, reductive explanation). The "primacy" here is of an epistemic, not an ontological, sort: it is a matter of deciding which properties are the more basic, if a science involving all the properties is to be formulated. Since not all explanations are reductive in character, it is not the case that every explanation in science involves a reductive (R) distinction between epistemically primary and secondary properties. The R-distinction is relative to a particular explanatory theory; there is no reason why another theory might not, for its own purposes, draw the distinction between "primary" and "secondary" differently. In particular, the R-distinction is not of itself unique; that is, there is no inherent reason why a domain might not be explained in alternative ways, by trying different sets of properties as "primary". The fact that one succeeds in explaining one

---

[21] Clearly, much more would have to be said about this notion of "dependence". There are two principal ways in which it can be defined; either in terms of dependence upon a perceiving organism, or in terms of dependence on *anything* outside the object (on the instrument used in measurement, for example). Dependence on an observer (though the kind that first drew attention to the problem) is not the only kind that would raise the issues of invariance and property-attribution. But this more general sort of relational dependence raises some very difficult issues and we shall thus use 'O-primary' in the first sense only, the sense that those who first stressed the "subjective" character of secondary quality always had in mind.

set of properties in terms of another by no means excludes the possibility that one might also invert the procedure and explain the former "primaries" in terms of the former "secondaries". As this distinction has often been drawn, the suggestion has been that R-primary properties are irreducible in a quite strong sense, this is that they are the *only* properties in terms of which the domain could be explained, that they *cannot* be reduced by the "secondary" properties or any subset of the "primary" and "secondary" properties together. This, of course, would be very difficult to prove in any actual instance because it would involve showing that no other theory yielding a different cut between "primary" and "secondary" could, in principle, succeed in explaining the domain.

There is a third sense of 'primary' which makes it refer to *essential* properties, properties which are necessarily present in all instances, by contrast with other properties which are "secondary" (i.e. accidental), ones which not all objects in the domain of discourse possess. (A slightly different form of this distinction would draw the line rather between properties one can *show* to be essential, and ones where no such proof is available, even though these "secondaries" *may*, in fact, belong to everything in the domain.) The grounds for claiming that a certain property is "primary-essential" (E-primary) can be of different sorts. They can be perceptual: if all the instances we have perceived possess the property, we may be disposed to generalize, and claim that all instances must possess it. Or they can lie in imagination: we cannot "imagine" an instance which does *not* possess this property. Or we may have built a theory on the assumption that every object in the domain *does* share this property, and the theory has proved a successful one.[22]

The interrelations between these different distinctions are quite complex. It is not the case that a property which is "primary" in any one of the three

---

[22] The two analogues in Aristotelian philosophy of a "primary-secondary" distinction do not, as it happens, suggest any other basic modes of distinction than the three we have discussed above. In the categories of accident, quality was always paired with quantity and was taken to be "primary" in both the O and R senses, for reasons that were not always clear. This thesis was exactly inverted by the mechanical philosophers who took quantity to be "primary" in all three of our senses, but broke it down into more specific measurables. The second analogue was the distinction between the "common sensibles" (properties capable of being perceived by more than one of the senses and therefore in some way "primary") and the "proper sensibles" (proper to one sense only). Although there was some suggestion of an R-primary status for the common sensibles, the analogy with the later primary-secondary distinctions cannot be carried very far, since the proper sensible corresponding to touch (resistance, solidity) was always claimed to be R-primary (as the mechanical philosophers also later made it).

senses defined above can on this account alone be asserted to be "primary" in one of the other senses. It is plausible to assume, for instance, that a property which is R-primary is likely to be O-primary, but it is by no means obvious that an R-secondary property would have to be O-secondary too. And there can obviously be both R-primary and O-primary properties which are not E-primary. But these fine points were of little concern to the mechanical philosophers, all of whom saw in the primary-secondary distinction a way of expressing the central tenet of their world view: the baffling variety of perceivable properties of physical things can be reduced to a few essential "primary" ones, the very ones which can be handled by the quantitative methods of mechanics. For mechanics to provide an adequate account of physical motions, the minimal properties it made use of had to be E-primary; for it to become the basis of a "mechanical philosophy" applicable to all motions, they had to be R-primary too. It was not evident that they had to be O-primary; the viability of a mechanical philosophy would not have been affected were there to be no O-primary properties at all. Nevertheless, if it could be argued that there *are* O-primary properties and that they are identical with the set of E- and R-primary properties already assumed, this would greatly strengthen the original claim that the properties truly are R-primary, something that in the nature of the case could never be more than a promissory note, and in those early years a pretty shaky one at that.

The notion of a primary-secondary distinction between properties was thus inseparably bound up with the origins of mechanics, and more especially of the philosophy of nature which the new mechanics prompted. In this perspective, "matter" came to be defined as the bearer of the "primary" properties. To be "material" was to possess these properties and thus fall within the scope of mechanical description and explanation. Depending on the emphasis given, the concept of matter thus served either to make an epistemological point about the minimal set of properties sufficient in principle to provide the basis for a complete account of the behavior of physical objects, or else to make an ontological point about which physical properties could be said to be "real" (absolute, intrinsic, independent of the observer, etc.). In either case, the suggestion was that there were certain quantitative characteristics common to all physical objects, and that these had to be accepted as the necessary and sufficient basis for mechanics (physics). Once again, then, as we have already seen in the discussion of Descartes above, matter becomes by definition exactly correlative with mechanics; mechanics is directed to the understanding of matter as matter, and matter is that which can be exhaustively comprehended by the methods of mechanics.

Bacon's tabular methods of discovering the "most real simple Natures" were intended to reveal the essential properties of matter; he also drew a clear O-distinction between "dispositions" that were intrinsic to the object, and "relative" features, such as smoothness or brightness. The latter depended upon our perception of them, and were "apparent", "exterior", "ordered to man", rather than "existent", "interior", "ordered to the universe", as Form (a rough analogue for the "primary property" of later writers) is. Galileo is more explicit in the famous passage in *Il Saggiatore* where he is discussing the nature of heat.[23] His argument is that the perceptual qualities of color, sound, taste, smell (not, however, touch, an exception he does not explain) are O-secondary in a very emphatic sense ("mere names"). Thus physics does not have to take account of them, since it can be assumed that they are R-secondary also. That is, a science of matter need deal only with the O-primary properties—size, shape, solidity, motion. The implication is that they (or a subset of them) must be R-primary also.

Boyle's treatment of this problem was, perhaps, the most detailed and most perceptive in the entire early empiricist tradition. He devoted a lengthy work, *The Origin of Forms and Qualities*, to the question: how can those features of natural bodies accounted for in Scholastic philosophy by the notions of *form* and *quality* be explained in the "corpuscularian" philosophy?

> I agree with the generality of philosophers so far as to allow that there is one catholic or universal matter common to all bodies, by which I mean a substance extended, divisible and impenetrable. But because this matter being in its nature but one, the diversity we see in bodies must necessarily arise from somewhat else than the matter they consist of.[24]

The "somewhat else" that is required to diversify bodies is first of all motion, and then size and shape, since the matter is divided into parts. These are "accidents" because even though they are altered, the "whole essence of matter" remains the same, but in another sense they are "essential" or "primary" because upon them (and upon them alone) depend what kind of a body something is, whether it is for example gold or water.[25] These "secondary qualities, if I may so call them" can be constituted from the primary properties because of the multitude of relations any given body can have in virtue of its size, shape, and motion alone, with all the other bodies in the universe:

---

[23] In translation in *Discoveries and Opinions of Galileo*, ed. S. Drake, New York, 1957, pp. 273–78.

[24] *The Origin of Forms and Qualities according to the Corpuscular Philosophy*, in vol. 3 of *The Works of the Honourable Robert Boyle*, London, 1772, p. 15.

[25] *Ibid.*, p. 24.

We shall not much wonder that a portion of matter that is indeed endowed with but a very few mechanical affections, as such a determinate texture and motion, but is placed among a multitude of other bodies that differ in those attributes from it and one another, should be capable of having a great number and variety of relations to those other bodies and consequently should be thought to have many distinct inherent qualities.[26]

In support of this, he notes that the heat of the sun can produce a wide variety of effects (melting, expanding, changing color . . .), and yet we would not wish to attribute these to separate qualities in the sun; their differences are ultimately due to differences of relation between the sun and the bodies it heats, as well as to the sizes, shapes and motions of the corpuscles of which the bodies are composed. Sensible qualities are to be explained in this way also; they are due to nothing more than the different ways in which the sense-organs can be affected by the mechanical properties of bodies. We tend to think of them as separate real qualities of objects:

whereas indeed (according to what we have largely shown above) there is in the body to which these sensible qualities are attributed nothing of real and physical but the size, shape, motion or rest, of its component particles together with their texture . . . nor it is necessary that they should have in them anything more, like to the ideas they cause in us.[27]

But this immediately raises a difficulty, "perhaps the chiefest that we shall meet with against the corpuscular hypothesis": how can sensible qualities be reduced to mechanical relations with sense-organs, when it is evident that "they have an absolute being irrelative to us"? Snow would be white, even though no man existed. Boyle's answer is a careful one:

I do not deny but that bodies may be said in a very favourable sense to have those qualities we call sensible, though there were no animals in the world. . . . [A body would still in such a case] have such a disposition of its constituent corpuscles that if it were to be applied to the sensory of an animal, it would produce such a sensible quality which a body of another texture would not. . . .[28]

So one body will have a "greater disposition" than another to reflect light of a particular color, and so forth. In other words, there is a *dispositional* sense in which the sensible qualities can be said to be really present in the bodies to which they are attributed; they are such that they *would* affect a particular sense-organ in a particular way, so that bodies are "dispositively endowed" with them. But Boyle adds that the quality "superadds nothing of real" to the specification of the object in terms of the primary properties,

[26] *Ibid.*, p. 20.
[27] *Ibid.*, p. 23.
[28] *Ibid.*, p. 24.

and concludes that since the qualities are no more than the effects of these properties they must be "deducible" from a sufficiently complete knowledge of them.[29]

On what does he rest his case? He frequently repeats phrases like "it would be hard to conceive". In a world without animals, "it is hard to say what could be attributed" to the matter-corpuscles except the properties of motion, shape and bulk. He also argues that since the interactions between sense-organ and environment are obviously mechanical, the qualities of objects *must* be reducible to, and explicable in terms of, simple mechanical properties. The "primary" properties are primary because they are non-relational, absolute; to alter them is to alter the nature of the body. Whereas the secondary qualities are relational, though rooted in a real disposition of the body. Thus his O-distinction rests on two rather different bases: absolute versus relational, or actual versus dispositional. No longer is it simply the real versus the "mere name" (as it had been for Galileo and much of the atomist tradition). But the weight of his argument clearly rests on the pervasive reductionism of the day, the assumption that all properties can in principle be reduced to the chosen few of the mechanical philosophy. They are the only ones that can be *conceived of* as universal and absolute; interactions fully describable in terms of motion, size, shape, are the only ones *conceivable* between material entities, and specifically between sense-organ and environment. Thus, he runs together all three ways of defining the primary-secondary distinction, assuming that matter is defined by the "primary" properties. Like the Greek Atomists, the original source of the primary-secondary distinction, he conferred primacy on the *spatial* characteristics of matter in the domains both of knowledge and of being.

But now an epistemological gulf was gradually beginning to open. If the qualities we perceive in bodies do not really belong to them but are somehow caused in us by the primary properties of the perceived objects, can we be said to perceive these properties themselves? And if we cannot, has not matter once again become a sort of unknowable principle, to whose existence and nature we have to infer via a causal analysis of sensation? The Locke-Berkeley dialectic was an exploration of the possible answers to this uncomfortable question. Locke maintained a causal theory of perception in which quantitative "modifications of matter" ("qualities") cause "ideas" in us. (He uses the word 'idea' indifferently for the "object of perception, thought or understanding"). All that the mind can know, then, are "ideas"; the qualities of objects are not directly known, but are

---

[29] Ibid., p. 26.

inferred. Nevertheless, we can distinguish between "ideas" of primary qualities and "ideas" of secondary qualities on the grounds that the former somehow "resemble" the qualities that cause them, whereas the latter do not, being caused by combinations of primary qualities in the object. Locke's primary-secondary distinction is thus an O-distinction, though he also uses it to support a reductionist claim for the universal scope of Newtonian mechanics.

A curious inversion of the traditional roles of substance and matter ought to be noted here. For Aristotle, matter was a constituent of substance; it was the indeterminate substratum of change, unknowable because lacking in any properties of its own, the ultimate subject of predication. Locke makes *substance* the substratum and the ultimate subject of predication, unknowable in itself, though he is understandably uneasy about the status of this "something, I know not what". Whereas matter (or body) for him, though not directly perceived, is still a determinate entity possessing spatial and inertial characteristics; it is now substance that is the "constituent" of matter, rather than the reverse. He has thus unloaded upon his concept of substance the ancient philosophical woes of the concept of primary matter. Or to put this more exactly, his concept of *substance* is close in some respects to that of matter in the Aristotelian sense, and quite remote from that of Aristotelian substance. Why, then, does he not use the traditional label 'matter' for his substratum-subject? The reason was simply that by his time the new usage of 'matter' as a general name for the objects of mechanics was too well established, especially in the Newtonian circles to which he belonged. So when he faced the metaphysical problems of predication and ontological constitution to which Aristotle had responded in terms of matter and form, he could no longer use either of these terms, and had to speak of "substance" and "property" when giving a solution to these problems not unlike that of Aristotle.

Berkeley called his philosophy "the immaterialist hypothesis" because its main theme was the rejection of Locke's notion of a "matter-cause" of sensation, itself unperceived and unperceivable. Berkeley strongly objected to the appearance-reality dichotomy implicit in the primary-secondary distinction of the new empiricist and mechanist philosophies. In his view, it led to philosophical absurdity and ultimately to religious unbelief because of the independent status assigned to matter. He argued against it along three lines. First, if material objects are not directly perceived, if all we know is our ideas, how can we be sure that matter exists at all? Must not Locke's brand of empiricism lead to complete skepticism and solipsism in the long run? Second, Locke's case rests implicitly on the adequacy of Newtonian mechanics as an explanatory system. But empirical science can never reach causes, it can never be more than descriptive; all that we know are the constant sequences of phenomena. Harking back to the nominalist philosophy of the fourteenth century, Berkeley argued that the intelligibility of the physical world does not come from any intrinsic causal structure but from a positive act of the Divine Will decreeing that the phenomenal sequence should be so, when it might just as well have been otherwise.

Finally, he rejects the primary-secondary distinction itself on the grounds that even in Locke's own system there is no real basis for it. To know which qualities are O-primary, we would need to know which qualities the object possesses independently of our knowledge of it. But this, on Locke's own admission, we can never know since the material object-in-itself cannot be known. Locke's quasi-mechanical argument for a "resemblance" between the *ideas* of primary quality like solidity (in the mind) and the primary qualities themselves (in the object) Berkeley finds completely confused. How can an idea in the mind be "like" a quality in an object? And how can we prove it by calling upon the sense of touch, as Locke does, to verify that our *idea* of solidity (arising from experiences of the resistance of ma-terial objects to change of motion) must be caused by a *property* of solidity in the object? Could we not as easily say the same for color? His argument is thus that the analysis of perception leaves no room for a privileged set of qualities. If *any* perceived qualities are "secondary", then *all* are.

Berkeley denied the existence of a "matter" which constitutes the re-ality of the physical world and is the cause of our sensations, though not *directly* knowable in itself. Yet he did assert that *some* cause for our per-ceptions is necessary, and this role he attributed to God. Thus, it could be said that in his system God replaces matter as the causal basis of perception and of the apparent continuities of the perceived world. The dualism be-tween what is perceived and the real cause of that perception is thus re-tained by him, despite its origins in empiricism; he assumes that the objections that can be raised against "matter" do not hold against God. His own objections to the empiricists' matter derived more from its being postulated as an inert "over-against-mind" than from its not being directly knowable in itself. But he had, nevertheless, put a finger on a crucial dif-ficulty for empiricists. There were, essentially, four alternatives open to them.

They might defend the primary-secondary distinction, and try to live with the skeptical problems of a matter that lies behind and beyond direct experience (though its properties could, of course, be discovered indirectly through mechanics). Such a "matter" would, further, suggest a very strong spirit-matter dualism, one that might challenge the Christian creationist view of the total dependence of being on the creative power of Spirit. Or else the primary-secondary distinction might be rejected, in which case three broad types of alternative offered themselves. One might defend an Aristotelian type of direct realism which would make all qualities equally "real" in the object and the object itself an autonomous substance. But the "mechanical philosophy" appeared to raise insuperable difficulties for the claim that qualities like color are intrinsic properties of the object. Or one might, with Berkeley, retain a mitigated dualism between the cause of perception (God) and ideas and percepts themselves. But this invoking of a Being transcending experience as the ground for the veridical character of experience, and the limiting of real existence to minds and their contents, went strongly against the grain of empiricism.

Phenomenalism: Finally, one might argue that there can be no justification for going beyond what is immediately given in perception ("sense-data", *"sensa"*) since on the one hand a "beyond" (whether of a Lockean or Berkeleyan sort) cannot be directly validated, and on the other hand, a claim to immediate knowledge of autonomous material objects, whose nature and existence are independent of our perceiving of them, runs into all the difficulties raised from both the Lockean and the Berkeleyan directions. This reduction of material objects to *sensa,* which has come to be called *phenomenalism,* is roughly equivalent to Berkeley's system without Berkeley's God. As the radical conclusion of a consistent empiricism, it has exercised a fascination over all discussions of empiricist epistemology since Berkeley's time.[30] Its central thesis may be stated as one about "matter"; it is simply a denial of the existence of a matter-cause of sensation of the sort postulated by classical empiricism and implicitly accepted by most scientists. This is one of the two major post-Cartesian philosophies for which a concept of "matter" serves as the defining theme; phenomenalism is an "immaterialism", and in that sense, at least, is the opposite pole of "materialism" (the other modern philosophy which hinges around a matter-thesis), even though both of these systems are of empiricist lineage. Yet since the stress in classical materialism was an ontological rather than an epistemological one (matter defined by its contrast with "spirit" and the existence of the latter denied), Berkeley's immaterialist affirmation of spirit would come closer than would an epistemologically-oriented phenomenalism to being the contrary of classical materialism.

Hume was the first true phenomenalist. Despite some obvious affinities between his thought and Greek skepticism and more especially later medieval nominalism, his dissolution of material objects into strings of sense-

---

[30] See K. Sayre's paper elsewhere in this volume.

data, lacking any intrinsic pattern of ordering, constituted such a radical denial of the major principles of Greek cosmological thought, matter and form, substance and nature, that it was in essence a new position. And an enigmatic one it was because of the many ambiguities concerning the nature and status of the "sense-data", which had replaced the commonsense, enduring material objects accepted by earlier philosophers. If the dependence of the "sense-data" upon the perceiver were to be stressed, it seemed as though one was not so far from Berkeley after all. If they were made into something non-mental, or at least partially independent of the consciousness of the subject, one would be back with the difficulties originally urged against the empiricist concept of matter. But the major difficulty with Hume's position lay in its inability to provide an even moderately plausible alternative to the concept of matter. We are quite sure that the things that we perceive exist even when they are not observed. It is essential to our reconstruction, both in common sense and in science, of the world in which we live that there should be an immense number of enduring objects that have never been directly observed by man and never will be. If sense-data are to be substituted for material objects, what is to provide the continuity ordinarily attributed to these objects when they are not being observed? The phenomenalist rejects Berkeley's answer to this question, but does not seem to have a more plausible replacement.

§7 *Newton*

Phenomenalism already carries our story beyond the seventeenth century. There is one gap still in the account of the transition from the classical Greek concept of matter to the modern mechanical concept. It was Newton more than anyone who shaped the schema of mechanical categories in terms of which we conceptualize motion. Instead of the broad and unspecific ideas, *matter* and *motion*, that had traditionally been used to describe moving bodies, Newton substituted systematic and quasi-operational concepts, *quantity of matter* (*mass*), and *quantity of motion* (*momentum*). In searching for a measure of the inertial response of a body to motion imposed on it, the most helpful metaphor was clearly that of a *stuff*, of which there could be more or less in a body. The body's response to impressed force would depend, then, on how much "stuff" there was.

But how was it to be measured? Descartes' attempt to equate it with volume had been a failure. Weight seemed a more promising correlate. But Galileo had shown that the speed of *in vacuo* fall is independent of weight, so that for analysis of fall, at least, the matter-factor did not seem as relevant as one would have expected. Still in the case of *impact*, it was obvious that

the degree to which the motion of a body is affected, as well as the degree
to which it could affect the motion of other bodies, are somehow dependent
upon its matter-content. But was there any reason to link this content with
weight? The analogy between the downward force we call weight and the
central force with which Newton was for the first time able to "explain"
Kepler's laws of planetary motion depended on making matter also somehow responsible for attraction. This was clearly pushing the metaphor
rather hard. Yet it *worked*!

The matter-measure (later to be called simply mass) was correlated with
change of motion. Such changes are brought about by "forces", the new
mechanical specification of the older notion of efficient cause. The amount
of the impulse-force is estimated by the change of momentum it produces
in a given body; in a simpler later version, this can be put by saying that
the force is proportional to the product of the mass of the body and the
acceleration the force brings about. This is intuitively plausible; it somehow seems an appropriate way of linking together, in a rather tighter array
than before, the commonsense notions of *force, matter*, and *motion*.[31] It
makes the "force" directly proportional both to the acceleration it produces
(twice the force produces twice the acceleration) and to the "resisting"
mass (twice the force will accelerate twice the quantity of matter to the
same velocity). Weight is a force, and consequently can now be split into
factors, an invariant mass-parameter proper to the body, and an acceleration
produced by gravity in that particular region. Since this latter depends on
the distance of the center of force, weight is now seen to be variable, and
Galileo's claim that the acceleration of falling bodies is constant and the
same for all bodies is seen to be an approximation holding only over short
distances, and for bodies at the same distance from the center of the earth.
The non-geometrical matter-factor *is* relevant after all, even for falling
motion, and the limited validity of a purely "Galilean" kinematics, of space
and time measures only, is revealed.

Newton was able to unify three types of motion that had long been regarded as disparate—terrestrial impact motions, terrestrial motions of fall,
and planetary motions—by means of unified concepts of force and mass
that could be applied to all three types of motion. The justification for doing
this was on the one hand the plausibility of such a generalization of the
force-inertia ideas, and on the other the success of the system based on
them. He could show the strict proportionality of weight to mass for all

---

[31] Of course, one has to beware of making this seem *too* obvious a precising of the
everyday notions. Aristotle, starting from these same notions, had come up with a
quite different way of relating force and motion, making the speed of fall proportional to the weight-force acting upon the body.

kinds of bodies by means of careful pendulum experiments. He could also show that falling motion and planetary motion are of the same type, and that to treat them as different would lead to a contradiction with observation.[32] He uses the expression 'quantity of matter' as though it were univocal. But the relationship of *force* and *mass* in the new system was, in fact, extremely complex, and was left unclarified in a number of important respects.

One of these unclarities is worth dwelling on, because of its significance for later physics. Newton assigned to matter three quite different roles, to quantity of matter (or mass) three different operational measures. He made mass a measure of the resistance of a body to change of motion imposed on it from the outside (inertial mass, to give it its later title, IM). Second, he made it a measure of the response of a body to a gravitational force acting on it (passive gravitational mass, PGM). Third, "mass" also became a measure of the ability a body has (or that at least is somehow connected with every body) of causing a gravitational response in other bodies (active gravitational mass, AGM). These concepts are by no means identical; it cannot be assumed without proof that they designate the same (or even related) factors of a body.[33]

Newton was concerned to show, by means of the results of careful pendulum experiments, the exact proportionality of weight and quantity of matter (IM) for bodies at the same distance from the center of force.[34] He wanted to show that they were all equally attracted, in proportion to their quantity of matter, which (given Law II) would mean that the same

[32] He shows that "the force by which the moon is retained in its orbit is that very same force which we commonly call gravity". *Principia*, Book III, Prop. 5; see also Prop. 6.

[33] At first sight, it might seem that IM and PGM are not really distinct; "resistance" sounds like a form of "response". But the PGM can be measured in cases where there is no motion, and thus IM is not involved at all, as in ordinary weight measurement where the net force is zero. Such a measurement will yield a value for the PGM of the body, if the gravitational acceleration characteristic of the locality be known. PGM is a tendency to move, whereas IM is a tendency to resist change of motion. See C. B. Mast, "Three concepts of Mass", in *The Concept of Matter*, 1963; also M. Jammer, *Concepts of Mass*, Cambridge, Mass., 1961, chap. 10; E. McMullin, "From matter to mass", *Boston Studies in the Philosophy of Science*, 2, 1965, 25–45.

[34] *Principia*, Book II, Prop. 24; Book III, Prop. 6. Even in this single case, both IM and PGM are involved, the former in calculating the motion of the pendulum, the latter in calculating the force acting upon it given the gravitational field (or potential, as it would now be expressed). The pendulum results would help to confirm the constant proportionality (not, of course, the equality or the identity) of IM and PGM.

acceleration is produced in all. This provided indirect confirmation for Law II, the law determining inertial mass. He did not, of course, address himself to the proportionality of IM and PGM (since he had not adverted explicitly to this distinction). Yet the accuracy of his calculation of planetary orbits did, in fact, implicitly support the assumption he had made that the two could be taken as identical.[35] The oddity that they *should* turn out to be the same (that "curious fact", as Weyl puts it,[36] does not appear to have struck him. The uneasiness he *did* feel over the attribution of an active role ("attraction") as well as a more conventional passive role ("inertia") to matter, reflected a tension that would endure to our own century, when it played an important part in the formulation of Einstein's general theory of relativity.

Another unclarity in Newton's notion of matter arose from a very different source. He often characterized matter as the dense, the impenetrable, by contrast with the vacuum of empty extension.[37] The body-void antithesis figured prominently in the Epicurean tradition that influenced him in his early years, through the writings of Gassendi and Charleton especially.[38]

---

[35] In the calculation of orbits, all three mass-factors are involved. The gravitational force between two bodies, A and B, is $Gm_a m_b/r^2$. If it be taken to act on B, then m_a will be AGM, and m_b will be PGM, and the acceleration produced will be Gm_a/r^2. The gravitational force is set equal to the centrifugal force, $m_b v^2/r$, where m_b is now IM. In applying the notions of force and impulse in the way he did, Newton implicitly took for granted the equality of m_b (PGM) and m_b (IM); if they had not been equal, the orbital calculations would have been erroneous. The fact that they did give correct results served to justify (though *not* to explain!) the validity of the implicit assumption. The relation of PGM and AGM was of less concern in the *Principia*, because AGM was absorbed into the value of the gravitational acceleration (Gm_a/r^2), and its absolute value was not needed. The equality of action and reaction asserted by Law III implied, if not the equality of the PGM and AGM measures, at least their constant proportionality over all bodies.

[36] H. Weyl, *Space, Time, Matter*, New York, 1952, p. 225: "The fact that a given gravitational field imparts the same acceleration to every mass that is brought into the field constitutes the real essence of the problem of gravitation. . . . [In the gravitational field] the force that acts on the particle is equal to gG, in which g, the 'gravitational charge' depends only on the particle, whereas G depends only on the field: the acceleration is determined here again [as in the case of an electrostatic field] by the equation $ma = gG$. The curious fact now manifests itself that the 'gravitational charge', or the 'gravitational mass', g, is equal to the 'inertial mass', m".

[37] See J. E. McGuire, "Body and void and Newton's *De Mundi Systemate:* some new sources", *Arch. Hist. Exact Sciences*, 3, 1967, 206–48.

[38] The antithetic terms here were 'body' and 'void', not 'body' and 'space', since space can be either full (of body) or void. The category of *space* is prior to *body* and *void*. There can be space without body, and (logically, at least) there can be space

Almost in his earliest works, one finds him regarding as likely the existence of a void, both in the interstices of matter and between the aetherial particles in the celestial space.[39] By the time the *Principia* appeared, he is quite certain: a *plenum* (because it would have no "pores") would be denser than quicksilver, and would thus quickly halt planetary motion. "Therefore a vacuum is necessarily granted".[40] When Leibniz criticized the doctrine of the void in his debate with Clarke, Newton began the drafting of an elaborate series of definitions of 'body', 'void', 'phenomenon', for possible interpolation in the third edition of the Principia. In these too, he linked corporeality with tangibility and *vis inertiae,* defining void simply as the absence of body.

But this definition became more and more questionable as his "void" began to fill up with various "active principles": magnetic effluvia, light, electric spirit, and finally an aether of dense but dubious character. The strains on his concept of materiality became apparent. It is instructive to notice how differently Leibniz and he responded to the same problem. To Leibniz, the causal agencies operating in otherwise empty space were "material" because they could affect physical bodies in a law-like way, making them capable of scientific treatment, and because they ensured that space was never really empty. For him, 'material' meant no more than: occupying space, physically real. That they were active in no way detracted from their materiality, since in his view, *all* matter must be active.[41] Whereas Newton denied their materiality in part because they *were* active principles (matter for him is ultimately passive), and even more because they lacked *vis inertiae.* Their presence in space still left it void, as far as he was concerned, because they offered no resistance to moving bodies. The "void" that he was affirming was thus not the "void" that Leibniz was denying. But the

without void. The existence of body, on the other hand, entails that of space but not of void. Since body is spatial, it cannot be defined in terms of an antithesis with space, only with *void* space, the burden of the definition thus falling on the notion of void, not that of space as such.

[39] See his untitled tract, beginning 'De gravitatione', in *Unpublished Scientific Papers of Isaac Newton,* ed. A. R. and M. B. Hall, Cambridge, 1962, pp. 90–121.

[40] *Principia,* Book III, Prop. 6, Cor. III. After a discussion with Cotes, he added a further corollary in the second edition, making his reasoning more explicit: "If all the solid particles are of the same density, and cannot be rarefied without pores, then a vacuum is granted." To admit a variable density for the primordial particles would risk (in his eyes) allowing the possibility that density might "diminish to infinity", thus leaving the particle without *vis inertiae.* This reasoning is reminiscent of that employed in the discussion of the intensity-invariance criterion of Rule III, a central theme in McMullin, *Newton on Matter and Activity,* Notre Dame, 1978.

[41] *Ibid.,* ch. 4.

dispute was by no means a merely verbal one; their verbal usages reflected irreducible differences between their underlying conceptual systems, one which in turn was expressive of their fundamental metaphysical disagreement.

Newton immediately noticed one fateful implication of his theöry of gravity. Since the dynamic effect of matter depends only on its quantity and not at all upon its volume, and since it operates from the center of mass and not from the surface, a very small volume of dense matter could produce the effect of a large body, if properly distributed. Thus it is possible that a body might have thousands of times more "pores" than parts, so that what appears to be a solid body might consist almost entirely of vacuities.[42] The plausibility of such a view derived, in Newton's mind, principally from its permitting all sorts of chemical and physiological processes to be understood as motions of the primordial unchanging particles, under the action of intense, short-range forces. The porosity of matter would also allow one to explain transparency, the transmission of electrical and magnetic influence and other puzzling phenomena.

The full impact of this doctrine upon the concept of matter would not be felt till later when forces replaced matter entirely in the dynamic theories of Boscovich and Kant. But for Newton, matter remained primary, even if exiguous, mainly perhaps because the nature of force always remained problematic for him. It was one thing to diminish the volume of matter in the universe a millionfold, but it would have been quite another to take away the remaining dense matter entirely. Newton could not forget how hard-earned an insight it had been to make force depend (in the multiple ways we have seen) on quantity of matter. Even though he later had recourse to active principles, electric spirit, and the like, the amount of the gravitational attraction exercised by a body was, after all, determined solely by its own matter-content.

§8 *The Parting of the Ways*

There is one further shift that occurred in this period that can easily escape attention. It culminates in the work of Newton, though its beginnings can be seen at least as far back as Galileo. It is at the level of method rather than content. We shall lead into it by asking how quantity of matter is to be known. Besides saying in Definition I that it is given by the product of density and volume, Newton also adds that it is "known by the weight

---

[42] See A. Thackray, *Atoms and Powers*, Cambridge, Mass., 1970. Most of Newton's speculation on this topic never saw publication; we find it abundantly illustrated in drafts he left for the *Queries* of the *Opticks*.

of each body, for it is proportional to the weight, as I have found by experiments with pendulums". But if the numerical measure of the "quantity of matter" is to be derived from a weight measurement, would this not make the Second Law into a definition, either of mass or of force, since only two of the three quantities mentioned in it would be independently measurable? Newton may have thought of density as an intensive, irreducible quality, as earlier natural philosophers had usually taken it to be. Or more likely, his corpuscularian view of matter may have led him to think of the "density" of bodies in terms of the number of constituent "primordial" particles per unit volume, these particles themselves being construed as of uniform density and size.[43] But this does not avoid circularity, since some way of knowing the particles to be of equal mass would have to be given in advance. From the operational standpoint, Newton's notion of mass seems to be prior to that of density, so that a definition of mass in terms of density inevitably runs into difficulties. The choice, therefore, seems to be between a circular definition of density and a definition in terms of weight or mutual acceleration that trivializes the Second Law. Yet Newton's mechanics obviously has an empirical grip on *quantitas materiae* somewhere.

Is the Second Law a definition or an empirical statement? Perhaps the problem lies here with the formulation of the question, the assumption that the "Law" must be either a definition of one term by means of others previously understood, or else an observation-report utilizing terms all of which are previously understood. The Aristotelian theories of knowledge current in the later medieval period suggested that our grasp of an empirical concept is acquired by an induction over (or intuition into) some experiential instances of the class of real things to which the concept applies. *Horse, rational,* even *point,* were to be understood by an abstraction over singular instances, an abstraction then to be expressed in language by means of a definition. The Greek theorists had long before pointed to the regress difficulties involved in the notion of definition itself. But their notions of *form* led them to assimilate the conceptual to the real order rather too readily and consequently to overlook the multiple contingencies of language. Connections between concepts were to be *discovered*, it seemed, not postulated; they existed, one way or another, prior to human science. Even in geometry, the key terms were assumed to be understood individually in advance; thus, the axioms were regarded as true statements about the objective relations holding between the concepts. Neither the inter-

---

[43] The "porosity" doctrine, alluded to above, would lend some support to this interpretation, particularly since Newton does assume in his discussions of density variations that the primordial particles all have the same density.

dependence of the terms used in geometrical axioms, nor the role played by these axioms in implicitly limiting the imprecision of the terms was grasped. There was, of course, from the beginning an appreciation of the "special" character of mathematics, which was reflected in a variety of views on mathematical objects and mathematical abstraction. But even during the later Renaissance, when quantities ever more remote from sensible experience were introduced, little attention was paid to problems of mathematical language.

Not surprisingly, the language of the natural sciences was even more taken for granted. The concepts of Aristotle's biology, of Eudoxus' astronomy, of the Mertonian or Parisian mechanics, were plausibly assumed to derive by abstraction from experience; neither the dependence of these sciences on the contingencies of the natural languages in which they were expressed, nor the considerable degree of stipulation needed to overcome the imprecisions of these languages, were adverted to because the sciences themselves did not as yet carry language much beyond the everyday level. The mechanics of the seventeenth century, and the *Principia* in particular, did so to a significantly new degree.

Newton made use of three classes of technical terms whose implicit modes of definition differed considerably. First were mathematical terms, a sophisticated algebra of changing quantities. Second were the descriptive kinematic terms of Galilean mechanics, like 'length', 'time', 'velocity'; they could be defined by means of specific direct measure-operations, without invoking (or so it seemed) any hypothetical elements. Third came the dynamic terms, 'force', 'momentum', 'inertia', 'action', 'matter', needed for a causal account of motion. The particular contribution of the *Principia* was to interlock these terms for the first time in a way which was not yet precise but which gave the promise of becoming so. They were not given separate operational definitions. Rather they were linked by way of empirically applicable "definitions" and "laws". The system as a whole is thus definition, description, and explanation, all at once; it is both postulated and empirically (approximately) true. A single element (concept or law) cannot be detached for separate analysis; one cannot ask of the Second Law whether it is definition or observation statement. It is both and neither when considered (as it must be) in the context of the system as a whole. Its warrant is neither a specific postulation nor a particular set of observations, but rather the warrant of the theory taken in its entirety. Of course, one could criticize the Second Law for ambiguity, or single it out for modification in the light of new evidence. But one's ultimate reason for accepting it could only be the adequacy of the theory of which it is an integral part.

The systematic character of the warrant for (as well as the understand-

ing of) the concepts and "laws" of the *Principia* was not, of course, an
entirely new phenomenon. Once language begins to tighten into the net
of "science", the interdependence of its elements becomes much more
marked. Greek philosophy had already made a major move in this direction, a move whose significance had not been fully appreciated; the postula-
tional and systemic aspects of the philosopher's language could too easily
be overlooked because of the persuasively familiar air of the terms used
and the ready availability of the experiences drawn upon. The example of
Euclidean geometry encouraged the hope that a science could be summed
up in a set of axioms, each capable of being evaluated separately and seen
to be true, once properly understood. Philosophers quarrelled about the
truth of individual axioms of their systems, often not realizing that what
almost certainly *ought* to be at issue between them was not the statement
itself but the entire conceptual framework from which its terms derived
their meaning, and by means of which the axiom itself came to seem obviously true or obviously false.[44] And this would demand a methodologically very different sort of discussion. When debates of the sort already
chronicled arose about the characteristics of "matter", they were usually
inconclusive, partly at least because the real source of the differences was
not clear, and also because criteria for adjudication between rival philosophical systems, taken as unitary wholes, have never been easy to find
or agree upon.

Newton did not invent new terms for his dynamics; he drew upon terms
that made intuitive sense of our ordinary mechanical experience. All of
them had appeared in earlier natural philosophy but the relations between
them there were almost as vague as in their occurrence in ordinary language.
He specified the relationships between them in a way which permitted
precise empirical questions to be formulated and answered. He was guided
in doing this by the senses the terms already possessed, and by the large
body of empirical knowledge of various sorts of motions that had been
building up since the days of Galileo and Kepler. But in the last analysis,
a leap of creative insight was needed in order to specify a concept-system
in terms of which all of this would fall into place. The immediate and last-
ing success of the *Principia* lay in the way in which it unified the science
of motion by means of a handful of partly interdependent concepts. The
precise predictions it made possible served as a further and highly visible
validation, not only of its postulated laws of force, but of the new ideal of

---

[44] See McMullin, "The nature of scientific knowledge: what makes it science?", in
Philosophy in a Technological Culture, ed. G. McLean, Washington D.C., 1964, pp.
28–54; J. Earman and M. Friedman, "The meaning and status of Newton's law of
inertia", *Philosophy of Science, 40*, 1973, pp. 329–59.

dynamic explanation implicit in the Newtonian concepts of *force* and *mass*. The validation was so immediate and so powerful that all other ways of understanding the main concepts of the new mechanics soon tended to be eclipsed.

The effect of this on the concept of matter was particularly striking. Matter in the *Principia* appears in a dual role, as that which resists change of motion and as that which is somehow responsible for change of motion in other bodies. In the Aristotelian tradition, matter had been identified with the *potential* aspect of physical things, that which allowed them to be acted on mechanically and which determined their mode of acting mechanically on others. Originally an indeterminate substratum, it became quantifiable (as we have seen) in later Scholastic philosophy; its quantitative conservation was then the mark of the body's continuing identity. Newton needed a measure-concept for the "response to force" parameter of a mechanical system, one that could be manipulated in algebraic terms. Why could he not have used 'matter' instead of the complex expression, 'quantity of matter'? After all, he did not speak of "quantity of length" or "quantity of force". But the mode of quantifying space, time and their derivatives, velocity and acceleration, was intuitively much more obvious. One could talk about the "length" of a body, and no one needed to be told that it was a quantitative measure. 'The matter of a body' carried no such connotation. 'Matter' was a substance-term and did not single out particular attributes like quantity. The qualifier 'quantity of' would have to be added in order to focus on what was for Newton only one aspect of matter. The same qualifier was used by him in defining momentum ("quantity of motion"), partly for the same reason. (The "speed" of a body was naturally something that could be estimated numerically, but what of the "motion"? Could one say that one body had twice as much "motion" as another?) Since it was not at all obvious how "matter" and "motion" were to be measured, the phrases 'quantity of matter' and 'quantity of motion' had to be expressly defined by him; these were, in fact, the opening definitions of the *Principia*.

Within a short time, these phrases had been shortened to the more workable 'mass' and 'momentum'.[45] Since the root terms 'matter' and 'motion' seemed to have no technical uses in mechanics other than in con-

---

[45] Newton himself ordinarily uses 'quantity of matter', and occasionally 'mass of matter'. The term 'mass' itself (meaning a lump of stuff) was in common use in the seventeenth century; it was used technically by Leibniz to mean "secondary matter", an aggregate of substances. Its technical use as an abbreviation for the "quantity of matter" of the Second Law seems to date back only to the work of J. Bernoulli. See Jammer, *Concepts of Mass*, pp. 74, 78.

junction with 'quantity of' and the new terms supplied for these latter uses, the older terms began to disappear from technical discussions of the behavior of matter in motion. What Newton had done was to separate off the inertial and gravitational aspects of bodies and provide a sharp systematic definition of a concept which would encompass these. *Mass* was an aspect of *matter*, but from now on the problematics to which the two concepts responded drew further and further apart. Philosophers would continue to discuss predication, individuation, continuity and the like in terms of matter, but scientists would predict and explain motions in terms of mass. In no other context are the beginnings of the separation between philosophy and empirical science more clearly revealed. Having a separate term for the "scientific" aspects of matter served quite effectively to detach, little by little, the physicists' working concept, *mass*, from its conceptual ground in natural philosophy.

We have focused above for the most part on the origins of the notion of mass, Newton's most immediate contribution to the story of the concept of matter. But Newton also struggled with a quite different set of problems, bearing on the concept of materiality from another quarter. They had to do with the underlying explanatory structure of the new mechanics: how is action transmitted? in what does it originate? is resistance to change of motion itself a kind of force? how is gravity related to bodies? These questions could not be answered by the *Principia* itself, within its own terms of reference. But they could not be avoided (as Newton's critics were quick to insist) if mechanics was to constitute a system of *explanation*. Newton took up the challenge, and devoted much of his creative effort to it after the appearance of the *Principia*. What was at issue was, at bottom, the relation between matter and agency; in this respect too, the new mechanics was to transform the traditional views almost entirely. But the story of this development is such a complex one that it must be reserved for detailed treatment elsewhere,[46]

§9 *Conclusion*

Newton bequeathed a host of problems to his successors, many of which could be described as a probing of the conditions of materiality.[47] In the following century, physicists, and to some extent chemists and biologists too, continued the work of clarification that Newton had begun. It would

---

[46] In *Newton on Matter and Activity.*

[47] See the very useful collection of primary sources edited by M. Crosland, *The Science of Matter: An Historical Survey,* London, 1971.

not have seemed to them to be ontology, and it certainly was not undertaken with any such goal in mind. Yet the "problem of incorporeals" which so beset eighteenth-century science was in many ways an ontological one.[48] How was one to regard air, magnetism, light, heat, aether and all of those other "effluvia" that did not appear to fall into the simple category of *body*? They are clearly real because they have observable effects. But are they material? What are the proper criteria of materiality? inertial mass? solidity? extension? subjection to physical law? And if they are not material, how are they related to spirit, to the causal agencies that bring about action in the case of living (and especially human) beings? What was needed was a map of the material realm, with the interconnections clearly marked. It was not just that more experimental data were needed; more important was an agreement on the conceptual system in terms of which the data would be described. This could be brought about only indirectly, by attempting a variety of theories with quite different structurings of the physical world implicit in them. The theory that proved most fruitful in opening up the way to an exact account of the multitude of new, carefully planned observations would be accepted, and with it a certain understanding of the implications of materiality. The efforts that went to the solution of such questions as: ought the *vis inertiae* be regarded as a force? can space sustain force in the absence of matter? do not show in the theories themselves, because once the theories were formulated, the conceptual articulations had already been made.

The problem of deciding on the boundaries of materiality was not limited to physics. The eighteenth century saw the beginnings of systematic chemistry. Two directions seemed to be open to it. Should it build on Newton's corpuscles and forces? Was not this much too hypothetical and too remote from the complexities of chemical change? Would it not be better to concentrate on the quantitative laws of chemical combination, on parameters that could actually be measured without worrying about the atomic-level processes underlying them? Both ways were followed, but the latter proved the more effective in the long run. Hales, Cavendish, Black, Priestley and Lavoisier managed to isolate the common gases and prove their materiality. The status of heat and of fire proved far more troublesome to decide. The earlier assurance that heat was no more than a form of motion began to waver in the face of new evidence about the thermal effects of boiling, freezing and various chemical changes. Many of the leading chemists, Lavoisier and Dalton among them, postulated an imponderable fluid, caloric, which was conserved throughout all thermal changes and ex-

---

[48] See chapter 9, "The problem of incorporeals" in Toulmin and Goodfield, *op. cit.*

plained thermal expansions and contractions. An even odder entity, phlo-giston, was suggested by Stahl to be a constituent of combustible bodies and metals. Burning and calcination were thought to occur when phlogiston was driven off. The trouble was that these effects were known to involve a *gain*, not a loss, of weight so that phlogiston seemed to have "negative" weight; it did not, therefore, fall within the bounds of "materiality" as these had been set by Newton. Lavoisier argued that gain of oxygen, rather than loss of phlogiston, provided a more satisfactory model; he could not produce altogether conclusive evidence of this—Priestley, for one, remained unconvinced—but he did show that oxygen provided a far more consistent and fruitful material postulate than did phlogiston.

Meantime, the debate about how *body*, *force* and *void* ought to be related continued among the physicists.[49] Boscovich carried one line of Newtonian thought to its logical conclusion by representing the ultimate constituents of matter as point-centers of force rather than as impenetrable corpuscles. He thus, in effect, filled space with force, and collapsed all three categories into one. By identifying matter with force, he reversed the atomist and neo-Platonic first principle that had had such an influence on the earlier stages of mechanics, according to which matter is basically inert, requiring action to be imposed on it from the outside. The point-centers of force could be mathematically defined in such a way as to give rise to precisely the same physical effects as a solid extended atom would. The disagreement between the atomist and the central-force models was not, therefore, one that could necessarily be decided on the grounds of observation. It was (as Faraday was later to note) basically a question of choosing an ontology, a way of relating *matter* and *space*, so that physical theory as a whole would appear most consistent with itself, and the progress of theoretical specula-tion would be most effectively aided. Much of nineteenth-century physics was concerned with the enlargement of mechanics to include electricity, magnetism and light. It was only as the century ended that it became clear that this could not be done without discarding some of the basic assump-tions of Newtonian mechanics, and modifying almost (but not quite) be-yond recognition the concepts of *matter, space* and *time* from which physics had begun more than two thousand years before.

In this recent discussion, the term 'matter' played very little part. When we speak of the evolution of the concept of matter in eighteenth- and nine-teenth-century science, it is not as though there is a term 'matter' to guide us to where the relevant speculation was. Rather, it is a much more indirect affair of studying the part played by considerations of conservation, of

---

[49] See M. B. Hesse's essay elsewhere in this volume.

transfer of action, of response to force, and so forth, the contexts in other words where the matter metaphor had in the past influenced the direction of thought. Among philosophers, now for the first time a separate group from "scientists", the concept of matter played a much more explicit and easily determined role. The phenomenalism of Hume, the materialism of La Mettrie and d'Holbach, the objective idealism of Kant, each centered around a characteristic concept of matter. Later, the idealism of Hegel was followed by the dialectical materialism of Marx and a variety of evolutionary philosophies, both naturalist and idealist. In all of these, different ideas were advanced as to what constitutes something as "material", and how matter relates to consciousness, to time, to space. In the essays that follow, an attempt will be made to unravel these tangled strands, so that the reader may come more fully to appreciate the fundamental part that the concept-complex of *matter* has had in the genesis of contemporary thought.

PART ONE

From Matter to Mass

DESCARTES' CONCEPT OF MATTER

Richard J. Blackwell

Occasionally one encounters in the history of philosophy a theory which is significant primarily because it represents a "pure position". Such a view consists in responding to a fundamental philosophical question by opting for one of the irreducible alternative answers which seem to be presented to human thought. Along with this, one usually also finds a relentless pursuit of the option taken into all its logical implications. Although such an inquiry may prove fruitless and although the subsequent history of thought may see fit to reject it, the undertaking itself is a valuable contribution to the exploration of the possibilities which are open to human knowledge.

A case in point is Descartes' identification of matter and extension. Faced with the question, What characteristic distinguishes a material from an immaterial entity? Descartes points to what at first sight is a simple and obvious answer. An entity is material if it is extended. Certainly a special characteristic of the objects of the material universe—as distinct, for example, from ideas, volitions and virtues—is that the former are spatially extended while the latter are not. But Descartes means to say much more than this. Extension is not merely a reliable indicator which can be used to identify material, as distinct from immaterial, entities. Rather, extension alone constitutes the essence of material bodies. In other words, not only must a material being be extended but its materiality is its extension. Or to put it in more contemporary language: for Descartes, extension is both a necessary and a sufficient condition for the materiality of bodies.

This is certainly a direct and unequivocal position. And Descartes has performed a major service to the history of thought by developing this view in its full consequences. To Descartes the prospect of working out this program must have appeared very appealing. If the properties other than extension which are usually associated with matter can be reduced to or derived from extension itself, then it seems that the material universe lies open to a purely geometrical description. Physics will become applied geometry, and the ideal of clear and distinct ideas will triumph in that area of reality—the material universe—which seemed to be the most recalcitrant to Descartes.

In an attempt to evaluate the Cartesian notion of matter, we will direct our attention to three central questions: (1) precisely what does Descartes mean by his identification of matter and extension?, (2) how does matter so understood function for Descartes as a principle in the formation of the physical universe?, and (3) are the consequences of the identification of matter and extension compatible with a viable physics? For Descartes to win an affirmative verdict on the third question, he must be able to show in response to the first two questions that the properties which are usually associated with physical objects either are erroneously attributed to them or are reducible to extension. As we shall see, Descartes' formulation of his theory of matter is expressly designed to prove this point.

In this study our approach to the Cartesian doctrine of matter will be limited more or less to the confines of his physics. However, in dealing with any specific issue in Cartesian thought, one must of course keep in mind the structure of his philosophy as a whole. In such a highly systematic thinker as Descartes, the components of philosophy are woven into an intricate, unified fabric, and the various concepts involved depend upon each other for their full meaning and support. Hence in viewing Descartes' theory of matter from the vantage point of his physics, we do not mean to deny that there are other levels of approach open to Descartes to bring out the full implications of his position.

§1 *Extension, the Void and Material Substance*

Our first task, then, is to determine what Descartes means by his identification of matter and extension. A convenient way to approach this question is to indicate that at first sight the Cartesian position is inconsistent. The extension that Descartes refers to is clearly spatial, not temporal, extension.[1] Hence, matter is somehow identical with extension in length, breadth and depth, or, in other words, spatial volume. Now the latter notion seems to be indifferent to the alternatives of being either filled or empty. That is, both an actual physical body and an empty space seem to possess the characteristics of length, breadth and depth. If so, the original equation becomes expanded to a threefold identification of matter, ex-

---

[1] It seems that it would not be meaningful to talk about temporal extension as derivative from the properties of motion in Cartesian physics. Although Descartes clearly grants that "no motion occurs at an instant", he also adds that "God conserves motion precisely as it is at the moment of time in which He conserves it and independently of the motion which chanced to occur a little bit earlier" (*Principia Philosophiae*, Book II, art. 39). He seems to conceive of time as a series of discrete instants united externally through divine conservation, but not bearing any internal continuous relation to each other. As a result, Koyré argues that motion for Descartes is fundamentally a-temporal, as when one refers to "geometrical motion". On this see A. Koyré, *Études Galiléennes*, Paris, 1939, *3*, 170 ff.

tension and space (both filled and empty). Thus, if matter is extension and *only* extension, as Descartes seems to be saying, then he has no way of distinguishing an actual physical body from empty space. The inconsistency referred to above lies in the fact that Descartes is very emphatic in denying any validity at all to the notion of empty space or a void. An existing, extended void is impossible. As Descartes puts it, "As regards a vacuum in the philosophic sense of the word, i.e. a space in which there is no substance, it is evident that such cannot exist, because the extension of space or internal place, is not different from that of body".[2]

As we have said, this is only an apparent inconsistency on Descartes' part. But the reason why this is so leads us to the central points of Descartes' theory of matter as extension. If Descartes himself were to answer the problem we have just raised, he would ask us first to recall that what is "really real" for him in the physical world are material substances. Hence, the question of defining matter must be considered in the light of the broader question of the status and constitution of material substances. This is precisely how Descartes himself approaches the definition of matter in the *Principia Philosophiae*, and if we hope to recapture his meaning, we must follow the same approach.

For Descartes, substance is defined as ". . . a thing which so exists that it needs no other thing in order to exist".[3] Strictly speaking, only God is a substance according to this definition, but if we add the notion of an existence conserved by divine power, then we can also speak of finite substances. Our knowledge of such substances, however, is through their attributes rather than the bare fact of their existence.

> But yet substance cannot be first discovered merely from the fact that it is a thing that exists, for that fact alone is not observed by us. We may, however, easily discover it by means of any one of its attributes because it is a common notion that nothing is possessed of no attributes, properties or qualities. For this reason, when we perceive any attribute, we therefore conclude that some existing thing or substance to which it may be attributed, is necessarily present.[4]

This text is of fundamental importance in understanding Descartes' theory of matter. Applied to material substances, it means that extension is epistemologically prior to (material) substance, while substance is ontologically prior to extension. That is, we know material substance through the attribute of extension which constitutes its essence, while at the same time extension is dependent upon and implies the substantiality of ma-

---

2 *Principia Philosophiae*, Book II, art. 16; tr., Haldane-Ross, *1*, p. 262.
3 *Ibid.*, Book I, art. 52; Haldane-Ross, *1*, p. 239.
4 *Ibid.*, Book I, art. 52; Haldane-Ross, *1*, p. 240.

terial things as its ground. To put this in the language of the *Regulae*,[5] extension is a simple nature which is bonded to (*vinculum*) or necessarily implies the prior simple nature of substantiality. (The reverse implication, of course, does not hold, for otherwise Descartes would be forced to admit that all substances are material.) If this be granted, then it is easy to see how Descartes would answer the difficulty raised above. Extension as identical with matter is an attribute of a substance which it necessarily implies.[6] For an attribute must always be an attribute of some subject. Hence, wherever there is extension, there must always be a substantial subject of that extension. An empty space, or an existing extension with no subject, or a void in the philosophical sense of that term, is an impossibility. Or to put it in the words of our original objection to Descartes, spatial volume is not indifferent to the alternatives of being either filled or empty. It must be filled because extension necessarily implies substantiality. As Descartes has said in the text quoted above: ". . . when we perceive any attribute, we therefore conclude that some existing thing or substance to which it may be attributed, is necessarily present".

This argument obviously either stands or falls with the success or failure of Descartes' insistence that extension is an attribute.[7] Extension is always

---

[5] For the theory of simple natures cf. *Regulae ad Directionem Ingenii*, sec. VI.

[6] An important consequence of this view is that the spatial extension of the physical universe should then be infinite, or as Descartes prefers to say, indefinite, since infinity can be properly predicated only of God. The question of whether the material universe is finite or infinite in extension was hotly debated in Descartes' day. Because of the lack of a visible parallax of the fixed stars, the Copernican system required a vastly larger universe than the Aristotelian-Ptolemaic tradition demanded. Near the end of the sixteenth century, Giordano Bruno (*On the Infinite Universe and Worlds*) had argued for the infinity of space comprising a plurality of different worlds. In contrast, Descartes argued for the indefinite extension of space and against the doctrine of a plurality of worlds. Contrary to the long-standing Aristotelian view, the matter of the heavens and the earth must be the same in a physics in which matter is simply extension (*Principia Philosophiae*, Book II, art. 22). This was one of the many attacks on the Aristotelian position that a separate physics is required for terrestrial as distinct from celestial phenomena. For an excellent discussion of Descartes' role in this controversy over the infinity of physical space, cf. A. Koyré, *From the Closed World to the Infinite Universe*, New York, 1958, chaps. 4-5.

[7] A subsidiary argument in support of this point is contained in Descartes' explanation of rarefaction. When a body expands, its parts become separated, and the intermediate spaces are filled with other bodies. The example he gives is a sponge soaking up water. The space originally occupied by the sponge itself, as distinct from the water it absorbs, is no larger than it was previously. Hence the extension of a body does not itself change. However all its other properties do change: the body is now hard, now soft; now red, now brown; now round, now square; and so on. Hence, these other properties must be modes because they are subject to change, while

an "extension of" something, and what comes after the "of" is the substantiality of material things, which rules out the existence of the void. Moreover, this position seems to imply that extension and material substance are not synonymous expressions for Descartes. Extension constitutes the essence or materiality of a material substance, but its substantiality is due to other attributes. If not, then every substance would have to be a material substance. Descartes is quite well aware of this and clearly states that there are a plurality of attributes which constitute any one type of substance.[8] One of these attributes, extension in the case of material substances, is preeminent in our knowledge of things, but the other attributes are still necessary to constitute a real being. In short, although extension is the essence of a material substance, the further attributes of unity, duration and per se existence are needed to ground its substantiality.

This raises a considerable difficulty when one turns to the texts of Descartes. We have argued that extension and material substance are not identical but that the former is only one attribute among several in the structure of the latter. In other words, extension in some sense is only a part, and not the whole, of a material substance. But Descartes frequently says that extension and material substance differ only in thought.[9] What kind of a distinction is there, if any, between extension and material substance? One thing that can be agreed upon is that this would be an instance of what he calls a distinction of reason, which is defined as follows. "Finally the *distinction of reason* is between substance and some one of its attributes without which it is not possible that we should have a distinct knowledge of it, or between two such attributes of the same substance".[10] Extension as an attribute of material substance clearly satisfies this definition. But the name "distinction of reason" is unfortunately very ambiguous here. Does

---

the unchangeability of extension indicates that it is an attribute and not a mode (*Principia Philosophiae*, Book II, arts. 5–7). However, such an appeal to sensory evidence is not really conclusive in Cartesian philosophy. The central argument which he must use to prove that extension is an attribute, and the only attribute which constitutes the essence of material substance, is the appeal to the institutions of the mind, that is, what the mind is forced to think about when it considers the notion of material substance.

8 "And because in God any variableness is incomprehensible, we cannot ascribe to Him modes or qualities, but simply attributes. And even in created things that which never exists in them in any diverse way, like existence and duration in the existing and enduring thing, should not be called qualities or modes, but attributes" (*Principia Philosophiae*, Book I, art. 56; Haldane-Ross, *1*, p. 242). Once again the characteristic of unchangeability seems to distinguish an attribute from a mode.

9 *Le Monde*, Part VI; *Descartes: Selections*, New York, p. 321; *Principia Philosophiae*, Book II, arts. 8–12.

10 *Principia Philosophiae*, Book I, art. 62; Haldane-Ross, *1*, p. 245.

this mean a distinction formulated by the mind alone without a parallel distinction in things, as the name imples, or does Descartes refer here to a special case of a real distinction in things independently of the mind, that is, a distinction between a substance and its attributes, which exists whether we happen to be thinking about it or not? If the former, then all attributes of a substance must be ontologically identical. How then can there be different kinds of substance since some attributes, e.g. *per se* existence and unity, are properties of all substances? Spinoza's placing of all the Cartesian attributes, including extension, in the one divine substance follows this first meaning of Descartes' "distinction of reason". But it seems truer to the intentions of Descartes to adopt the second alternative and conclude that the expression "distinction of reason" is unfortunate and that Descartes means to refer to a special type of real difference in things, i.e. between a substance and its attributes. Some texts seem to support our interpretation; others do not. This is not as clear and distinct as one might hope.

Be that as it may, we are now in a position to evaluate more directly Descartes' theory of matter. The frequently quoted but often insufficiently analyzed expression "matter is extension" is equally ambiguous. If this phrase is interpreted to mean "extension is identical with the materiality of material substance", then this is an accurate account of what Descartes is saying. And this is equivalent to Descartes' much more common expression that extension is the essence of material substance. But if the phrase "matter is extension" is interpreted to mean "extension is identical with material substance", then Descartes has been misunderstood. The substantiality of material substance is constituted by factors other than extension. If this were not the case, Descartes would have no defense against (1) Spinoza's collapsing of all substances into the one divine substance and (2) the charge that he has no means of distinguishing a material body from empty space. There is a considerable difference between saying "extension is identical with material substance" and "extension is identical with the materiality of material substance". Descartes is clearly aware of this difference and wisely chooses the latter meaning. To say simply that for Descartes matter is extension is to say both too little and too much.

Up to this point we have considered only the relation between extension and material substance. As Descartes insists, extension and *only* extension constitutes the essence of material substance. What then becomes of the properties of material substance other than extension? In a philosophy in which extension is given an exclusive position of pre-eminence, all other properties of material things must somehow be subordinated to extension. How does Descartes bring about this reduction?

The first step is to make use of the old distinction between primary and secondary sense qualities, originally suggested by Democritus and re-introduced by Galileo. The secondary sense qualities, which the common man attributes to material bodies, are really creatures of the mind. As such they are removed from the problem at hand, for they are merely the furnishings of human sensory consciousness. This leaves only the primary sense qualities, perhaps better designated as quantities, as the properties of material things that need to be subordinated to extension. It is not to our purpose here to look into the involved epistemological difficulties and consequences implied in the use of the primary-secondary quality distinction.[11] Our point simply is that by limiting the problem to the consideration of the primary qualities only, Descartes is left with the far simpler question of reducing the quantitative aspects of material things to the basic attribute of extension. Although this program is itself faced with formidable difficulties, it is at least hopefully manageable in a physics which is structured as an applied geometry.

Descartes' position on the status of sense qualities is indicative of another important aspect of his theory of matter. Sensory evidence plays a very small role in the discussion of this question. The senses are needed to ascertain the existence, but not the essence, of the bodies in the physical universe. If God is not a deceiver, and if the overwhelming impression of sense experience is that a material world exists as independent from both the self and God, then sensory evidence does establish the existence of such a world. But that is all. The question of the nature of material bodies must be answered in terms of an intellectual intuition of the mind. This is the point of the famous wax example in the *Meditations*.[12] Knowing what material things are, as distinct from the bare fact of their existence, is not the business of the senses but of reason analyzing what the mind is forced to think about in considering the notion of material substance. In the light of this there can be little wonder that Descartes found the earlier doctrine of the subjectivity of the secondary sense qualities very congenial to his methodology. Matter is not to be defined in terms of the deliverances of sense experience but rather in terms of the intuitions of the mind.

It is this point which enables Descartes to go far beyond Galileo's view of sense qualities and which also provides him with the key to the reduction of all the primary qualities to extension. In the *Assayer*, while discussing whether or not heat is a property which actually resides in bodies, Galileo reasons:

---

[11] For a contemporary critique of this distinction and its consequences in regard to our understanding of the nature of the physical world, cf. A. N. Whitehead, *The Concept of Nature*, Ann Arbor, 1957, esp. chap. 2.

[12] *Meditations on First Philosophy*, II; Haldane-Ross, *1*, pp. 154-55.

Now I say that whenever I conceive any material or corporeal substance, I immediately feel the need to think of it as bounded, and as having this or that shape; as being large or small in relation to other things, and in some specific place at any given time; as being in motion or at rest; as touching or not touching some other body; and as being one in number, or few, or many. From these conditions I cannot separate such a substance by any stretch of my imagination. But that it must be white or red, bitter or sweet, noisy or silent, and of sweet or foul odor, my mind does not feel compelled to bring in as necessary accompaniments. Without the senses as our guides, reason or imagination unaided would probably never arrive at qualities like these.[13]

Now compare this with the following remarks from Descartes.

In this way [understanding alone] we shall ascertain that the nature of matter or of body in its universal aspect, does not consist in its being hard, or heavy, or coloured, or one that affects our senses in some other way, but solely in the fact that it is a substance extended in length, breadth, and depth. For as regards hardness we do not know anything of it by sense, excepting that the portions of the hard bodies resist the motion of our hands when they come in contact with them; but if, whenever we moved our hands in some direction, all the bodies in that part retreated with the same velocity as our hands approached them, we should never feel hardness; and yet we have no reason to believe that the bodies which recede in this way on this account lose what makes them bodies. It follows from this that the nature of body does not consist in hardness. The same reason shows us that weight, colour, and all the other qualities of the kind that is perceived in corporeal matter, may be taken from it, it remaining meanwhile entire: it thus follows that the nature of body depends on none of these.[14]

The point of Galileo's argument is to prove that the primary, but not the secondary, qualities actually reside in physical bodies. Ultimately Descartes agrees on this, but he is more consistent and carries the logic of the argument to its full conclusions. If reason alone, unaided by the senses, is to settle the question of the constitution of matter, then no sensory evidence of any kind is really relevant to the discussion. Galileo does not seem to be fully aware of this. But Descartes does not miss the point. The question is, What am I forced to think about when I consider the notion of a material substance? Galileo's answer is that he must think of a whole group of primary qualities, apparently all on an equal footing. Descartes' answer is that he is forced to think only of extension. The remaining primary qualities are of a secondary status as modes of extension and do not form part of the essence of a material substance. By rejecting sensory evidence as irrelevant and by appealing instead to the intuitions of reason, Descartes

---

[13] *Discoveries and Opinions of Galileo,* tr. S. Drake, New York, 1957, p. 274.
[14] *Principia Philosophiae,* Book II, art. 4; Haldane-Ross, *1,* pp. 255–56.

feels that he has established a hierarchy among the primary qualities, all of which really reside in physical bodies but not all on the same level. Galileo would find his argument carried to an unexpected conclusion.

The second stage, then, in Descartes' reduction of all the properties of material bodies to extension is to show how the primary qualities other than extension are really just modes of extension. To say the least, this is a difficult undertaking, and much of the later criticisms of Cartesian physics were brought to bear on the details of this reduction. In what sense are motion, rest, plurality, weight, resistance, shape, figure and a host of other apparently real properties of physical bodies explainable as modes of extension? If Descartes can successfully answer this question, then his project of formulating physics in the geometrical mode will be accomplished. The doctrine that matter is extension, or better that extension and only extension is the essence of material substance, stands in the balance.

§2 *Matter as a Cosmic Principle*

"Give me extension and motion, and I will construct the world." In this famous remark from *Le Monde*, Descartes indicates, among other things, that he realizes that extension alone is inadequate as a principle for the explanation of the formation of the physical universe. Extension pure and simple is inert, lifeless, undifferentiated, the same at all places and at all times. It does not contain within itself any factors capable of explaining the tremendous variety of different types of material bodies, their interrelations and interactions upon each other, or the dynamism and novelty which emerge in the course of the history of the universe. He must introduce a second factor, motion, as an original divine given needed for the structuring of the physical world. Although motion, strictly speaking, is a mode of extension for Descartes, it functions in Cartesian physics as a principle co-equal to extension in the process of cosmic genesis. Descartes argues that if he is granted these two principles, he can explain all the aspects of the physical world, including all the other primary qualities. Hence in this way all the properties which actually reside in physical bodies, including even motion itself, are ultimately reduced to extension. The details of this argument are the main supports which Descartes offers for his claim that extension is an attribute and the only attribute which is needed to constitute the essence of material substance. As we have seen, the Cartesian doctrine that matter is extension depends upon Descartes' showing that extension is an attribute rather than a mode of material substance and that all other physical properties of bodies are modes, not attributes. This program is carried out in his theory of cosmic genesis.

The central point in this argument is Descartes' attempt to define motion as a mode of extension. If we turn to the texts of Descartes, we find that the type of motion admitted into Cartesian physics is a specially formulated doctrine designed to fit the needs of a geometrically embodied universe. Koyré has shown that this motion is a geometrical idealization of motion, that is, motion for Descartes is the trajectory described by a moving body rather than a progressive, ongoing process.[15] In other words, motion is a static representation of the path traversed and is not understood as a dynamic process in its own right. And this is precisely what is to be expected if motion is to be defined as a mode of extension. Since this point is so fundamental, let us allow Descartes to speak for himself. The definition of motion is presented in the *Principia*:

> . . . we may say, in order to attribute a determinate nature to it, that it [motion] is the *transference of one part of matter or one body from the vicinity of those bodies that are in immediate contact with it, and which we regard as in repose, into the vicinity of others*. By *one body* or by a *part of matter* I understand all that which is transported together, although it may be composed of many parts which in themselves have other motions. And I say that it is the *transportation* and not either the force or the action which transports, in order to show that the motion is always in the mobile thing, not in that which moves; for these two do not seem to me to be accurately enough distinguished. Further I understand that it is a mode of the mobile thing and not a substance, just as figure is a mode of the figured thing, and repose of that which is at rest. [Italics in the original text.][16]

This definition contains many of the commitments which Descartes needs to make in order to generate the physical universe and to carry out his universal reduction of all physical properties to extension. First of all the only type of motion to be granted is local motion.[17] Changes of substance, quality and quantity, to use the remaining Aristotelian categories, must be looked upon as special cases of local motion. In the context of the seventeenth-century search for mechanical explanations, however, this point would receive rather little challenge from Descartes' contemporaries. Local motion itself is understood as the translation from one neighborhood of contiguous bodies to another. In a *plenum* universe, all bodies must be

---

[15] A. Koyré, *Études Galiléennes*, Paris, 1939, *3*, pp. 162 ff.

[16] *Principia Philosophiae*, Book II, art. 25; Haldane-Ross, *1*, pp. 266. For a detailed analysis of Descartes' theory of motion, cf. M. Gueroult, "Métaphysique et physique de la force chez Descartes et chez Malebranche", *Revue Metaphys. Morale*, *59*, 1954, 1–37, 113–34.

[17] Just prior to his definition of motion, Descartes explicitly states that he can conceive of no motion other than local motion. See *Principia Philosophiae*, Book II, art. 24.

contiguous, and when a body moves, the vacated place must be refilled by other bodies, as when a fish swims through water, to use Descartes' own example. As a consequence, all actual physical motion must be circular, and thus the definition of motion already contains the seeds of the Cartesian theory of the vortices. Moreover, the definition also implies that the designation of which body is moving is a relative consideration. Depending on what we "regard as in repose", we can say either that A moves in relation to B or that B moves in relation to A. Descartes will find this to be a very convenient safeguard in the dispute between Galileo and the Church over whether the sun or the earth is in motion. Each is at rest in its own vortex, and yet each is in motion in respect to others.

Unfortunately, however, Descartes does not make full use of this doctrine of the relativity of local motion when he considers other physical questions. For example, his explanation of the impact between two colliding bodies might not have been so clearly erroneous if he had argued that either body can be considered to be in motion in respect to the other.[18] Furthermore the definition quoted above also implies that the unity and individuality of a material body is constituted by its parts sharing in a unity of motion. This again is a consequence of Descartes' theory of extension. In itself extension is continuous and contains no actual parts at all. If God had not applied motion to extension at the moment of creation, then the universe would consist of one large, inert material substance. The breaking up of this original extension into ever-divisible parts (hence Descartes' denial of atomism in the sense of indivisible units of matter)[19] is the result of motion. Descartes cannot even have a plurality of distinct bodies in the world by appealing to extension alone. Motion working on

---

[18] The seven rules of impact appear in *Principia Philosophiae*, Book II, art. 46–52, as corollaries to Descartes' third law of nature. All but the first of these seven rules are quite clearly contradicted by factual evidence. Descartes' failure to use this notion of the relativity of motion in this context is not the sole source of the difficulties. Consistent with his doctrine that matter is extension, he understands the "quantity of matter" of a body as equal to its spatial volume. Later in Newton this same concept is defined as volume times density, or mass. Moreover Descartes' "quantity of motion" is equal to volume times scalar speed in contrast to Newton's definition of "quantity of motion" or momentum as mass times vector velocity. The fundamental problem for Descartes was his inability to conceive of the concept of mass in a physics in which matter is extension. The transition from Cartesian to Newtonian physics is in large part the story of the transition from matter as extension to matter as mass. For an analysis of the seven rules of impact, cf. D. Dubarle, O.P., "Remarques sur les régles du choc chez Descartes", *Cartesio*, Milan, 1937, pp. 325–34; P. Mouy, *Le développement de la physique Cartésienne*, Paris, 1934, pp. 22–23; D. M. Clarke "The impact rules of Descartes' physics", *Isis*, *68*, 1977, 55–66.

[19] *Principia Philosophiae*, Book II, art 20.

extension is the principle of diversification and unity of individual physical bodies.[20] Here we have a clear case of one of the traditional primary qualities—plurality—reduced to extension and motion.

Is is also interesting to note in the text quoted above that Descartes attempts to define motion without any reference to the external force or action which produces it. Motion is a translation which resides solely in the mobile object itself. If we emphasize the point that motion is a *mode* of the moved body, motion seems to be an internally possessed property of the body, very similar, for example, to its size or figure. So understood, motion does not seem to be an essentially relational reality. It is very difficult to see how this is consistent with the relativity of motion which is also expressed in the definition unless one is willing to grant that at least some of the Cartesian modes are relational entities, a point which Descartes does not explicitly develop here. On the other hand Descartes does not deny, of course, that an external force or action is needed for the production of motion. But what is this force? Originally it is God's creative and conserving power. If it is granted that there is no genuine secondary causality in the Cartesian physical world, a point which is certainly open to debate, then the bodies in the universe are purely passive and Cartesian matter is totally inert and inactive. It is this aspect of the Cartesian theory of matter which was especially objectionable to Leibniz.

The central point which emerges from all this is that motion has been expressly defined so that it can serve its role as the chief instrument in the genesis and history of the universe. The origin of the universe is seen as the result of the divine act of the creation of extension and motion and the imposition of the latter upon the former. In this way, motion is constituted as the first and primary mode of extension. And the interaction between these two principles progressively builds up the structure of the universe. It is not our intention here to follow through this entire analysis, but the description of the first stages of this genesis may help to illustrate how extension and motion are used by Descartes as cosmic principles. The first result of the mixing of matter and motion is that the originally given homogeneous extension is broken up into a vast number of parts having all conceivable shapes, sizes and velocities. In a *plenum* universe, these parts of matter must be all contiguous to one another, and hence their respective motions bring about innumerable collisions. The largest bodies are impervious to further division and come to rotate around their respective vortices at various distances from the center. The medium-sized bodies are gradually worn down as fine particles of matter are broken off. This

---

[20] *Ibid.*, Book II, art. 23.

.subtle matter, the smallest, fastest and most easily divisible of all, gravitates to the center of the various vortices to form luminous bodies. The remaining matter constitutes the *plenum* of the heavens.

From this results Descartes' famous doctrine of the three types of matter. The first element, consisting of the smallest and fastest-moving particles, forms the sun and the stars at the center of the vortices. The second element, comprising larger and slower-moving matter, fills interstellar space. The third element, which contains the largest, slowest and least divisible matter, constitutes the earth, the planets and the comets. The first element is luminous, the second translucent, the third opaque. From this, Descartes moves to progressively more specific physical phenomena: the motion of the planets and comets, sunspots, the transmission of light, the structure of the earth, chemical reactions of matter, heat, magnetism and so on.[21] As we have said, we do not intend to go into the details of these explanations. The main point of interest to us is that the various theories worked out for these problems involve fundamentally a successive re-application of the principles of extension and motion. And Descartes is willing to grant, as he moves further down the ladder to increasingly more specific physical phenomena, that his projected method of explanation grows more difficult to realize.[22] The need to appeal to experimentation and sensory evidence becomes more pronounced. There are several reasons for this. The actual universe before us is only one of an infinite number of worlds that God could have freely chosen to create. The principles of philosophic explanation are too general for us to clearly see their deductive application to very specific phenomena. The tremendous variety and detail in the physical world seem to escape from the limited powers of the human mind and are apprehended only by an appeal to carefully controlled sensory observation and experimentation. One is reminded here of Galileo's distinction between intensive and extensive knowledge.[23] What the human mind can know is grasped with as full a degree of certitude as is found in the divine mind itself. The difference is that God knows everything in this plenary fashion, while this is prevented by the finitude of the human mind. The point is that if the Cartesian God were a divine physicist, everything would lie open to the divine mind in full clarity and certitude. And the principles which such a physicist would find adequate would be matter and motion.

At this point it should be clear how Descartes goes about reducing all the

---

[21] The whole of *Le Monde* and Book III of the *Principia Philosophiae* are devoted to working out the details of this theory of cosmic genesis.

[22] *Discourse on Method*, Part VI.

[23] Galileo Galilei, *Dialogues concerning the Two Chief World Systems*, tr. S. Drake, Berkeley, 1962, p. 103.

primary qualities, which actually reside in bodies, to extension. Such properties are the result of various types of matter experiencing various types of motion. And motion has already been reduced to extension as one of its modes, indeed its primary mode. Space does not permit us to discuss Descartes' treatment of all the primary qualities from this point of view, but perhaps if we look at one such analysis the general line of argumentation will be illustrated. A central point of importance in the origins of modern science is the problem of understanding the nature of gravity. Descartes presents his definition of gravity in the context of the effects of the relatively fast-moving celestial matter. He argues that a drop of a liquid is made round by the innumerable impacts of celestial matter from all sides. He then continues:

> The force of gravity is not much different than this third action of celestial matter. For just as these celestial particles, moving indiscriminately in all directions by their own motion, drive all the particles of a drop of liquid toward its center and thus make it round, likewise by this same motion they impel all the parts of the earth toward its center when they collide with the bulk of the earth and are deflected. And in this consists the weight of terrestrial bodies.[24]

There is no notion here of mass as a distinct, possessed property of material bodies, or indeed of the now traditional distinction between weight and mass. Extension is not mass, and weight must be explained in other terms. Gravity is the result of the types of matter and comparative velocities of motion in the Cartesian universe. A body is lighter or heavier depending on its type of matter and motion and also the types of matter and motion which fill the spatial region in its general vicinity. And since the types of matter in turn are not discriminated by extension itself, but by the motion imposed on extension, in the last analysis gravity for Descartes is a function of the comparative velocities of motion.

A bold theory to be sure—but one which is quite consistent with the principles of Cartesian physics. This illustration also brings out another important feature of Descartes' explanation of the physical universe. The theory is formulated on the basis of only two principles: extension and motion. But when one looks into the details of the Cartesian physics, one finds a very unequal division of labor between these two principles. Motion is forced to do yeoman's service for extension. Consider again the question of defining gravity. Extension alone does not bear within itself any actual

---

[24] *Principia Philosophiae*, Book IV, art. 20. For a slightly different account of gravity, see *Le Monde*, XI. However, in both versions gravity is explained in terms of the different velocities possessed by the different types of matter in the Cartesian universe.

division of parts, much less any further structural differences which might account for the physical differences of weight. How would two volumes of extension alone be any different as regards weight? Because one is larger than the other? Hardly. Irrespective of size, no element of spatial volume pure and simple can be meaningfully designated as either heavy or ·light. It is rather the motion of the various parts of matter to which Descartes must turn. And this is true not only of his analysis of gravity. The Cartesian explanation of all the primary qualities, including even shape and size, must appeal more to the effects of motion than to extension. For as we have seen, there cannot even be a plurality of material bodies in the Cartesian universe without the assistance of motion. This unequal division of labor is itself an indictment of the Cartesian enterprise. Extension simply does not contain within itself the characteristics which are needed to constitute the complexities and richness of material being.

§3 *Concluding Remarks*

During the latter half of the seventeenth century, Cartesian physics won a rather widespread acceptance. For the first time the forces generating the revolution in science had succeeded in producing a thoroughly systematized physics which was every bit as comprehensive and total in its perspective as the old Aristotelian system which had been rejected. Others must have been attracted to the Cartesian physics bcause of its uncompromising mechanism in which the age placed such high hopes. Still others were well disposed to the new physics because of their prior convictions of the values of the Cartesian epistemology and metaphysics, which provided so much of the background for Descartes' teachings on the physical world. There could be no doubt that the Cartesian physics was a system which no serious investigator in this area could afford to overlook.

But as the century drew to a close, the attacks against Descartes' physics mounted in intensity, largely as a result of the promises offered by the Newtonian synthesis. This, of course, is an extremely involved development, and much more than the notion of matter as extension is involved. But the overall effect is unmistakable. Matter as extension was replaced by matter as mass. Descartes' "quantity of motion" (volume times scalar speed) was replaced by Newton's "quantity of motion" (mass times velocity, or momentum). Although Newton perhaps did not define the nature of matter as explicitly as did Descartes, he was much more successful in formulating a quantitative description of material bodies in terms of the measurement of mass. The entire problematic was undergoing a major shift of approach. The question was no longer the philosophic meaning of matter, but rather the scientific measurement of mass. The new tools forged by Newton successfully solved many fundamental problems and conveniently sidestepped others. During the first few decades of the eighteenth century the center of intellectual attention had clearly shifted to the new problematic introduced by Newton.

But perhaps the most pertinent criticisms of the Cartesian doctrine of matter were those of Leibniz. He mounted a two-pronged attack directly against the identification of matter and extension. First of all, he argued that extension is not an ultimate attribute of material substance. Rather, extension is a derivative property of the material universe arising from the plurality, continuity and co-existence of substances. As such, extension cannot be the defining attribute of material substance, since it is further analyzable into more fundamental factors. As Leibniz says: "Extension itself is for me an attribute which results from a plurality of co-existing and constantly interdependent substances."[25] For Leibniz, it is force, not extension, which constitutes the essence of substance. This is the point behind his second objection. Extension alone is unable to account for the natural inertia of bodies, that is, their resistance to motion. As evidence Leibniz points to the Cartesian laws of impact, which, as he says, are irreconcilable with experience. He then continues:

> I still agree that naturally every body is extended and that there is no extension without body. None the less we must not confound the notions of place, space, or of pure extension with the notion of substance which, besides extension, includes resistance, that is to say, action and passivity.[26]

In this text, Leibniz has hit upon the Achilles' heel of the Cartesian notion of matter. If extension and only extension constitutes the essence of material substance, then matter must be totally inert. It can do nothing, and nothing can happen to it. For Leibniz, it is inconceivable how motion could arise from, or even be imposed upon, such an unreceptive subject. And it is this latter point which is expressly required by Descartes for the genesis of the universe. In short, if matter is not open to the initiation and reception of motion, then such a matter must fail in giving an account of the physical world. A much more dynamic notion of matter is in order.

Even if we step back from the details of Cartesian physics, the force of Leibniz' second objection cannot be avoided. By this I mean that one might

---

[25] *Leibniz Selections*, ed. P. P. Wiener, New York, 1951, p. 166.

[26] *Ibid.*, p. 102. For an overall account of Leibniz' critique of Cartesianism and the role that this played in the development of Leibniz' own philosophical position, see Yvon Belaval, *Leibniz critique de Descartes*, Paris, 1960. See also John W. Nason, "Leibniz's attack on the Cartesian doctrine of extension", *Journ. Hist. Ideas*, 7, 1946, 447–83.

conceivably work out a physics which identifies matter and extension otherwise than did Descartes. For example, the metaphysical language of substances, attributes and modes might be dropped and be replaced by other
categories. And instead of trying to reduce everything to extension through
the medium of motion as the primary mode, one might try to bring this
reduction about in some other way. Still, if matter is extension and ultimately only extension, and if extension is understood in the usual sense
of a three-dimensional spatial spread, then such a matter must be inert. It
cannot contain within itself the structures and the properties which are
needed to explain the variety and especially the dynamic character of the
world in which we live. Descartes himself struggled valiantly with this
problem. But the dual status of motion as a mode of extended substance
and as a functionally co-equal principle with extension in the genesis of
the universe is a sign of his uneasiness. The problem still remains. Matter
as extension is intrinsically inert. Unless this objection can be answered,
it is not possible to build a viable physics on the principle that matter is
extension.[27]

---

[27] Further studies of the Cartesian doctrine on matter not referred to earlier in this
paper include: E. Bréhier, "Matière cartésienne et création", *Revue Metaphy. Morale*,
44, 1937, 21–34; J. Laporte, "La connaissance de l'étendue chez Descartes", *Revue
Philos France et de l'Etranger, 123*, 1937, 257–89; B. M. Laing, "Descartes on Material
Things", *16*, 1941, 398–411; J. Prost, *Essai sur l'atomisme et l'occasionalisme dans la
philosophie cartésienne*, Paris, 1907; J. Moreau, "La réalite de l'étendue chez Descartes",
Études Philos., 5, 1950, 185–200; A. Franceschi, "El concepto de 'materia sutil' en Descartes", *Descartes*, Homenaje en el tercer centenario, no. 40, Buenos Aires, 2, pp.
11–41.

MATTER IN

SEVENTEENTH CENTURY SCIENCE

Marie Boas Hall

§1 *Introduction*

The new theories of matter promulgated by seventeenth century scientists were many, varied and complex; they all shared a profound conviction that the previously accepted Aristotelian view of matter was philosophically false and scientifically sterile. Natural philosophers sought for new concepts that could aid them in their attempts to understand nature. Although in the end there never was any firm agreement as to what, exactly, matter was, there was universal agreement about what a theory of matter ought to be able to do in explaining the world, and in what terms this explanation should be expressed. The mechanical philosophy gradually came to be an integral part of seventeenth century scientific thought, susceptible of many interpretations but widely agreed to be a necessary ingredient in any scientific theory or explanation.

Early in the seventeenth century it became apparent that, in fact, there must be more involved in a useful theory of matter than a mere definition of what the stuff which made up the physical world might be. There was indeed much useful discussion of the construction of matter in terms not entirely alien to pre-Socratic thought, and a universal conclusion that matter was not continuous, but physically discontinuous; that there was no *Aristotelian* plenum, though there might be no true vacuum; and that the world was made up of very small, discrete, individual bodies. Had this been all that was required there would be little to discuss here, for the conclusion is not difficult to reach or to understand, nor need it alter the pattern of one's general scientific thought, as many early proponents of particulate theories (Daniel Sennert[1] for example) unconsciously or consciously demonstrated. But in fact such new theories of matter only assumed scientific importance when use was made of the particles they envisaged in trying to explain natural phenomena and when the Aristotelian theory of forms (developed

---

[1] Sennert (1572–1637) in *De Chymicorum cum Aristotelicis et Galenicis Consensu ac Dissensu*, (1619).

over the centuries) was replaced by the mechanical philosophy. For thereby the properties of bodies were subjected to a new and rigid examination which utterly changed men's attitude towards the external world and profoundly affected the development of science. Though it was the philosophers who gave this point of view real sophistication, it was the scientists who introduced it into general discussion and proved its usefulness.

§2 *The Rise and Fall of Ancient Atomism*

It is tempting to assume that scientists, rebelling against accepted Aristotelian doctrines, would naturally turn to anti-Aristotelian notions already in existence: the atomic theories of Graeco-Roman antiquity. In fact, this did happen—had already happened before the seventeenth century—but it provided no real solution to the problems that theories of matter were now to be called upon to face. Democritus, Epicurus and Lucretius were much read, as innumerable editions of *De rerûm natura* testify, but they had relatively little to offer seventeenth century natural philosophers, whatever they offered thinkers like Bruno or Gassendi. Partly because every seventeenth century scientist had read Aristotle, who was the only real source for Democritean atomism, the arguments against ancient atomism prejudiced men in advance. Added to this were the repeated discussions of many generations' disputing and redisputing about minima, least particles, infinite divisibility and so on, as well as the incisive discussions by mathematicians, all of which tended to show that the concept of an indivisible atom was very nearly untenable.

More important was the unfortunate fact that ancient atomic theories were scientifically out of date. Democritus's views on the physical universe were more primitive than those of Aristotle; Epicurus was no scientist and did not even understand the nature of sound; Lucretius had not attempted to subsume later Greek physics in his cosmology. Something could be salvaged from the views of Hero of Alexandria and the Alexandrian school of physicists and physicians on whom later Greek commentators on Aristotle had leaned heavily, but this was more suggestive than constructive and provided no complete system. No wonder then that the works and even the names of those who tried to rescue ancient atomism and transfer it unchanged to the seventeenth century world have an unfamiliar ring. Nicholas Hill's *Philosophia Epicurea* (1601), Sebastian Basso's *Philosophia naturalis adversus Aristotelem* (1621); J. C. Magnen's *Democritus reviviscens sive de atomis* (1646) all are deservedly forgotten, and merely served to prove that neither Epicurus nor Plato nor Democritus could live again in the seventeenth century.[2] Even Gassendi's great work on Epicureanism is scientifically a magnificent failure which survived because of its literary and philosophic influence. No scientist could take seriously Gassendi's elaborate, turgid and prolix attempt to introduce Democritean atoms into the physics of the mid-seventeenth century—though his success in Christianizing atomist cosmology deprived anti-atomists of one of their most fatal weapons. It is impossible to find any real influence of Gassendi's work on later scientists who already had other, more cogent theories of matter to choose from; for it must not be forgotten that Gassendi's *Philosophiae Epicuri Syntagma* of 1649 was only a small collection of atomic texts, and the real doctrine of Gassendi's interpretation had to wait for posthumous publication (as *Syntagma philosophicum*) until 1658, too late to be of importance to science.[3]

The reason why atomic theories of antiquity found themselves too out of date for acceptance was not only because of the positive advance of physical science, though this was naturally important, since it is difficult for a scientist to take seriously a theory propounded by someone ignorant of what to him are "simple facts". Sixteenth century science had, as it happens, not added many new discoveries which needed to be interpreted by a theory of matter; on the other hand, the experimental and mystic sciences of natural magic and alchemy had added many new phenomena, and scientists were increasingly inclined to feel that science should have something to say in these matters. Natural magic included magnetism, which Gilbert investigated in an experimental fashion, though he was unable to offer any useful or interesting explanation of the origin of the phenomena he discovered and extended. The use of lenses in telescopes and microscopes outdid the illusions of the conjurer; the suction pump was as mysterious in its operation as any pneumatic toy described in della Porta's *Natural Magick*; surely these lay in the province of natural philosophy, which ought to be able to explain their origin. Alchemy, rapidly becoming either technological, or medical

---

[2] Of course, the names, and even the existence, of ancient atomists were frequently invoked to give sanction to contemporary acceptance of scientific corpuscularianism, and "Moschus the Phoenician" (mentioned by Strabo) became Moses to give theological support. But this kind of reasoning was commoner to philosophers and divines than to scientists. (Cf., e.g., Cudworth's *True Intellectual System of the Universe*). Walter Charlton's *Physiologia Epicuro-Gassendo-Charltoniana* (1654) was Charlton's own work, much modified from Epicureanism by adjustment to contemporary science. Charlton was a physician, and later F.R.S.

[3] The publication of Bernier's French abridgement in 1675 introduced Gassendi's ideas to the literary and general public; it could not reach the scientist. Even Robert Boyle, still a young man in 1658, never read Gassendi's work in its entirety; he refers several times to Gassendi's *Syntagma*, but always as a "little" work. This could only mean that he knew the 1649 work, not that of 1658.

chemistry, offered a host of mysterious phenomena which also needed explanation: some substances when mixed together gave off heat; certain salts added to water produced cold; some colorless solutions turned from blue to either green or red, others to yellow or white when appropriate colorless substances were added; some liquids mixed to produce solids; some solids dissolved in liquids and permanently lost their solidity; shapes, odors, tastes, colors, solids, liquids, heat, cold, came and went in bewildering confusion. The philosophically minded chemist could offer no explanation but that of forms: through the chemical art he was able to impress or withdraw forms and qualities to any extent. He did not yet quite know how this was done; but one day the secret of complete command would appear. Then the forms of yellowness, density, malleability, incorruptibleness, perfection and so on would be mastered by the chemist, and then it would be easy to impress them on a base metal and prepare true gold. Though the natural philosopher could hardly concern himself with the dream, the properties of heat, cold, color, fluidity and so on were real, and the natural philosopher could not ignore the need to include them in his concept of the physical world.

Ancient atomism could not help very much. It could explain density, solidity, transparency and coherence by arguing that substances varied in the amount of vacuum between the atoms,[4] but even so it was incapable of coping with the host of forms and qualities now forcing their attention on the seventeenth century natural philosopher. A new approach was needed: and this was supplied by a new kind of theory of matter, at once non-atomic and anti-Aristotelian, which offered immense possibilities for scientific explanation and investigation. It derives from ancient atomism in a very limited sense only, for the actual concept *atom,* is irrelevant to its development. Yet ancient atomism had some influence; inadequate though it was to deal with seventeenth century science, it appeared to be closer to the truth than the Aristotelian way. Ancient atomists had "explained" the observed properties of matter by attributing appropriate characteristics to individual atoms, which was very much simpler than assuming that properties arose from extraneous "forms". Its disadvantage was that it merely pushed the explanation back a stage, from gross matter to its component atoms. (And indeed the ancient atomists had never gone so far as to consider general properties like heat, dryness or color). The atomic theories of antiquity suggested to seventeenth century thinkers that matter might have certain simple and fundamental characteristics which would account for a large number of observed phenomena. That is, every macroscopic property might not require an individual microscopic cause; but variations in some state or configura-

---

[4] This was particularly stressed by Hero, in the preface to his *Pneumatica.*

tion of the small parts of matter might produce different effects in gross matter.[5] More specifically, the new theories were to suggest that not only were the size and shape of the component units of matter relevant, but their *motion* might play an important—even a critical—role. It should be noted that in this approach motion was no longer related to form: velocity, impetus, acceleration, momentum, inertia were all totally divorced from the concepts of fourteenth century mathematical physics and were regarded as states subject to mathematical and quantitative treatment and requiring no analysis of their nature.

§3 *Matter and Motion*

This typically seventeenth century attention to dynamics appears almost simultaneously about 1620 in the work of three very different men: an English lawyer, an Italian scientist and a Dutch schoolmaster, Bacon, Galileo and Isaac Beeckmann. Beeckmann's theory, developed mainly between 1616 and 1620, comes closest to ancient theories, for he believed in real atoms. In fact he asserted the existence of four kinds or shapes of atoms, one corresponding to each of the four elements. The essences or properties of substances derived, he held, directly from the atoms. But motion, as well as shape and size of the atoms determined the properties of the component body. Motion, shape and size were to him equally important and all must be considered for a satisfactory explanation. For example, fire particles, small, subtle, pervasive and most apt for motion, caused bodies to become hot by setting the component atoms in motion; and cold was mere absence of heat.[6] Though Beeckmann's theory is now known only through his diary, his influence upon Descartes is obvious, and well known.

In 1620 appeared Bacon's *Novum Organum*, a work which has often baffled philosophers by the conflict between its expressed aim and its achievement. Bacon's persistent attempt to illustrate his inductive method by its application to what he called "the discovery of forms" has aroused much criticism, not least from his nineteenth century editors. Seventeenth century English scientists found Bacon's discussion both plain and suggestive, because they found it easy to divest it of its incrustation of logic and consider it only in relation to theory of matter. Unlike Beeckmann, Bacon

---

[5] This suggests a certain indefiniteness in explanation which was, indeed, characteristic of seventeenth century mechanical philosophies. It was an inevitable result of the method of explanation employed, but one worth risking for the sake of manifold advantages.

[6] *Journal*, ed. Cornelis de Waard, 3 vols., La Haye, 1939–45, *1*, 25, 152–53, 216; *2*, 96, 198; *3*, 138 *et passim*.

was not an atomist; on the contrary, he thought men could waste as much time mentally analyzing matter into atoms as they could in spinning webs of metaphysics.[7] Yet the atomic point of view was, he thought, far superior to the Aristotelian doctrine of forms because it was more useful.[8] It fell short to chiefly not repudiating the notion that matter was inert; for motion, so he believed, was the clue to the understanding of forms. It may be worth repeating here as a reminder Bacon's quasi-legal definition:

> For though in nature nothing really exists besides individual bodies, performing pure individual acts according to a fixed law, yet in philosophy this very law, and the investigation, discovery, and explanation of it, is the foundation as well of knowledge as of operation. And it is this law, with its clauses, that I mean when I speak of *Forms*.[9]

Surely Bacon was trying, clumsily, to express the need for general principles to explain the properties of bodies; and indeed he managed to be at once more general and simpler than Beeckmann. Bacon's theory needed no atoms or special kinds of matter; only particles—literally small bits of matter—and motion. Bacon stressed motion even more than did Beeckmann, insisting that it was the real key to the understanding of the properties of matter. His definition of heat, almost too well known to quote, amply illustrates this: "Heat is a motion, expansive, restrained, and acting in its strife upon the smaller particles of bodies."[10] Less familiar is his prediction that many other properties, including specifically color, whiteness and chemical action, would also prove to be the result of the motion of "the smaller particles of bodies", a guess that was to prove influential later.

It is surprising to find Bacon so deeply influenced by contemporary mechanical thought; he is indeed the only early protagonist of the mechanical philosophy who had not worked on dynamical problems. It is, on the other hand, only to be expected that Galileo should have considered the role of motion in the small parts of matter, as well as in gross matter. More especially since he was well read in the classics of ancient atomism and seems to have adopted the theory espoused by Hero of Alexandria as a basis. (As, for example, in his discussion of cohesion in the First Day of *The Two New Sciences*, and its application to the problem of the strange limitation on the ability of a suction pump to raise water). Galileo's most complete and important discussion—on heat—arose from the exigencies of polemic in *The*

---

[7] *Novum Organum*, Book I, Aph. lxvi; cf. Book II, aph. viii.
[8] Cf. *Novum Organum*, Book I, aph. lxiii; also *Cogitationes de natura rerum*, and the Fable of Cupid.
[9] *Novum Organum*, Book II, aph. ii.
[10] *Ibid*, aph. xx.

Assayer (*Il Saggiatore*).[11] He wished to give a philosophic basis to his contention that comets (even if they had been solid bodies, which he thought they were not) could no more grow hot as a result of rapid motion through the atmosphere than an egg could be cooked by anyone (except an ancient Babylonian) by whirling it on a sling. Digressing briefly to consider the nature of heat, Galileo placed heat in the same category as taste, odor, color and other "qualities" which only became manifest through the action of the senses. These he described as "mere names . . . residing only in the consciousness". Yet this was not the case with all properties; as he put it:

> Now I say that whenever I conceive any material or corporeal substance, I immediately feel the need to think of it as bounded, and as having this or that shape; as being large or small in relation to other things, and in some specific place at any given time; as being in motion or at rest; as touching or not touching some other body; and as being one in number, or few, or many. From these conditions I cannot separate such a substance by any stretch of my imagination.

Heat was emphatically not to be classed with these fundamental properties, but rather with those which required the action of the senses to acquire full reality: it was the result of "nothing . . . except shapes, numbers, and slow or rapid movements" (of the small parts of matter). So odor arises when small particles impinge upon our noses, taste when small particles strike our tongues. In each case three things are requisite (for obviously not all bodies can produce each of these sensations) : particles of a certain kind (fire, in the case of heat), motion of these particles, and an organ of sensation.

§4 *Descartes*

Galileo, so much at home in macroscopic dynamics, never tried to apply the same methods to microscopic dynamics, no doubt realizing the magnitude of the task. Gravity thus remained a property whose action he could describe, but for which he never tried to find an explanation. It was thus that he laid himself open to Descartes's rebuke:

> All that he says of the celerity of bodies falling in a vacuum &c is without foundation; for he ought to have determined beforehand what weight is; and

---

[11] For the passages cited see *Discoveries and Opinions of Galileo*, tr. Stillman Drake (New York, 1957), pp. 272–78, or *The Controversy on the Comets of 1618* (Philadelphia, 1960), tr. Drake and C. D. O'Malley, pp. 300–314. Attention was first drawn to these in a classic discussion by E. A. Burtt, *The Metaphysical Foundations of Modern Physical Science*, chapter III, sec. C (pp. 73–80 of the revised edition, London, 1949).

if he had truly understood its nature, he would have known that there is none in a vacuum.[12]

Descartes's rebuke was perfectly fair: for he himself did not dream of ignoring the necessity of explaining such a fundamental property of matter. Where previous discussions of the mechanical philosophy had dealt with description and possibilities, Descartes made his version of it an integral part of his philosophy, and labored to render it perfectly general. He showed none of the vagueness of the first progenitors of the mechanical philosophy, but was complete and precise in every detail of his theory of matter, which he discussed in connection with every branch of science.

The general outlines of his theory are too well known for detailed exposition here. That matter and extension are one, so that there is no empty space; that motion has divided matter into particles, mathematically though not physically divisible to an indefinite extent; that these particles are of three kinds, distinct yet mutually transmutable; that motion (necessarily by impact) is imparted to terrestrial matter by the "subtle matter" or "aether" —to say this is to outline his theory without explaining either its purpose or its importance. For what was important about Descartes's theory is less its details than the conviction that this theory was adequate to explain the whole physical universe, since from matter so conceived and from motion one could mentally construct the whole universe. And the universe so conceived would at least approximate the world around us, which demonstrated the correctness of the approach. As he wrote at the end of the *Principles of Philosophy*:

> Being certain that every body which we perceive is composed of other bodies so small that we cannot perceive them: I think no one who uses his reason can deny that it is better philosophy to judge what happens among these little bodies (which only their minuteness prevents us from perceiving) in the light of what we see happen among those bodies which we do perceive, and to account by these means for everything in Nature—as I have tried to do in this treatise—than to explain the same things by inventing I don't know what others which bear no relation to what we perceive—first matter, substantial forms, and all that grand apparatus of qualities which many are in the habit of imagining, each one of which may be more difficult to know, than the things which are to be explained by them.[13]

___

[12] Letter 146 to Mersenne, 11 October 1638 (*Oeuvres,* Adam-Tannery edition, II, 385).

[13] *Principia Philosophiae,* Book IV, art. 101. I have taken this and subsequent quotations from the French translation, using the edition of 1659. The Latin version, from which many English translations are derived, is less explicit, but as Descartes saw and approved the French text, it seems reasonable to assume that this represents his meaning accurately.

An important characteristic of Descartes's theory was that relatively few properties of gross matter were dependent on the nature of the particles themselves; what mattered was rather motion. The size, shape, and so on of particles determined the nature of the substance of which they were parts; the properties exhibited by this substance resulted from variations in the motion of the particles caused by the action of the subtle matter or aether. Thus, it was the characteristics of the particles of water which determined the fact that the substance they composed was water, rather than some other fluid; but its fluidity was caused by the fact that the particles were perpetually moving among themselves; if their relative motion stopped, the substance of which they were parts ceased to be a fluid, and became a solid; which is what happened when water solidified to ice. This was a great advance, because it reduced the number of causes required for known properties, and at the same time simplified them. Not that Descartes entirely escaped the excessive elaboration of detail characteristic of ancient atomism; for example, he wrote: "I suppose that the little particles of which water is composed are long, smooth and slippery, like little needles."[14] It was true that to him fluidity resulted from motion; but he still felt it essential to explain why the particles of water could move so easily under the action of the aether. The idea of considering the relative rest or motion of particles was of great generalizing value, for it related the various states of matter in such a way as to explain how it was possible for solids and liquids to be converted one into another; what the peculiar fluidity was which characterized both air and fire; how rarefaction and condensation were possible. Further, the action of the aether in producing motion of the particles caused the body to grow hot; light was the result of nothing but motion in the aether; gravity arose from the peculiar motion characteristic of aether. And from the nature of light Descartes was able to explain such phenomena as the tails of comets.[15] Descartes even tried—at length, but not very clearly—to explain such mysterious and occult problems as the nature of certain chemicals and the cause of magnetism (which he found to consist of the action of the aether on specially shaped particles).

As is so often true of Descartes, the scientific details fall far short of the philosophic concept. The notion of the subtle matter as responsible for all motion, and of motion in turn as responsible for very many of the secondary qualities of matter, was well conceived. It offered a simple and satisfactory explanation for many diverse phenomena, and effected enormous simplification of detail. For the sake of this advantage, many Cartesians were prepared to overlook or minimize such naive and contradictory details as the

14 "Meteores" in *Oeuvres*, Adam-Tannery ed., *6*, 233.
15 *Principia Philosophiae*, Book III, arts. 133, 135.

"screwed" magnetic particles, the transmutation of the elements (now beginning to be regarded as an "occult" concept), and the strange small, flexible, shapeless particles which existed to fill up the interstices between other particles and thereby prevent the otherwise inevitable yet impossible vacuum. The enormous logic of the Cartesian system overbore such relatively minor considerations (as they were felt to be), more especially since the system was so conceived that it could be called "mathematical" in an age when that was the pinnacle of rationalist praise—though one did not need to understand geometry in order to read Descartes's *Principles of Philosophy*.

§5 *Cartesianism*

Even for experimental scientists the Cartesian system was useful and acceptable, for it had sufficient flexibility to permit the inclusion of facts unknown to its inventor. New discoveries added to its strength, when it appeared that Descartes had left principles whereby the new discoveries could be explained. The puzzling "vacuum" in the Torricellian tube or the receiver of Boyle's air pump was difficult for true atomists to explain, for it was not a real vacuum; at least, one could hardly call it "empty space" when it admitted light, magnetism and perhaps fire. The Cartesian had no such difficulty: it was empty only of air, and still filled with the aether which naturally transmitted the observed light, magnetism and so on, quite unaffected by the presence or absence of air. New discoveries in optics—diffraction, the double refraction of Iceland spar, the color of thin films—called forth new theories on the nature of light by, for example, such scientists as Hooke and Huygens. Both were Cartesians in that they accepted the general premises of Descartes's theory of matter; both evolved their own theories of light, different from that of Descartes but conceived within the framework of the Cartesian system. Hooke's theory of pulsed vibrations was intended to explain colors—which it did, though not in such a way as to withstand the assault of Newton's experimental evidence and alternative, non-Cartesian theory. Huygen's theory—precise, detailed and truly mathematical—had more stamina. Light, he held, certainly "consists in the motion of some sort of matter" since it is produced by fire which has the property of disuniting the particles of bodies, which "is assuredly the mark of motion, at least in the true Philosophy, in which one conceives the causes of all natural effects in terms of mechanical motions".[16] In fact, Huygens maintained, light consists in the interaction between particles of luminous bodies and the aether (supposedly an elastic fluid) which in turn sets up wave fronts similar to those of sound in air. A whole series of wave fronts is generated,

---

[16] *Treatise on Light*, tr. Sylvanus P. Thompson, Chicago, 1945, Chapter I.

whose generation and propagation are susceptible of mathematical treatment. By this concept Huygens was able to explain reflection, refraction, diffraction and (in part) the double refraction of Iceland spar. There was indeed only one drawback to his theory: he could not explain the production of colors; but he thought himself excused because that was a difficult subject "in which no one until now can boast of having succeeded".[17]

§6 *Experimental Corpuscularianism*

Huygens's work showed once again how far one could proceed in explaining the properties of bodies on purely rational grounds. This indeed seemed to Cartesians the only reasonable way, especially when dealing with invisible particles; to such thinkers it seemed absurd to try to "prove" details of structure by means of experiment. Experiment might find out new phenomena to explain, but it could never demonstrate anything, nor was it really necessary to confirm obvious (that is, ratiocinative) truths. Hence the ambiguous attitude displayed by many Cartesians towards the work of Robert Boyle. They rightly took him to be the exemplar of the ideals of the Royal Society; yet while valuing highly his pneumatic discoveries—which none could forestall, though many could enjoy repeating—most Cartesians thought the very notion of an experimentally based corpuscular philosophy childishly and unnecessarily empirical. For what, they argued, could Boyle hope to achieve, except to demonstrate with immense labor what they knew to be true already by the surer and swifter method of reason? And in addition there was Boyle's strange addiction to chemistry, a science most physicists, especially on the Continent, neither understood nor wished to understand.

In actual fact, Boyle's "corpuscular philosophy" (as he preferred to call it) was original in conception and in the method he chose for its demonstration. In spite of an informal style and a pretense of eclecticism, Boyle had developed a carefully thought out and well constructed theory of matter. He was not an atomist, recognizing with the best scientists of his time that any particle was, as he put it, "mentally, and by divine Omnipotence divisible"; but he believed that the ultimate or least particles were sufficiently small and solid so that "nature doth scarce ever actually divide (them) . . . and these may in this sense be called *minima or prima naturalia*". These least particles in turn combined into "primitive concretions or clusters" which "though not absolutely indivisible by nature into the *prima naturalia* . . . yet . . . they very rarely happen to be actually dissolved or broken, but remain entire in great variety of sensible bodies, and under various

---

[17] *Ibid.*, Preface.

forms or disguises".[18] These were the smallest particles detectable, the corpuscles directly responsible for stimulating the senses. The size and shape of both *prima naturalia* and the corpuscles formed by their association affected the nature and properties of the matter they composed. Even more important, however, was the motion of these two kinds of corpuscles, for motion alone was responsible for the innumerable changes in properties which every theory of matter had to explain. Motion "altered the texture" as Boyle said, and thereby altered the properties. The source of the motion did not make any difference; its existence was enough. So Boyle instanced the change from water to ice as an example of loss of motion; the rubbing of two sticks of wood together to produce, successively, heat, flame and charcoal, as an example of gain in motion; and the turning of milk into butter and whey, the fermentation of apples after bruising and similar domestic mechanical actions, as examples of what should be explored scientifically and carefully.

Granted that God had made the universe out of what Boyle called "those two grand and most Catholic principles, matter and motion", how was one to discover the relations between the structure of matter, its changes of motion and the properties which were the result of this structure and motion? Boyle was no Cartesian, to frame what he regarded as *a priori* hypotheses; at least he would frame no more than need be. He took his inspiration direct from Bacon, and insisted that the key to an understanding of nature must be experiment. And a more profound innovation was his belief that for the "discovery of forms" no kind of experiment was better than chemical experiment. For it was, Boyle thought, through chemistry that the corpuscles composed of *prima naturalia* were revealed as having independent and permanent existence. Thus he instanced the fact that liquid mercury

> may be turned into a red powder (mercuric oxide) or a fusible and malleable body (a metallic alloy) or a fugitive smoke (the vapor), and disguised I know not how many other ways, and yet remain true and recoverable mercury.[19]

One of his most famous early experiments to demonstrate that changes of properties need not mean the loss of existence for a well-defined corpuscle was that on the "redintegration of nitre" (saltpeter).[20] He decomposed

---

[18] This view of the structure of matter is most clearly expressed in *The Origin of Forms and Qualities* (*Works*, ed. T. Birch, 6 vols., London, 1772, *3*, 29–31). There is a very similar discussion in the "Propositions" of the *Sceptical Chymist* (*ibid.*, *1*, 474–75).

[19] *Origin of Forms and Qualities, loc. cit.*, p. 29.

[20] "A Physico-Chymical Essay, containing an Experiment with some Considerations touching the differing Parts and Redintegration of Salt-Petre", in *Certain Physiological Essays.*

saltpeter (potassium nitrate) with a glowing piece of charcoal (which he thought to be pure fire) and found he had volatile niter, which could be dissolved in water to give spirit of niter (nitric acid) and fixed niter (actually potassium carbonate because the potassium combined with the carbon of the coal and oxygen from the air); since these could be recombined to form saltpeter again, Boyle argued that he had shown that this familiar salt was composed of two permanent corpuscles, temporarily joined. This was to have applications in chemistry; his primary purpose here was to further the corpuscular philosophy. In the preface he declared that his purpose was in large part "to beget a good understanding betwixt the chemists and the corpuscular philosophers"[21] both by showing chemists that natural philosophy could teach them to understand the processes which they undertook in the laboratory and by showing the natural philosophers that chemistry, which dealt with finely divided matter and its properties, could be of assistance in demonstrating their theories of the structure of matter. But he also intended, as he told Spinoza,

> to explain how the common doctrine of substantial forms and qualities, accepted in the schools, rests on a weak foundation, and that what they call the specific differences of things can be reduced to magnitude, motion, rest and position of the particles.[22]

If he could successfully demonstrate this, he thought he could convince men that the explanation by means of forms was sterile, and should be replaced by an explanation in terms of matter and motion.

This, in fact, was what he set himself to do; and there poured from the press a long series of works dealing with various physical and chemical properties and their mechanical explanation. The elasticity of air, the nature of fluidity and firmness, the origin of colors, the nature of cold, the mechanical origin of heat, taste, odor, volatility, fixedness, corrosiveness, magnetism, acidity were nearly all explored in individual essays or treatises, along with others. The list is almost oppressively long, especially since it was interspersed with general treatises on the corpuscular philosophy.

To understand the use Boyle made of his theory of matter it is necessary to consider some specific examples of his method and success. One of his

---

[21] "Some Specimens of an Attempt to make Chymical Experiments Usefull to Illustrate the Notions of the Corpuscular Philosophy".

[22] Oldenburg to Spinoza, 3 April 1663, printed in A. Wolf, *The Correspondence of Spinoza*, (New York, n.d.), pp. 110–115. This is a reply to a very long letter of criticism which Spinoza wrote in the spring of 1662, after reading the Latin edition of *Certain Physiological Essays*, sent by Oldenburg. Spinoza certainly misunderstood both Boyle's purpose and his argument; in particular he understood nothing of chemistry, insisting that niter and spirit of niter differed only as ice and water.

most interesting works is *The Experimental History of Colour* (1664) which is crammed with novel and fascinating experiments, and in which he tried, for the first time, to explain color mechanically. He was convinced that color was a secondary quality and therefore amenable to explanation in terms of matter and motion. That it *was* a secondary quality he argued from the fact that one can see color without light (just as one can feel heat without fire): that is, if the organ of sight is stimulated either by a blow, or by the imagination (as in dreams) one believes that one sees color. *The Experimental History of Colour* begins with a brilliant investigation of whiteness and blackness. Boyle concluded that whiteness and blackness were the result of the arrangement of the particles on the surface of the body: a white body's particles were so arranged as to act almost like a mirror to reflect a very high proportion of the incident light; in a black body, on the other hand, the particles were so arranged as to absorb almost all the incident light, reflecting very little. To demonstrate this he pointed out how it is easy to set black paper alight with a burning glass, difficult to set white paper alight in this way; that if a tile painted half white and half black were exposed to the sun, the black half became very hot while the white was still cool; that a room hung with black hangings was darker than one hung with white hangings and so on.[23] A century later Benjamin Franklin could do no more.[24] Armed with these successful and accurate conclusions, Boyle naturally proceeded to investigate color along the same lines; in this (more difficult) case with mixed success. He made some interesting discoveries: he noted the colors of soap bubbles (Newton's rings); he explored the changing colors of a solution of *lignum nephriticum,* a yellow solution with a blue opalescence which it loses, as Boyle found, upon the addition of acid—the first detailed discussion of its optical properties; he investigated the manner in which blue vegetable solutions were turned red by acids, green by alkalies, and showed that this was a reliable test for acidity and alkalinity (which therefore were the result of the texture and motion of the particles); he tried endless combinations of chemical solutions to produce an enormous variety of color changes. All this was intended to demonstrate that the "forms" of colors were the result of nothing but the action of matter and motion. Aside from his own success, Boyle by his work provided inspiration for the investigations of Hooke (trained in Boyle's own laboratory) and Newton (an indefatigable student of Boyle's work); both approached the problem of color in quite a different way from Boyle, considering changes in color as a result of

---

[23] *Works,* I, 721–22.
[24] I. Bernard Cohen, "Franklin, Boerhaave, Newton, Boyle & the Absorption of Heat in Relation to Color", *Isis, 46,* 1955, 99–104.

changes in light itself, rather than in the structure of the colored body. Yet without Boyle's work, neither might have been familiar with the facts of color which their theories had to explain.

To us it seems only natural to consider "light and colors" together, as Newton was to do ten years after the publication of Boyle's work. To Boyle it was more reasonable to consider light in connection with fire, and hence with heat. Like all mechanical philosophers, he took heat to be the effect of the agitation of the particles; he pointed out many times that it could be produced by mechanical action (like hammering a piece of metal) as well as by chemical activity and, of course, by fire.[25] Indeed, if a hammer struck a nail on the head, the resultant blow could produce *either* motion of the whole nail (into a block of wood) *or*, if this were impossible, motion of the particles of the nailhead, which would then grow hot. Even *in vacuo* friction alone would produce heat, though not flame.[26] Yet flame was closely related to heat; Boyle at first wondered whether it might not be merely a state of matter, that is, matter in a special kind of motion; he finally concluded:

> flame is little or nothing else than an aggregate of those corpuscles, which before lay upon the upper superficies of the candle, and by the violent heat were divided into minute particles, vehemently agitated and brought from lying as it were upon a flat, to beat off one another.[27]

Flame (and fire) then consisted of bodies heated hot enough to shine. Hence fire was certainly material, a conclusion with two important consequences: first, that no occult forms were required to explain either its existence or its action on other bodies (both the result of vehement agitation of the particles); and second, that it ought to be possible to demonstrate this by experiment. Hence Boyle's *New Experiments to Make Fire and Flame Stable and Ponderable,* in which he described the elaborate and quantitative series of calcination experiments on metals and various "minerals" like sulphur, whose invariable gain in weight he attributed to the addition of fire particles: hardly a surprising conclusion in view of the state of knowledge about the chemical structure of air. Boyle's experiments appeared to prove his point amply: fire was manifestly a mechanical agent, not an occult force. Light appeared to Boyle closely related to fire; he planned a series of experiments on calcination by means of a burning glass, but was defeated by

---

[25] Cf. "A Discourse about the Absolute Rest in Bodies" (*Works, 1,* 446), and "The Mechanical Origin and Production of Heat" (*Works, 4,* 223, 249–50).

[26] *A Continuation of New Experiments Physico-Mechanical* (1669), *Works, 3,* 265–66.

[27] *A Defense against Hobbes and Linus* (part II, ch. 2), *Works, 1,* 142.

the English weather. But he had once more opened up a new area of scientific investigation; Hooke, for one, pursued it by examining under a microscope the sparks struck off flint and steel, and showing them to be glowing bits of metal.[28]

Boyle's boldest attempt at a mechanical explanation of occult properties arose from his investigations into the causes of magnetism and electricity. Seizing on Gilbert's observation that bars of iron left standing upright over a period of time became magnetized, and that a bar of iron laid north and south on an anvil could be magnetized by hammering, Boyle concluded that magnetism must result merely from a peculiar arrangement of the iron corpuscles. As he put it:

> if there be introduced a fit disposition into the internal parts of the metal by the action of the loadstone, the metal, continuing of the same species it was before, will need nothing, save the continuance of that acquired disposition, to be capable of performing magnetical operations.[29]

That it *was* merely a rearrangement of the corpuscles which caused the change from non-magnetic to magnetic iron was further demonstrated by the fact that if a bar of iron were heated before standing it upright, or hammering it, the resultant motion of the particles made it receive "the magnetic disposition" more quickly. This was a truly mechanical explanation; unfortunately, it failed in this case to stimulate further investigation. About electricity Boyle was less clear, though he hazarded a guess that it too had a mechanical origin, arising from "electric effluvia": he thought that the friction necessary to render an electrical body attracting rubbed off material, non-aetherial particles. Not enough work had yet been done on electricity to provide him with experiments to suggest other sorts of mechanism; he intended to investigate the matter himself, as a brief paper entitled "Enquiry and Experiments about Electricall Bodys" shows, but never did so.

From the examples above it hould be clear that Boyle found no physical property too complex for experimental investigation and mechanical explanation. The most important aspect of his explanation was that it not merely supplanted prevailing explanations which relied on forms and occult forces, but proved beneficial to the advance of science by suggesting new fields in need of experimental exploration and mechanical interpretation. This stimulus was equally active in his use of the corpuscular philosophy in chemistry, though there the problem was far more complex, since, before Boyle, few physicists or chemists thought of chemistry as a physical sci-

[28] *Micrographia*, observation VIII; facsimile in R. T. Gunther, *Early Science in Oxford*, *13*.

[29] "Mechanical Origin and Production of Magnetism", *Works*, *4*, 340–45.

ence.[30] Boyle, having made original contributions to natural philosophy, could afford to dabble in the quasi-esoteric science of chemistry; at the same time interest in chemical experiment never led him to abandon physical, mechanical reasoning. He believed, as indicated above, that complex corpuscles were the units of chemical reaction, since they, and they alone, could survive unchanged through a long series of reactions, to reappear unchanged at the end of the process. Only these corpuscles, and the more complex substances made by joining corpuscles together, appeared to Boyle to have chemical validity. As a consequence he utterly rejected the doctrine of elements or chemical principles, denying the existence of any set of substances out of which all other substances were made and into which all other substances could be decomposed or analyzed. For both logically and experimentally, he insisted, corpuscles and *prima naturalia,* and they alone, were the true elements.[31]

Having rejected the concept of elements and principles—of earth, air, fire and water, or salt, sulphur and mercury, or oil, phlegm, spirit, &c.—as the chemical basis of matter, Boyle was free to treat all chemistry from a corpuscularian point of view. This was important less for itself—it is not certain that chemistry always benefits from being treated like physics— than because it led Boyle to useful conclusions. He concentrated on what he believed to be permanent corpuscles, that is to say, on those which could undergo transformation in the course of a chemical reaction and yet, after further reaction, be recovered again unchanged. As a result he built up a large group of simple chemical substances whose properties he studied carefully and whose identities he was sure of. He was the first to devise detailed identification tests for individual substances: indicator tests for acids and alkalies; crystal shape for salts; flame tests for copper; characteristic reactions for various substances. Thereby he simplified the chemical picture,

---

[30] Thus Fontenelle, comparing Boyle (after his death) with the now rightly forgotten chemist Samuel Du Clos, who spent all his time as a member of the *Académie royale des Sciences* analyzing plants into salts, spirits, phlegms and oils, found Boyle's attempt to apply the corpuscular philosophy to chemistry too simple and clear to be true chemistry. (Section for *Chymi,* 1669, in *Histoire de l'Académie Royale des Sciences,* 1666–1698).

[31] This is the conclusion of the *Sceptical Chymist.* Misled by a partial and hasty reading of Boyle's definition of an element (and ignoring the fact that he said he was trying to give a definition acceptable to all contemporary chemists), modern historians have blamed Boyle for being too much of a physicist and for neglecting the now-useful concept of an element. But the "element" of the 17th century was not just a substance impossible to analyze further; it was a substance universally present in all bodies and hence *always* a product of analysis. This misinterpretation is a conspicuous defect in the otherwise admirable paper by Thomas S. Kuhn, "Robert Boyle and Structural Chemistry in the Seventeenth Century", *Isis, 43,* 1952, 12–36.

showing the identity of many common substances known under a wide assortment of names, and gaining at least partial knowledge of the composition of many more substances. This showed chemists a new side to their subject, and directed their major activities away from the futile and repetitious analysis of organic materials (which obligingly yielded the supposed elements of each chemist) into more fruitful lines towards the understanding of the nature, composition and preparation of inorganic substances.[32] It is possible to imagine that Boyle's corpuscular emphasis was not necessary for this kind of chemical thinking; it is however a fact that Boyle, who always kept the corpuscularian philosophy in mind and applied it to every chemical operation, was the only chemist to achieve this changed point of view. The success of his approach is witnessed by the usefulness of the chemical experiments he described in stimulating the work of other chemists, and the fact that particles remained a part of chemical explanation, though not so obtrusively as with Boyle.

§7 *Newton*

Boyle showed that matter and motion alone could explain a vast number of chemical and physical properties, when such an explanation was combined with experimental investigation and confirmation. Newton, influenced strongly by Boyle, showed that an experimentally based theory of matter could explain the world, and thereby outdid Descartes. Newton was stirred to emulate Descartes precisely because he found Descartes's theory of matter totally unpalatable, both in principle and detail. To Newton, Descartes's famous first assumption about matter, that it was identical with extension, was unacceptable, so that he could never be a Cartesian. To equate matter with extension was philosophical nonsense as far as Newton was concerned, since he had no difficulty in postulating a vacuum; worse, it was theologically unsound, since it placed an unwarranted limitation upon God's omnipotence and was nearly atheistic.

> If we say with Descartes that extension is body, do we not offer a path to Atheism, both because extension is not created but has existed eternally, and because we have an absolute idea of it without any relationship to God?[33]

Matter differed from extension precisely as it had attributes impressed upon it by God. (Newton's theory of matter ineluctably led him to a discussion of

---

[32] For a fuller development of these points, see Marie Boas, *Robert Boyle and Seventeenth Century Chemistry*, Cambridge, 1958.

[33] *De Gravitatione et aequipondio fluidorum*, Cambridge University Library Ms. Add. 4003.

the relation of God to His universe, because "the main business of natural Philosophy is to . . . deduce Causes from Effects, till we come to the very first cause, which is certainly not mechanical".[34]) Boyle's corpuscular theory was, basically, more acceptable to Newton than Descartes's theory of matter; but it failed in not being mathematical. Boyle could explain changes in properties as being the result of changes in motion, but he could not characterize the changes in motion in any rigorous way. No one as yet had devised a dynamics of particles comparable to that developed for the macroscopic world. Only Newton seriously attempted it.[35]

A clear understanding of what Newton really thought about the nature of matter and the proper explanation of its properties has always been (in his own phrase) "pressed with difficulties." Is one to accept the apparent mathematical austerity of the *Principia* with its refusal to say what the cause of gravity might be (though there was more to Newton's theory of matter than an explanation of gravity) and its insistence on *"hypotheses non fingo"* (though, as Professor Koyré has shown, Newton made use of hypotheses here as well as elsewhere)?[36] Or should one lay more stress on the speculative, experimental attitude manifested in the Quaeries to *Opticks?* What should one make of the aetherial hypotheses discussed in Newton's early papers and letters, like those on light and colors, and the famous letter to Boyle?[37] Did Newton, as he hinted, really frame these hypotheses only to please his correspondents? Did he, indeed, care at all about a mechanical philosophy? If he did (and in fact there is abundant evidence that he did) was he a firm proponent of an aetherial theory or not? Fortunately, Newton's unpublished manuscripts, of which he was a zealous hoarder, provide a clearer picture, and show that the connection which one might suppose to exist between the discussions of matter scattered through the *Principia* and the more speculative discussions in *Opticks* is a real connection. The manuscripts reveal as well both why Newton devoted so much time to aetherial speculations and why in the end he remained too unsatisfied to proclaim them as doctrine.[38]

To understand thoroughly the nature of matter and to derive from

----

[34] *Opticks*, Quaery 28.

[35] For a fuller discussion of this point, see Marie Boas and Rupert Hall, "Newton's Mechanical Principles", *Journal of the History of Ideas, 20*, 1959, 167–78, and A. Rupert Hall and Marie Hall, "Newton's Theory of Matter", *Isis, 51,* 1960, 131–44.

[36] A Koyré in *Revue d'Histoire des Sciences, 8,* 1955, 19–37, and in *Bulletin de la Société Francaise de Philosophie,* Avril-Juin 1956, 59–79.

[37] These are all conveniently available in *Isaac Newton's Papers & Letters on Natural Philosophy,* ed. I. B. Cohen, Cambridge, Mass., 1958.

[38] The relevant documents are to be published in A.R. & M.B. Hall, *Unpublished Scientific Papers of Sir Isaac Newton,* Cambridge, 1962.

thence the properties of bodies was an early ambition of Newton's. Steeped as he was in the "new philosophy" of the 1660's, he could hardly fail to be aware of the importance of the mechanical philosophy. And he quickly found such contemporary accounts as those of Descartes, Hooke and Boyle deficient in the extent of their explanation. They could and did explain much; but not all that Newton thought could and should be explained. And, unlike his predecessors, he sought for explanation which should belong to mathematical as well as to experimental physics. Very early in his work he began to grapple with attraction and repulsion which he recognized as omnipresent forces, tiresomely occult, yet at the same time susceptible of mathematical expression. Descartes had explained apparent cases of attraction as the effect of impulse of the aether; Boyle, without an aether, had to assume that only the fortuitous concourse of corpuscles was involved, as when he defined hygroscopic salts as those which "happened" to have pores the right size and shape to admit particles of water casually brought into contact with the salt particles by the ordinary motion of the atmosphere in which the water particles were dissolved. Newton found this cumbersome, and too remote from the world of mathematical law. After his early investigations of gravity he seems to have become convinced that attraction and repulsion were such common forces in nature that it was impossible to ignore them; it was imperative to try to fit them into mathematical physics. Newton's discussions of his theory of matter, therefore, nearly all are extensively concerned with the problem of how these forces, once their universal existence has been demonstrated, may be characterized and explained.

Underlying all Newton's writing on science is a profound belief in the particulate structure of matter: it is embedded in the structure of the *Principia* no less than that of his writings on optics. As he wrote in the Preface to the first edition of the *Principia:* "the whole burden of philosophy seems to consist in this—from the phenomena of motions to investigate the forces of nature, and then from these forces to demonstrate the other phenomena." Further, the phenomena of motions should be universally applicable; it was obviously applicable to larger bodies, but it equally certainly was applicable as well to the lesser particles of which bodies were composed. It is indeed the purpose of the third of the "Rules of Reasoning in Philosophy" with which Book III opens to relate the behavior and motion of particles to those of large bodies.[39] So

---

39 "The Qualities of bodies, which admit neither intensification nor remission of degrees, and which are found to belong to all bodies within the reach of our experiments, are to be esteemed the universal qualities of all bodies whatsoever."

the extension, hardness, impenetrability, mobility, and inertia of the whole, result from the extension, hardness, impenetrability, mobility, and inertia of the parts; and hence we conclude the least particles of all bodies to be also extended, and hard and impenetrable, and movable, and endowed with their proper inertia. And this is the foundation of all philosophy.

As for the parts themselves, Newton was very nearly an atomist—more so certainly than was Descartes. He readily admitted that the least parts could be divided mathematically and imaginatively; whether they could be divided "by the powers of Nature" was a more difficult question. On the whole he thought not; in *Opticks* he wrote,

All these things being consider'd, it seems probable to me, that God in the Beginning form'd Matter in solid, massy, hard, impenetrable, moveable Particles, of such Sizes and Figures, and with such other Properties, and in such Proportion to Space, as most conduced to the End for which he form'd them; and that these primitive Particles being Solids, are incomparably harder than any porous Bodies compounded of them; even so very hard, as never to wear or break in pieces; no ordinary Power being able to divide what God himself made one in the first Creation.[40]

With all mechanical philosophers, Newton assumed that the properties of matter—heat, magnetism, color, gravity—resulted from the motion of the least particles. The peculiarity of the Newtonian explanation lies in the fact that he so often explained the motion as caused by forces of attraction or repulsion. All mechanical explanations involve, ultimately, an arbitrary limit to causative argument; at some point the mechanical philosopher must stop, and assume that he has explained far enough. So Descartes found that the motion of ordinary particles needed explanation, but not the motion of the aetherial particles. Boyle, seeing that this was no great advantage, inquired no further than the changes in motion of the particles themselves; this eliminated the suppositious aether, but left many forces (including those of chemical action) unexplained except in terms of chance. Newton was obsessed with the need to find out the laws governing the universe, which he felt must exist even in the regulation of the properties of matter. He was able to treat gravity mathematically, by assuming a force of attraction; no wonder he extended the concept of a force of attraction (with its converse, repulsion) to other properties. So he argued that attraction and repulsion operated on light to produce reflection, refraction and diffraction; on metals, acids and other chemicals to produce observed chemical reaction; on rubbed bodies to produce electricity. It was an enormous economy to be able to interpret so many properties of bodies as being the result of a single force (for repulsion was only negative attraction); and it eliminated

---

[40] Quaery 31.

the loose action of chance necessitated by Boyle's explanations. Further, it was potentially subject to treatment in mathematical terms.[41]

But Newton was too much a product of his age not to wonder whether forces of attraction and repulsion were suitable ingredients of a mechanical philosophy. They sounded suspiciously occult. Sometimes, however, New - ton was content to treat these forces as self-sufficient, without trying to offer a causal explanation of their existence. In some of his early papers, in the first edition of the *Principia,* in the *Conclusion* which he planned as part of that edition, but later suppressed, in the Preface (in several versions) he was content to develop and explore various examples of the connection be- tween attractive and repelling forces and the particles of bodies as the cause of observed properties. This remained always the heart of his doctrine; for if this much were granted his explanations were self-sufficient. Yet he could not help speculating on the possible cause of these forces, which he was reluctant to regard as inherent in matter.

The only possible cause of these forces lay, to seventeenth century thought, in some kind of aetherial hypothesis. Newton firmly rejected the dense aether of Descartes, which he proved to be cosmologically impossible. But he could not resist trying to find a possible aether, which would explain the forces of attraction and repulsion and avoid the difficulties of occult forces. In many (but not all) of his optical papers, in his Letter to Boyle, in some manuscript versions of the General Scholium (introduced into the second edition of the *Principia*), and in many of the Quaeries to *Opticks,* Newton tried out various versions of a possible aetherial hypothesis. He had no difficulty in devising such an hypothesis, one that would (as he hinted in the final paragraph of the printed version of the General Scholium) ex- plain the action of light, gravity, chemical action, sensation, electricity, all by means of an electric "spirit" (elastic fluid). The difficulty was, as he ruefully concluded:

> these are things that cannot be explained in few words, nor are we furnished with that sufficiency of experiment which is required to an accurate determina- tion and demonstration of the laws by which this electric and elastic spirit operates.

So it remained an hypothesis, not to be treated on the same level as his mathematical and experimental demonstrations.

But because Newton failed to find an explanation for his forces of at-

---

[41] It is fairly clear that this was what Newton had in mind from the textual evi- dence; a suppressed version of the General Scholium, and some annotations by Newton have made it quite plain. See A. Rupert and M. B. Hall, "Newton's Electric Spirit: Four Oddities", *Isis, 50,* 1959, 473–76, and A. Koyré and I. B. Cohen, "Newton's "Elec- tric & Elastic Spirit" ", *Isis, 51,* 1960, 337.

traction and repulsion this by no means makes his theory of matter a failure. The mechanical philosophy as he applied it had led him to many grand discoveries: to his understanding of light and color, to his optical theories, to an understanding of gravity, to the Newtonian system of the world. Hence his conviction that this, and this alone, was the proper method of reasoning in natural science. As he wrote in the Preface to the *Principia*

> I wish we could derive the rest of the phenomena of nature by the same kind of reasoning from mechanical principles, for I am induced by many reasons to suspect that they may all depend upon certain forces by which the particles of bodies, by some causes hitherto unknown, are either mutually impelled towards one another, and cohere in regular figures, or are repelled and recede from one another. These forces being unknown, philosophers have hitherto attempted the search of nature in vain; but I hope the principles here laid down will afford some light either to this or some truer method of philosophy.

It was a justifiable boast; and Newton had excellent precepts to leave to his disciples. His true disciples were not those who devised elaborate theories of matter (like Boscovich) but those who tried to find out more about the properties of matter, and to explain them rigorously.

§8 *Conclusion*

The seventeenth century approach to the study of matter differs markedly from the modern approach. Its most striking difference lies in its strict utilitarianism. Convinced that there was no way of finding out what the least particles of matter were really like, mechanical philosophers concentrated on explaining the properties of matter in the most economical manner possible, with the general assumption of the existence of mobile particles. While the original progenitors of the mechanical philosophy all stressed motion as an important cause of properties, they all retained the need to consider the size and shape of the particles for a complete explanation, even though these could only be discovered imaginatively. Descartes's introduction of an aetherial mechanism to account for motion solved certain philosophic problems, and introduced some simplicity, but did not eliminate all need for speculation about size and shape of particles, and added a necessarily inexplicable aether. Boyle, rejecting the aether, achieved considerable simplicity, but at the cost of denying the existence of forces, supposed occult, and handing back the world to chance, except for the omnipresent supervision of God, which could be perceived rationally, but hardly experimentally. Newton's introduction of the forces of attraction and repulsion further simplified the number of explanations required to account for the properties of matter: the size and shape of the particles was even less important to him than to Boyle, and properties were now considered subject to mathematical law. Boyle had showed how corpuscularian notions might be used in chemistry; Newton showed how the force of attraction between particles was similar, perhaps identical with that between gross bodies, so that chemical action was related to the now non-mysterious gravity. It is perhaps not surprising that the chemist's atom was the first to be stabilized on an empirical basis by Dalton's ingenious conception of *weight* (an experimentally detectable property) as the key to the differences between particles of different kinds of matter. Physics had to wait longer to see the introduction of an atom which could be treated with the mathematical rigor which evaded Newton, and which could be explored directly through experiment.

COMMENT

In the deft and masterful manner we have come to expect from her, Dr. Hall has given us in this paper a superb summary of the "corpuscular philosophy" of matter proposed in the seventeenth century. To this summary, I can add nothing of significance. If I tried to add anything new, I would only find that she has already mentioned it elsewhere in her authoritative writings on the corpuscular philosophy. It need not trouble us that she does not discuss Leibniz or other non-corpuscular views. The corpuscular philosophy in some form or other was, indeed, the prevailing philosophical interpretation of nature in the seventeeth century.

I have only two short comments to make on the paper. The second is more provocative. The first is simply that her consistent distinction between philosophers and scientists in the seventeeth century seems to me to be terribly anachronistic. Of course, she is not the only one to assume such a distinction. It is to be found in our current histories of science and in all our histories of philosophy since the nineteenth century. Today such a distinction is taken for granted, even when we talk about the relation between philosophy and science. But is there any reason for perpetuating this distinction when we talk about the seventeeth century? Why should we go on calling Galileo, Huygens, Boyle and Newton mere "scientists", when they considered themselves to be philosophers cultivating the "true philosophy"? And why should we insist on calling Bacon, Descartes and Leibniz "philosophers", when they thought of themselves as true scientists? The seventeenth century thinkers did distinguish mere technicians, artisans and "an astronomer merely arithmetical", as Galileo put it, on the one hand, and natural philosophers, scientists and "the astronomer philosophical" on the other. But they did not distinguish between true philosophy and true science. Such a distinction was not envisaged by them, and I for one do not see any advantage in imposing such a distinction anachronistically upon them.

My second comment concerns interpretation. First, we can all admit, I think, that the new philosophies of matter in the seventeenth century were pointedly anti-Aristotelian, at least before the time of Newton. They were, as Dr. Hall points out, a reaction to such "occult qualities" as the Aristotelian first matter, substantial forms, accidental forms, final causes and the like, all of which were still taught in the schools with varying accuracy and seriousness. But the new, reactionary theories of matter could not be, according to her, a simple return to the atomic theories of Graeco-Roman antiquity (i) because every scientist of the seventeenth century knew Aristotle's objections to Democritus' atomism, and (ii) because those Graeco-Roman theories were scientifically out of date. Obviously there could be no *simple* return to Greek atomism in the sense of accepting the world as Democritus or Lucretius knew it. Too much had happened since then, as Bruno and Gassendi themselves fully realized. But that the new theories of matter—the corpuscular philosophy of Boyle as well as the atomic theory of Beeckman—were in *fact* a return to the atomism rejected by Aristotle can, I think, be validly and profitably shown.

Dr. Hall is certainly correct in saying that the new theory of matter, despite minor differences of detail among authors, was fundamentally mechanistic. Descartes, Huygens, Boyle and others recognized no other criterion for "the true philosophy, in which", to use the words of Huygens, "one conceives the causes of all natural effects in terms of mechanical motions". This mechanical philosophy tried to explain all observable properties by the motion, shape, and size of the particles of matter, as Dr. Hall has clearly shown. She is also correct, as I have said, in calling this approach anti-Aristotelian. But when she calls this approach "non-atomic", the word 'atom' must be taken in its narrow etymological sense of 'indivisible'. Possibly this sense of the term is sufficient to classify seventeenth century scientists into atomists and anti-atomists. But there were many who were uncertain about the natural divisibility of ultimate particles, and even the best scientists recognized that any material particle was, in the words of Boyle, "mentally, and by divine Omnipotence divisible".

In the last analysis it is really irrelevant to seventeenth century thought whether the particles of matter are absolutely indivisible, as Beeckman, Gassendi and Democritus would have it, or not. It is even irrelevant whether the vacuum exists in nature, or whether space is really empty, or whether the universe is a plenum. By "irrelevant", I do not mean that seventeenth century scientists were indifferent to these questions. Indeed there was much controversy over atoms and the vacuum, and this must be recognized in order to appreciate the differences of approach. Nevertheless, whatever those particles may be, whether atoms, little bodies, corpuscles, small parts of matter, or *prima naturalia*, they all had this in common: their motions were strictly mechanical, and they themselves were ingenerable and incorruptible. It is this basic assumption which brings seventeenth century theories of matter back to the view of Democritus and the objections of Aristotle.

It is true that Democritus and Leucippus proposed to explain all natural phenomena in terms of atoms and the void. But their attempt must be understood as an answer to the Eleatic problem. Parmenides of Elea had denied the reality of change (*genesis*) because for him change is inexplicable. As he saw it, the admission of change landed one in the following dilemma:

> What comes to be must do so either from what is or from what is not, both of which are impossible. For what *is* cannot come to be, because it *is* already; and from what *is not* nothing could have come to be, because what *is not* cannot exist as a substratum. (*Phys.*, I, 8; 191a 29–32)

In other words, since the non-real cannot exist, nothing can come from it; and since the real already exists, there can be no coming-to-be, or *genesis* of the real.

Further, since the non-real (nothingness) cannot exist, there cannot even be locomotion, for every place is full. Therefore Parmenides had to deny the multiplicity and mutability obvious to the senses. In other words, for Parmenides the truth of the matter is that all reality is one solid mass devoid of change, although the opinion of mankind maintains multiplicity and change. It was in order to avoid the Eleatic dilemma and at the same time to account for multiplicity and mutability that Leucippus, the teacher of Democritus, said "that *what is* is no more real than *what is not*, and that both are alike causes of the things that come into being." (Theophrastus, *Physic. Opiniones*, I, fr. 8. 19–21) Thus Leucippus conceived the non-real, or emptiness (the void) to be as real as the solid, hard realities in nature, which he called atoms. This little concession of admitting the reality of non-being made all the difference in the world, for Leucippus could explain multiplicity by recognizing a void between individuals. He could also explain locomotion by recognizing a void into which individuals could move. This void, of course, was not the atmospheric air, for Empedocles had shown air to be a corporeal substance. It was simply non-being, emptiness, nothingness,. which Leucippus and Democritus maintained existed just as much as physical bodies. This conception of non-being having physical existence is philosophically very strange, but it is very similar to our conception of empty space existing in reality.

There are two important points, it seems to me, to notice about this atomism of antiquity. First, it was unable to answer the objection of Parmenides to the reality of *genesis*, or the coming-to-be of atoms. The void, indeed, made movement in place intelligible, but it could not account for any intrinsic mutability of the atom. If an atom really changed intrinsically, the new atom would have to come from atom, which is no change at all, or it would have to come from nothing, which is impossible. Consequently Leucippus had to concede the impossibility of intrinsic change, or *genesis* and *olethros*. In fact Leucippus gave his atoms the character of the Parmenidean One, except that he allowed for many such ones. The second point to notice is that the only motion provided for was mechanical motion, i.e., an external force exerted on an atom allowed it to move spatially from one place to another. In such a view the appearance of becoming is nothing but the composition of atoms, and perishing is nothing but the separation of atoms. Growth is really nothing but addition of more atoms, while so-called qualitative changes are really nothing but the locomotion of atoms. There is no need here to mention anything about the size, shape or density of the various particles recognized by the ancients. Nor is there any need to mention the eternal vortex (*diné*) in which all bodies are whirled (Diog. IX, 31–34). We already have the two essential elements of a purely mechanistic philosophy of matter. These two are the denial of intrinsic *genesis*, or *fieri*, and the exclusive recognition of locomotion mechanistically produced as the only real motion in nature. It was precisely against these two assumptions that Aristotle directed his strongest objections, not only in the *De generatione et corruptione*, but also in the *Physics*.

If the seventeenth century scientists knew Aristotle's objections to Democritus' atomism, as Dr. Hall claims, then we must go on to say that they paid little attention to them. In fact, the seventeenth century scientists accepted all the consequences of Democritean atomism, viz. denial of the substantial unity of things, the intrinsic mutability and spontaneity of nature, the reality of şensible qualities such as color, heat, odor and sweetness, the reality of teleological operations, the stability of species, and the reality of all movements other than locomotion. Thus, the seventeenth century "mechanical philosophy" was in fact a return to the mechanistic theories of antiquity. However, the seventeenth century scientists did succeed in bringing the anti-Aristotelian mechanical theory "up to date". In this sense the new theory of matter was not *simply* a return to the atomic theories of Graeco-Roman antiquity.

James A. Weisheipl O.P.

THE REJECTION OF
NEWTON'S CONCEPT OF MATTER
IN THE EIGHTEENTH CENTURY

J. E. McGuire and P. M. Heimann

§1 *Matter and Powers*

Primary and secondary quality doctrines were fundamental to seventeenth-century discussions of matter. Natural philosophers and philosophers such as Descartes, Boyle, Newton and Locke were concerned to establish certain essential or primary characteristics of matter, properties such as extension and solidity, which they held to exist independently of human perception. Qualities such as sensations of sounds and colors, on the other hand, were held to arise as a result of a relation between the perceiving mind and the primary qualities of matter which were held to provide their causal nexus. Newton enunciated a classic formulation of this doctrine in his third Rule of Philosophizing, where he declared that qualities such as solidity and extension were essential qualities of all material entities, and were known to be so through experience. Arguing that the primary particles of matter, the atoms, were in principle unobservable, he justified the ascription to unobservable atoms of essential qualities, which were known from experience of gross bodies, by means of an argument based on the "analogy of Nature". Seventeenth-century thinkers accepted the distinction between absolute qualities (the primary qualities) and relational qualities. A number of eighteenth-century thinkers came to reject the validity of this distinction with respect to theories of matter, and with it the validity of the theory of matter as atoms possessed of primary or essential qualities. For thinkers such as Joseph Priestley and James Hutton, the essential characteristics of matter were entirely to be found in the seventeenth-century domain of secondary qualities, in the realm of mind-dependent and relational properties. These thinkers denied that qualities such as extension and solidity defined the essence of matter and argued that such qualities were the effects of certain powers. They also rejected Newton's doctrine of essential qualities and with it his atomic theory of matter, for these "powers" defined the essence of matter.

Though these thinkers are not entirely unambiguous in their use of the term 'powers', in general they held that to ascribe a power to a material thing is to assert what it can or cannot do in virtue of its intrinsic nature in relation to specifiable extrinsic circumstances. They tend to leave open a complete characterization of an entity's constitution in virtue of which it is held to be endowed with such powers, arguing that the characteristics of matter could only be established by means of the ways in which sensations were aroused. Thus, Priestley refused to discuss the question of the way in which powers could be held to be inherent in matter, for the powers were inherent properties which by their action gave the appearance of impenetrability and solidity to matter.

This notion of powers as inherent properties of matter is of crucial importance. Though thinkers such as Descartes, Hobbes, Boyle and Locke argued that bodies merely have a disposition to produce sensations of colors and tastes under certain conditions of perception, they denied that there was anything in the bodies themselves which corresponded to our ideas of these secondary qualities. These qualities resulted from primary qualities which were held to exist absolutely in material entities. For these thinkers, bodies were not ontologically causative in nature, and in arguing that an entity possessed the power or disposition to produce a sensation of color in relation to certain extrinsic circumstances, they did not suppose that the entity possessed that power as part of its intrinsic nature. Only primary qualities were intrinsic to the essence of bodies. However, in denying the seventeenth-century distinction between primary and secondary qualities, thinkers such as Priestley and Hutton argued that the essential characteristics of matter were to be characterized in terms of its power to give rise to certain effects. In their view, for an entity to have the power or disposition to give rise to a certain quality implied that the entity possessed that *power* as part of its intrinsic nature. They therefore defined the essence of matter in terms of such powers.

This conception of matter involved the notion of activity in nature, which contrasts with the general seventeenth-century emphasis on the passivity of material entities. Moreover, it was developed in the context of the rejection, by Berkeley and Hume, of the primary-secondary distinction; their discussion of Locke's notion of "power"; and lastly of Thomas Reid's defense of the notion of power in reply to Hume's rejection of the view that we have such an idea.

As in the seventeenth century, theories of matter were developed in close connection with new systems of philosophy, and we will therefore be concerned to analyze an important debate between philosophers and natural philosophers in the post-Newtonian period.

§2 *The Empiricist Debate*

In his *Essay concerning Human Understanding* (1690), Locke advanced the view that bodies have the disposition to produce sensations in the mind which do not correspond to anything in the bodies themselves. Thus, besides having "original" or primary properties, which defined the essence of material entities, Locke held that such entities possessed "imputed" and relational properties, which he called "powers":

> For the power in fire to produce a new colour, or consistency, in *wax* or *clay* by its primary qualities, is as much a quality in fire, as the power it has to produce in *me* a new idea or sensation of warmth or burning, which I felt not before by the same primary qualities, viz. the bulk, texture, and motion of its insensible parts.[1]

Locke argued that secondary qualities were "powers barely, and nothing but powers",[2] designating secondary qualities as relational qualities. Locke employed the term 'power' in an active causal sense. According to this conception, the primary or original qualities give rise to causal powers which were either actually efficacious or were able so to act. To say of matter that it possesses a secondary quality is to say that it is capable of producing an idea of that quality in our awareness under specifiable conditions; but to have a causal power to produce an idea of color in the mind is not the same as to have a property such as color. Nevertheless, primary qualities existed absolutely in objects and also had the power to produce ideas of primary qualities, which in the mind resemble those qualities themselves. Though Locke tended to use the term 'power' contextually in close association with the notion of causal efficacy, he denied that powers inherent in physical objects are active:

> A body at rest affords us no idea of any active power to move; and when it is set in motion itself, that motion is rather a passion than an action in it. For, when the ball obeys the motion of a billiard-stick, it is not any action of the ball, but bare passion. Also when by impulse it sets another ball in motion that lay in its way, it only communicates the motion it had received from another, and loses in itself so much as the other received; which gives us but a very obscure idea of an *active* power of moving in body, whilst we observe it only to *transfer*, but not *produce* any motion.[3]

For Locke, therefore, bodies are not ontologically causative in nature. Moreover, he admitted that "we have, from the observation of bodies by our

---

[1] John Locke, *An Essay concerning Human Understanding*, ed. Fraser, 2 vols., New York, 1959, *1*, 171.

[2] *Ibid., 1*, 179.

[3] *Ibid., 1*, 312.

senses, but a very imperfect obscure idea of *active* power; since they afford us not any idea in themselves of the power to begin any action, either motion or thought",[4] but claimed that the clearest idea of this notion came from "a consideration of God and spirits"[5] and the mind "from reflection on its own operations",[6] never doubting that it was possible to have an idea of natural power.

This analysis was to be contested by Berkeley and Hume. Rejecting the distinction between primary and secondary qualities, they were concerned to reduce primary qualities to the perceptible domain of the secondary. Moreover, they both held that the idea of power was not established either from direct sensory experience or reflection. In the *Principles of Human Knowledge* (1710), Berkeley opposed Locke's doctrine of abstract general terms (held to correspond to abstract ideas), when criticizing the Lockean distinction between things and ideas. He also rejected the notion of a substratum of invisible particles. Berkeley furthermore argued that all that can be known of external reality is what is perceivable or can perceive. And so the insensible corpuscles of the new science existed only as reified concepts. In the same way he rejected the distinction between primary and secondary qualities:

> For my own part, I see evidently that it is not in my power to frame an idea of a body extended and moved, but I must withal give it some colour or other sensible quality which is acknowledged to exist only in the mind. In short, extension, figure, and motion, abstracted from all other qualities, are inconceivable. Where therefore the other sensible qualities are, there must these be also, to wit, in the mind and nowhere else.[7]

Both sorts of quality were therefore held to be on the same ontological footing, and he applied the same argument to Locke's notion of powers:

> All our ideas, sensations, or the things which we perceive by whatsoever names they may be distinguished, are visibly inactive, there is nothing of power or agency included in them . . . since they [ideas] and every part of them exist only in the mind, it follows that there is nothing in them but what is perceived. But whoever shall attend to his ideas, whether of sense or reflexion, will not perceive in them any power or activity; there is therefore no such thing contained in them.[8]

Thus, Berkeley concluded that an idea of power cannot be obtained from experience; ideas are passive and can reveal neither activity nor power. It

---

[4] *Ibid.*

[5] *Ibid.*, *1*, 310.

[6] *Ibid.*, *1*, 313.

[7] George Berkeley, *Principles of Human Knowledge*, in *The Works of George Berkeley, Bishop of Cloyne*, ed. Luce and Jessop, 9 vols., London, 1948–57, *2*, 45.

[8] *Ibid.*, *2*, 51.

follows that the primary qualities can neither be the cause of sensations nor "the effects of powers" resulting from the configuration of corpuscles. For Berkeley, agency and change were to be located "in an incorporeal active substance or spirit",[9] to accord with his conception of divine causation.

In the *Treatise of Human Nature* (1739), Hume presented a devastating attack on the principles of the new science, especially those relating to the doctrine of primary and secondary qualities. Having accepted that secondary qualities have no real independent existence, Hume argued further that we have no idea of solidity. The primary qualities, therefore, do not afford us a distinct idea of the nature of matter:

> The idea of solidity is that of two objects, which being impell'd by the utmost force, cannot penetrate each other; but still maintain a separate and distinct existence. Solidity, therefore, is perfectly incomprehensible alone, and without the conception of some bodies, which are solid, and maintain this separate and distinct existence. Now what idea have we of these bodies? The ideas of colours, sounds, and other secondary qualities are excluded. The idea of motion depends on that of extension, and the idea of extension on that of solidity. 'Tis impossible, therefore, that the idea of solidity can depend on either of them. For that woul'd be to run in a circle, and make one idea depend on another, while at the same time the latter depends on the former. Our modern philosophy, therefore, leaves us no just nor satisfactory idea of solidity; nor consequently of matter.[10]

Drawing on his argument that activity is not a necessary characteristic of a physical thing's continuing in existence, Hume rejected the notion that we have an idea of power, force or efficacy. Instances of power in bodies are not observed under any specifiable conditions: "All ideas are deriv'd from, and represent impressions. We never therefore have any idea of power".[11] Even if force or power did exist as an independent entity, it could not be a characteristic or substitute for matter, the existence of which Hume also doubted: "When we reason from cause and effect, we conclude, that neither colour, sound, taste, nor smell have a continuous and independent existence. When we exclude these sensible qualities there remains nothing in the universe, which has such an existence".[12]

Though the natural philosophers of the eighteenth century could accept the reduction of primary qualities to the perceptible domain of the secon-

---

[9] *Ibid.*, 2, 52.

[10] David Hume, *A Treatise of Human Nature*, ed. Selby-Bigge, Oxford, 1888, pp. 228f.

[11] *Ibid.*, p. 161.

[12] *Ibid.*, p. 231.

dary, they could not accept such a skeptical conclusion, which threatened the very possibility of scientific knowledge. The Scottish Common-Sense school of philosophy, in particular, reacted strongly against it. Thomas Reid attempted a repudiation of what he characterized as Hume's uncompromising skepticism.

§3 *Thomas Reid*

Reid's appeal to common sense appears very clearly in his examination of the concept *power* in the thought of Locke, Hume and, by implication, Berkeley. In the *Essays on the Active Powers of the Human Mind* (1788), Reid considers terms like 'agency', 'efficacy', 'action', 'cause' and 'change', which in everyday language are closely associated with power.[13] He argues that the idea of power cannot be derived from experience, and though he agrees with Hume that we have no idea of power either from sense or reflection, he holds, nevertheless, that we have such a concept and moreover that it is meaningful, clear and distinct:

> The only distinct conception I can form of active power is, that it is an attribute in a being by which he can do certain things if he wills. This, after all, is only a relative conception. It is relative to the effect, and to the will of producing it. Take away these, and the conception vanishes.[14]

Active power is thus known only relative to its effects. Reid denied that we can know power in the Lockean sense of arising from collocations of primary qualities, but argued that we derive the idea of power from an instinctive disposition to see nature as uniform with respect to change and from attention to the operations of the mind, though "we neither perceive the agent nor the power, but the change only".[15] Accordingly, though the concept of power is not found in sensory experience, like causation and the uniformity of nature it anticipates and structures our experiences. In this way Reid hoped to provide a justification of the concept of power, since he locates the genesis of the idea in an area of human nature not treated explicitly by Hume's philosophical principles. This is clearly stated in the following passage from his *Essays on the Intellectual Powers of Man* (1785):

---

[13] Thomas Reid, *Essays on the Active Powers of the Human Mind*, intro. B. Brody, Cambridge, Mass., 1969, p. 28.

[14] *Ibid.*, pp. 38f.

[15] *Ibid.*, p. 33.

It is not easy to say where we first get the notion or idea of power. It is neither an object of sense nor of consciousness. We all see events one succeeding another; but we see not the power by which they are produced. We are conscious of the operations of our minds; but power is not an operation of mind. If we had no notions but such as are furnished by the extreme senses and by consciousness, it seems to be impossible that we should ever have any conception of power. Accordingly Mr. Hume, who has reasoned the most accurately upon this hypothesis, denies that we have any idea of power, and clearly refutes the account given by Mr. Locke of the origin of this idea.

But it is in vain to reason from an hypothesis against a fact, the truth of which every man may see by attending to his own thoughts. It is evident that all men, very early in life, not only have an idea of power, but a conviction that they have some degree in themselves, . . . without which no man can act the part of a reasonable being.[16]

In his *Essay on the Active Powers of the Human Mind*, Reid went on to argue that for Newton, science was restricted to establishing laws connecting antecedent and consequent conditions. In this view a law can be construed as a cause. For this reason, it was not the concern of natural philosophy to discover causes having powers, and so natural philosophy can avoid the ambiguities inherent in terms like 'cause', 'agency' and 'active power'.[17] The notion of active power is nevertheless meaningful, and efficient causes actually exist in reality even though the mind can never know "what their nature, their number, and their different offices may be".[18] Despite this, Reid held that bodies could be conceived as sets of powers, and stated, in his *Inquiry into the Human Mind* (1764) : "I conclude, then, that colour is not a sensation, but a secondary quality of bodies, in the sense we have already explained; that it is a certain power or virtue in bodies, that in fair daylight exhibits to the eye an appearance, which is very familiar to us, although it hath no name".[19]

The similarity to Locke's analysis of powers giving rise to secondary qualities is apparent, and it is possible that Reid is indebted to the former for his dispositional theory of material objects. In this way, Reid can be seen to have provided a justification for the ascription of powers to matter. It is this doctrine that is fundamental to the thought of Priestley and Hutton.

---

[16] Thomas Reid, *Essays on the Intellectual Powers of Man*, ed. Woozley, London, 1941, pp. 382f.

[17] Reid, *Active Powers of the Human Mind*, p. 41.

[18] *Ibid.*, p. 47.

[19] Thomas Reid, *Inquiry into the Human Mind on the Principles of Common Sense*, ed. Duggan, London, 1970, p. 101.

§4 *Joseph Priestley*

Priestley rejected many of the specific doctrines of the Common-Sense school—for example, the doctrine of instinctive principles of the mind and their analysis of sensation—but in accepting Locke's doctrines of power and causation, the influence of Reid on his thought is apparent. In an effort to combat the immaterialism and skepticism of Berkeley, Priestley propounded a philosophical monism in which mind was held to be a modification of matter: this he called "materialism". His conception of matter, therefore, is central to his general philosophy of nature and man. Adhering to a theory of knowledge related to the contents of sensory experience, he rejected the traditional theories of matter which insisted on the absolute existence of primary qualities such as solidity. He argued, moreover, that our notion of solidity was occasioned by experience of resistance which was itself occasioned by powers of repulsion. In his *Disquisitions Relating to Matter and Spirit* (1777), he argued that all that could be gathered about external reality was that "all resistance can differ only in degree, this circumstance . . . [leading] to the supposition of a greater or less repulsive power, but never to the supposition of a cause of resistance entirely different from such a power".[20] Newton's third Rule of Philosophizing is here implicitly denied, since an invisible realm beyond the experience of the senses, though it may exist, is not a possible object of knowledge. This conception of matter, Priestley held, was not incompatible with the characteristics of the mind, since on his view being "destitute of what has hitherto been called *solidity* matter was no more incompatible with sensation and thought, than that substance, which, without knowing anything farther about it, we have been used to call *immaterial*".[21]

Fundamental to Priestley's philosophy of matter was his denial that there were two distinct kinds of substance—matter and spirit. While matter had been held to be endowed with primary qualities and to be devoid of powers, spirit had been supposed destitute of extension but possessed of active powers. In rejecting this dichotomy, Priestley argued that two substances could not be "capable of intimate connection and mutual action unless they had common properties"[22] and claimed that matter could not be defined as separate from its powers of attraction and repulsion:

---

[20] Joseph Priestley, *Disquisitions Relating to Matter and Spirit*, in *The Miscellaneous and Theological Works of Joseph Priestley*, ed. Rutt, 25 vols., London, 1817–31, 3, 227.

[21] *Ibid.*, 3, 230.

[22] *Ibid.*, 3, 218f.

> I therefore, define it [matter] to be a substance possessed of the property
> of *extension*, and of *powers of attraction or repulsion*. And since it has never
> yet been asserted, that the powers of *sensation* and *thought* are incompatible
> with these (*solidity*, or *impenetrability* only, having been thought to be re-
> pugnant to them) I therefore maintain, that we have no reason to suppose
> that there are in man two substances so distinct from each other, as have
> been represented.[23]

The essence of matter was therefore extension together with inherent
powers of attraction and repulsion. Resistance was not due to the impenetrability and solidity of matter but to a power of repulsion; powers were
essential to the actual existence of matter and so he rejected Locke's argument that solidity constituted the essence of matter:

> I by no means suppose that these powers, which I make to be essential to
> the being of matter, and without which it cannot exist as a material sub-
> stance at all, are *self-existent* in it. All that my argument amounts to, is, that
> from whatever source these powers are derived, or by whatever being they
> are communicated, matter cannot exist without them. . . . Whatever *solidity*
> any body has, it is possessed of it only in consequence of being endued with
> certain *powers*.[24]

Thus solidity and substance were the mere effects of the powers, so that
matter was such in that it possessed extension and powers of attraction and
repulsion. Without these powers it would be nothing except vacuous extension. Priestley did not explain the way in which powers could "inhere
in" or "belong to" matter,[25] and he regarded the powers of attraction and
repulsion as active properties which by their action gave the appearance of
solidity and impenetrability to matter. Thus, in effect, Priestley was arguing
that matter was a set of substantive powers with respect to extension.

Though in general Priestley was critical of Reid's philosophy, his view
of the origin of the concept *power* was similar. In his *Introductory Essays
To Hartley's Theory of the Human Mind* (1775), Priestley stated that:

> The idea of *power* seems at first sight to be a very simple one; but it is in
> fact exceedingly complex. A child pushes at an obstacle, it gives away, . . .
> in like manner he practises a variety of other bodily and mental exercises, in
> which he finds that it only *depends upon himself* whether he perform them
> or not; and at length he calls that general feeling, which is the result of a
> thousand different impressions, by the name of *power*. . . . Even inanimate
> things have certain invariable *effects*, when applied in a particular manner.
> Thus a rope sustains a weight, a magnet attracts iron, a charged electrical
> jar gives a shock, etc. From these and other similar observations, we get the

---

[23] *Ibid.*, 3, 219.
[24] *Ibid.*, 3, 224f.
[25] *Ibid.*, 3, 238.

idea of *power, universally and abstractedly considered;* so that in fact, the idea of power is acquired by the very same mental process by which we acquire the idea of any other property belonging to a number of bodies, viz. by leaving out what is peculiar to each, and appropriating the term to that particular circumstance or appearance, in which they all agree.[26]

Priestley then went on to claim that the "idea of *solidity,* or *impenetrability* [is] what could not be deduced from *sense,* but must have its origin in the understanding".[27] In his *Letters to a Philosophical Unbeliever* (1780), he argued that "there is nothing in the idea of *power* or *causation* (which is only the same idea differently modified) that is not derived from the impressions to which we are subject, this being to be ranked in the class of *abstract ideas,* where it does not appear that Mr. Hume ever thought of looking for it".[28] The influence of Locke is evident here.

It is significant that Priestley developed his theory of matter as powers in terms of the Newtonian theory of the primacy of force and the paucity of matter in the world. In arguing that the powers and forces in nature were all that constituted matter, he referred to the Newtonian doctrine that space contained very little solid matter:

> The principles of the Newtonian philosophy were no sooner known, than it was seen how few, in comparison, of the phenomena of nature were owing to *solid matter,* and how much to *powers,* which were only supposed to accompany and surround the solid parts of matter. It has been asserted . . . that all the solid matter in the solar system might be contained within a nutshell, there is so great a proportion of *void space* within the substance of the most solid bodies. Now, when solidity had apparently so very little to do in the system, it is really a wonder that it did not occur to philosophers sooner, that perhaps there might be nothing for it to do at all, and that there might be no such thing in nature.[29]

Thus, Priestley connected his denial of solidity and his theory that matter was a set of powers with the Newtonian doctrine of the paucity of matter and vacuity of the universe; powers and forces thus became the primary agents in nature. Priestley extended the Newtonian theory of the paucity of matter and the primacy of force in the context of the non-Newtonian doctrine of matter as defined by powers, rather than by atoms characterized by essential qualities.

---

[26] Priestley, *Introductory Essays to Hartley's Theory of the Human Mind,* in *Miscellaneous and Theological Works,* 3, 191.

[27] *Ibid.*

[28] Priestley, *Letters to a Philosophical Unbeliever,* in *Miscellaneous and Theological Works,* 4, 398.

[29] Priestley, *Disquisitions Relating to Matter and Spirit,* in *Miscellaneous and Theological Works,* 3, 230.

§5 *James Hutton*

For Hutton the proper business of natural philosophy was to "investigate the powers or laws of action", as he put it in his *Dissertations on Different Subjects in Natural Philosophy* (1792).[30] But these could only be investigated on the basis of a theory of matter resting on established principles of knowledge. Hutton made a distinction between sensible and perceptible qualities, the first arising from the immediate effect of sensation and the second carrying an existential import by the action of the mind. Moreover, he distinguished these qualities from judged or inferred qualities which "proceed in reason from those sensible and perceptible qualities".[31] For Hutton, sensible and perceptible qualities only characterized observable natural phenomena—which Hutton called "body"—whereas the constituents of body—which Hutton called "matter"—"denoted something inferred, or judged of, from things which appear".[32] Hutton argued that all qualities were relational, so that properties such as extension and resistance did not "arise from the absolute nature of the thing",[33] but were dependent on the resisting powers from which they arose. Thus knowledge of bodies was relational, that is relative to conditions obtaining in the powers external to the mind. As Hutton put it, powers and the sensible qualities to which they gave rise were "conditional":

> Instead then of saying that matter, of which natural bodies are composed, is perfectly hard and impenetrable, which is the received opinion of philosophers, we would affirm, that there were no permanent properties of this kind in a material thing: but that there were certain resisting powers in bodies, by which their volume and figures are presented to us in the actual information, which powers, however, might be overcome . . . [thus] the extension of a body, would be considered only as a conditional thing.[34]

The teleological cast of Hutton's thought is very important, for he aimed "to see the general order that is established among the different species of events, by which the whole of nature, and the wisdom of the system is to be perceived".[35] He denied that final causes could not be discovered, asserting that they were "the proper object of our knowledge". It was only when final causes were discovered that "we may be said to understand those things, when we have seen the end for which they are

---

[30] James Hutton, *Dissertations on Different Subjects in Natural Philosophy*, Edinburgh, 1972, p. ix.

[31] *Ibid.*, p. 287.

[32] *Ibid.*, pp. 278f.

[33] *Ibid.*, p. 292.

[34] *Ibid.*, p. 290.

[35] *Ibid.*, p. 262.

intended in the system of this world, and perceive the means by which, in the wisdom of nature, the end is certainly effected".[36] This teleological approach is of special importance for Hutton's theory of knowledge and matter. He drew an analogy between the intentions of the mind and the structure of nature, arguing that the order and design constitutive of the mind was reflected in the structure of nature teleologically oriented: "when a regular order is observed in those changing things, whereby a certain end is always attained, there is necessarily inferred an operation somewhere, an operation similar to that of our mind, which often premeditates the exertion of a power and is conscious of design".[37] Thus the operations of powers producing the sensible qualities of bodies could be premediated in judgment, and for Hutton the characteristics of such powers were to be inferred from these sensible qualities.

The notion of *power* is crucial to Hutton's philosophy of matter. In discussing the nature of substance in his *Investigation of the Principles of Philosophy* (1794), he argued that Newton's view of solidity as the proper idea of substance was untenable, and stated that "magnitude and figure have no other existence than in the conceiving faculty of our mind" for "these qualities are truly ideas formed upon certain occasions and according to certain established rules". He concluded that "this philosophical idea of substance falls to the ground, and, together with it, all the material system built thereon".[38] Like Priestley, Hutton denied that solidity was the essence of substance, but he went further than Priestley in holding that power alone characterized matter. For Hutton "body" possessed extension and figure, while "matter", which constituted "body", was to be "considered as the substance, essence, or principles of external things".[39] On his view, "matter . . . will appear to be a thing absolutely different from that external thing which is perceived by our mind; and the proper attribute of matter will be, having the power to affect our mind in making us to know".[40] However, with respect to experience, "power, the cause of our sensation, is to be considered as a first cause".[41]

Hutton made an important distinction between "the physical and metaphysical ideas of matter".[42] The physical idea of matter was derived from

[36] *Ibid.*, p. 624.
[37] *Ibid.*, p. 285.
[38] James Hutton, *Investigation of the Principles of Knowledge, from Sense to Science and Philosophy*, 3 vols., Edinburgh, 1794, 2, 393.
[39] *Ibid.*, 2, 399.
[40] *Ibid.*, 2, 407.
[41] *Ibid.*, 2, 387.
[42] *Ibid.*, 2, 407.

our experience of powers, and on the level of experience "power and matter are found to mean the same thing".[43] Nevertheless, he pointed out that when power was exerted, there must be "a substance existing, in which that power should reside";[44] this was matter in its metaphysical aspect, "matter being properly the thing, and power the attribute thereof".[45] Though he argued that we cannot literally ascribe sensible qualities to invisible matter, there being no absolutes in reality, in his *Dissertations* he stated that it is from these sensible qualities that we are to *infer* the characteristics of powers and the metaphysical idea of matter: "nothing is to be allowed, as belonging to matter, that is not authorized in the strictest examination of actual things".[46] Thus, like Priestley, Hutton rejected Newton's third Rule of Philosophizing; for to characterize solid extended "body" by means of solid, extended "matter" was to assume as the principle of bodies "nothing but the bodies themselves under the pedantic designation of atoms or corpuscles".[47] Nevertheless, the nature of the unobservable substratum was to be inferred from the sensible qualities which were characteristic of bodies.

Hutton's theory of material existence has three distinct levels: body, matter in its physical aspect (as manifested by powers) and matter in its metaphysical aspect (a non-spatial substance, powers being its attribute). It is at this third level, which was only indirectly manifested through the action of powers, that the qualities of matter as a substratum can only be characterized negatively:

> But whatever matter is of itself, it must be considered as the cause of motion and resistance in natural bodies; and this is all that we are permitted to judge of in the science of physics. We are never able to learn to know what matter is in itself; nor have we any occasion for that knowledge. . . . But though we know not what matter truly is, we certainly may know what it is not.[48]

Thus "matter" is destitute of bodily form, cannot change place in space, is non-solid, non-spatial and incessantly active. Though its ontological status is somewhat unclear, it seems that for Hutton the substratum is non-material in nature. That there is a harmony between the constitution of the mind and the processes of nature assures us that we may have some idea of matter in itself through the design of its operations manifested by powers.

---

[43] *Ibid.*, 2, 403.
[44] *Ibid.*, 2, 389.
[45] *Ibid.*, 2, 403.
[46] Hutton, *Dissertations on Different Subjects*, p. 300.
[47] *Ibid.*, p. 669.
[48] *Ibid.*, p. 315.

Dugald Stewart supported Hutton's theory by arguing that Locke had implicitly conceived the essence of matter in terms of repulsive powers. Stewart gave a clear statement in support of the conception of matter advanced by Priestley and Hutton:

> The effects . . . which are vulgarly ascribed to actual contact, are all produced by repulsive forces, occupying those parts of space where *bodies* are perceived by our senses, and, therefore, the correct idea that we ought to annex to *matter*, considered as an object of perception, is merely that of a *power of resistance*, sufficient to counteract the compressing power which our physical strength enables us to exert.[49]

Referring to Book II, chapter IV of Locke's *Essay* ("Of Solidity"), Stewart maintained that Locke, in analyzing cohesion and compressibility, implicitly held that the essence of matter could be conceived in terms of repulsive powers. Thus Stewart consciously connected Locke's theory of powers with the conceptions of matter of Priestley and Hutton. Stewart concluded that the views of Priestley and Hutton "with respect to *matter*, so far as hardness or relative incompressibility is concerned, offer no violence to the common judgement of mankind, but aim only at a more correct and statement of *the fact* than is apt to occur to our first hasty apprehension".[50]

The view that the essence of matter is power led, in Scotland, to a critique of contact action. The argument was clearly put by John Playfair in his "Biographical Account of James Hutton":

> But if this be granted, and if it be true that in the material world every phenomenon can be explained by the existence of power, the supposition of extended particles as a *substratum* or residence for such power, is a mere hypothesis, without any countenance from the matter of fact. For if these solid particles are never in contact with one another, what part can they have in the production of natural appearances, or in what sense can they be called the residence of a force which never acts at the point where they are present? Such particles, therefore, ought to be entirely discarded from any theory that proposes to explain the phenomena of the material world.
> Thus, it appears, that power is the essence of matter, and that none of our perceptions warrant us in considering even body as involving anything more than force, subjected to various laws and combinations.[51]

Playfair went on to state that matter conceived in this way was "indefinitely extended" through all space as "is proved by the universality of gravitation". As the passage from Playfair shows, if interstitial particles can be conceived

---

[49] Dugald Stewart, *Philosophical Essays*, 3d ed., Edinburgh, 1818, p. 123.

[50] *Ibid.*, p. 133.

[51] John Playfair, "Biographical Account of James Hutton," in *The Works of John Playfair*, 4 vols., Edinburgh, 1822, *4*, 85.

as being unable to come into contact, they serve no important role in explaining the properties of things, since these can be shown to arise from intrinsic powers. He made the point clearly with respect to gravitation. It is not "according to this system . . . the action of two distant bodies upon one another, but it is the action of certain powers, diffused through all space, which may be transmitted to any distance".[52] Thus, conceiving the essence of matter as powers was seen to be not only incompatible with contact action, but also with the theory of invisible particles.[53]

---

[52] *Ibid.*, *4*, 86.

[53] The themes of this paper are discussed in greater detail, and with a wider reference to the natural philosophy of the seventeenth and eighteenth centuries, in our paper on "Newtonian forces and Lockean powers: Concepts of matter in eighteenth-century thought", *Historical Studies in the Physical Sciences*, ed. R. McCormmach, *3*, 1971, pp. 233–306.

ACTION AT A DISTANCE

Mary B. Hesse

The question whether matter can act at a distance across empty space or whether some medium of contact, or ether, is required, is one which has constantly recurred in physics, and has been thought to raise philosophical as well as purely scientific issues. Stated thus baldly, the problem lacks definiteness, and requires careful analysis of what is meant by 'matter', by 'act', by 'empty space', and by 'ether'. Many would now hold that the answer to the question of action at a distance depends essentially on our concept of matter and is relative to the existing state of science, rather than that our concept of matter must be framed in accordance with an *a priori* decision about action at a distance. But this has not always been so, and it is clear that, historically at least, metaphysical convictions about possible types of interaction have often been related to the acceptability of certain kinds of physical theory. This essay attempts to trace the historical relation between the action-at-a-distance problem and the progress of science, and to assess the permanent conceptual issues that seem to be involved.

§1 *Definition of the Problem*

Things may influence or act upon other things in many different ways: they may collide with each other; they may react chemically; they may attract or repel like two magnets, or on the other hand like two human beings; they may communicate at a distance by shooting particles at each other, or by transmitting a disturbance in a medium; they may, if they are human beings, speak to one another, or even (perhaps) communicate telepathically. All these and many other processes have been used as models for fundamental interactions in the history of physics. It will be noticed that those which involve only non-intelligent bodies, tend to be actions by contact (with the significant exception of the magnet), while those which are, superficially at least, actions at a distance, usually involve human beings. This general feature of physical interaction is already enough to explain the widespread preference for contact-action theories in physics, and the intuitive suspicion that action at a distance must retain overtones of anthropomorphism.

Two questions have to be distinguished here. There is first the conceptual or methodological question of whether interaction at a distance can be understood or explained at all without postulating a medium of *some kind;* and there is secondly the empirical question of whether, if a medium is postulated or "discovered", it is a substance similar to or different from ordinary observable matter. The first question can be made independent of whatever concept of matter may be accepted, so long as one is sufficiently liberal in one's understanding of what a "medium" can be (even empty space could be said in a sense to be a medium), but the second question cannot even be formulated until the concept of matter is more or less clearly defined. As a matter of historical fact, it does not seem that the first question has ever been put in the absence of *some* conception of what would count as a "material" medium, but sometimes a negative answer to it has forced natural philosophers to conceive of "immaterial" media as conceptual necessities.

Since it was the Greeks who first explicitly studied the concept of matter, it is to them that we must look for the first statements of the problem of action at a distance. Two streams of thought in Greek philosophy combined in the view that there is no action at a distance, and that all action is by contact of parts of matter. The first was Atomism, which attempted to solve the dilemma of the Parmenidean changeless "Being" by postulating atoms moving in the "Not-Being", or void. This meant that the atoms were, like Parmenides' Being, imperishable, homogeneous, and without intrinsic qualities, but the possibility of motion and change was saved by distinguishing atoms from void, and hence by endowing atoms with geometrical shape. Thus, chiefly as a consequence of the metaphysical impasse created by Parmenides, the notion of primary and secondary qualities of matter first entered philosophy; the primary qualities being mechanical and geometrical, and the secondary qualities being definable or constructible in terms of these. There was no suggestion that the atoms could act at a distance, for this would have meant that they were endowed with powers or qualities other than those necessary to define pure Being: moreover any such action would have to take place, not merely across empty space in any modern sense, but across "non-existence". Being could affect Being only by coming into contact with it.

A second characteristic of Greek thought, more practical in application, led to the same conclusion. This was the widespread use of mechanical analogues in scientific explanations, both among the Presocratic cosmologists, and among the Hippocratic medical writers before and during the time of Aristotle. Increased familiarity with mechanical techniques and inventions led to comparisons between the mechanisms of the body and such gadgets as pumps, bellows and levers, and between the movements of bodily fluids and irrigation channels, boiling liquids, differential deposition of solid matter in centrifuges, and so on. Some such mechanical processes appear to involve action at a distance, or at least attractive as opposed to impact forces: for example the "suction" of a vacuum, and the behaviour of magnets. But it is clear even in Plato and Aristotle that such apparent exceptions to the general rule of mechanical action by impact or pressure were not regarded as fundamental, and attempts were made to explain them in terms of the circular motions of the air or other "emanations", exerting pressure rather than attraction. According to Plato and Aristotle and their schools, there are no "pulls" but only "pushes" in nature.

It is not clear that this conviction was based on any consistent metaphysics, and even Aristotle's remarks about it read like rationalizations after the event. But Aristotle, unlike the Atomists, left open the possibility of diversity of kinds of material substance. Since there is no void for Aristotle, his "matter" is ubiquitous, and in itself qualityless, and therefore when Aristotle insists that every motion and change has a mover in contact with it, he may be understood to be saying no more than that matter is informed by power of *some* kind at places adjacent to a body in forced motion. This would not commit him to any particular mechanism for the transmission of this power from agent to distant patient, and the many kinds of immaterial emanation and substantial form which were later postulated to account for it could rightly be claimed to be consistent with Aristotelian teaching. There is here the ambiguity which runs right through the history of concepts of action at a distance, namely that if the medium of transmission is not defined too closely, the assertion that action must be continuous through the medium may be no more than a directive to look for some continuity of qualities, no matter what, and the search is almost always successful, although the result may be quite unlike matter with the qualities we know from experience of gross bodies.

§2 *The Mechanisation of Physics*

For this reason, the problem of action at a distance became much sharper with the successful mechanisation of physics in the seventeenth century. Cartesianism and Atomism, though to some extent rivals, both presupposed a concept of matter involving only the "primary", geometrical and mechanical, qualities. One of the most revolutionary features of this period of scientific revolution was the elimination from physical theory of the quasi-material, quasi-animistic, substances which had been postulated in Renaissance philosophy to account for the transmission of physical action. It was not so much Aristotle's mechanism that was rejected in the seventeenth

century as the self-moving souls, substantial forms, sympathies and antip-
athies, *horror vacui,* multiplication of species, and the rest, which pervaded
later Aristotelianism. Thus it was not enough for Descartes to seek to
explain, for example, magnetic attraction, by some kind of emanation "lay-
ing hold of" pieces of iron and drawing them to the magnet, like, as Gassendi put it, a chameleon darting out its tongue to catch a fly. Descartes
rather sought a vortex mechanism in terms of subtle particles forced
through the pores of the magnet, and drawing particles towards the magnet
by the pressure of circular thrust. As with Aristotle, there are no pulls but
only pushes in nature.

The programme of the corpuscular philosophy, accepted by almost all
seventeenth-century philosophers, was to explain all physical, and even
biological, processes in terms of mechanical particles in motion. A subsidiary point of disagreement between Atomists and Cartesians was the question whether or not there is void, but both parties agreed that there is in
any case no action across void—for the Atomists, particles can only affect
one another by colliding. The mechanistic programme, however, caused
scientific difficulties. How could it be made to account for those obvious
phenomena which seem to require forces of attraction, and even forces
acting at a distance—for example, the cohesion of solid bodies, the rise of
liquids in thin tubes, the expansion and contraction of gases, the motion of
heavy bodies towards the earth, the action of the moon on the tides, the
behaviour of magnets? Cartesian vortex mechanisms invented to account
for these were ingenious but unconvincing, especially as the behaviour of
bodies came to be described in more detail and in quantitative and mathematical terms. The stage was set for Newton's fundamental physical synthesis, and at the level of the concept of matter, this synthesis implied one
significant addition to the furniture of the world allowed by the mechanical
philosophers: namely the mutual attraction and repulsion of bodies at a
distance.

The first significant appearance of attractive forces was in Newton's
theory of gravitation, although Newton himself was reluctant to regard
this attractive power of bodies as an ultimate property, and sought some
mechanical explanation in terms of pressures in a fluid ether pervading all
space. But he would not allow such speculations to affect his mathematical
presentation of the theory in the *Principia Mathematica,* and insisted there
that a merely *phenomenal* description of the motions of the heavens implied that bodies have an acceleration towards each other defined by the
inverse-square law, no matter what *cause* of this acceleration may be
imagined. In his later discussions of other physical phenomena, however,
chiefly in the *Optics,* he seems to accept as fundamental, at least provi-

sionally, attractive and repulsive forces acting at a distance in order to account for cohesion, elasticity of gases, the behaviour of magnets and electric bodies, certain chemical reactions, and reflection and refraction of light corpuscles, as well as the phenomena of gravitation.

For the more conservative mechanists, this seemed like a reversion to all the immaterial influences, sympathies, and occult qualities which had been banished from physics so recently and with so much difficulty. One of the best accounts of the points at issue between the Newtonians and their critics is to be found in the Leibniz-Clarke correspondence. This was initiated by Leibniz in 1716 in a letter in which he asserted that the philosophy of Newton was contributing to the decay of natural religion in England. Clarke replied on behalf of Newton, and the correspondence continued until Leibniz's death, ranging widely over Leibniz's metaphysical principles and over scientific topics such as the nature of space and time, the existence of vacuum, and the theory of gravitation.

Leibniz argues on grounds of continuity and sufficient reason that there is no void, and asserts that the vacua apparently produced in the barometer tube and in air pumps are void only of gross matter and not absolutely empty. The space contains rays of light "which are not devoid of some subtle matter", and "the effluvia of the load-stone, and other very thin fluids may go through" the glass. Clarke objects that if these were material they would exert resistance, and there is no resistance to bodies moving in a vacuum. But he is prepared to admit that immaterial substances may be present in void space: "God is certainly present, and possibly many other substances which are not matter; being neither tangible, nor objects of any of our senses". As for action at a distance, Clarke agrees with Leibniz that it is impossible. In reply to Leibniz's assertion that an attraction causing a free body to move in a curved line would be a miracle, he says:

> That one body should attract another without any intermediate means is indeed not a miracle, but a contradiction: for 'tis supposing something to act where it is not. But the means by which two bodies attract each other, may be invisible and intangible, and of a different nature from mechanism; and yet, acting regularly and constantly, may well be called natural; being much less wonderful than animal-motion; which is never called a miracle.[1]

But Leibniz will not have any talk of

> . . . some immaterial substances, or some spiritual rays, or some accident without a substance, or some kind of *species intentionalis*, or some other I know not what . . . That means of communication (says [Clarke]) is invisi-

---

[1] *The Leibniz-Clarke Correspondence,* ed. H. G. Alexander, Manchester, 1956, p. 53.

ble, intangible, not mechanical. He might as well have added, inexplicable, unintelligible, precarious, groundless and unexampled.[2]

Clarke replies that attraction is only the name of an empirical fact, whatever may be its explanation, and therefore it should not be called occult, even if its efficient cause is not yet discovered. If Leibniz or anyone else can explain these phenomena by the laws of mechanism he will "have the abundant thanks of the learned world". But elsewhere Clarke is doubtful of the possibility of explaining gravitation by impulse, since it depends on quantity and not surface area of matter; hence he thinks it must be due to something "immaterial" which penetrates matter.

The argument is the same as that which runs through the whole history of theories of interaction. If there is continuity of causes transmitting actions from one part of space to another, what is the nature of these causes? Leibniz holds the comparatively "modern" view of Cartesian mechanism, asserting that matter can act upon matter only by contact and according to the laws of mechanics, and that soul or spirit cannot act upon matter at all, for if it did it would violate the natural conservation of *vis viva*. Clarke holds the traditional view that immaterial spirit may act upon body, remarking that this is as easy to conceive as cohesion "which no mechanism can account for", and he maintains in particular that God is substantially present everywhere and may intervene by acting directly upon matter. For Leibniz the activity of God is of another order, and from the scientific point of view he was no doubt right, in 1716, to question Clarke's immaterial substances and non-mechanical explanations, for the time had not come when these could be made sufficiently precise to be acceptable in a scientific theory. There is no disagreement between them about the facts, the disagreement lies in the question of what kinds of theory are to be admitted in explanation of the facts, and Leibniz shows a less empirical spirit than Clarke in insisting that theories must conform to certain *a priori* conceptions of interaction.

Leibniz's arguments were in fact ignored by most eighteenth-century physicists, who accepted Newton's matter endowed with attractive and repulsive powers, and used it in particular in various fluid theories to account for the phenomena of heat, electricity and magnetism, and chemical reactions. Leibniz had, however, one professed disciple: Boscovich, who published in 1758 a universal physics, the *Theoria Philosophiae Naturalis,* ostensibly based on Leibnizian principles. But in fact Boscovich had turned Leibniz inside out—instead of rejecting the void in the interests of continuity, he effectively rejected the extension of matter on the same grounds.

---

[2] *Ibid.,* p. 94.

For Boscovich "matter" exists in the form of point particles carrying inertia, but interacting by means of distance-forces of attraction and repulsion whose magnitude depends on the distance between the particles. It was this theory which later influenced Faraday at a crucial point in nineteenth-century physics, and led to the notion of force-fields rather than "matter" as the fundamental physical entities. To this development we now turn.

§3 *Field Theories*

In the early nineteenth century three modes of transmission of physical action were recognized:

(i) Action by impact of ultimate atomic particles, as described by Newton's mechanics.

(ii) Action at a distance by attractive and repulsive forces, including repulsive forces exerted at short distance between parts of gases and other elastic fluids, and gravitational attraction apparently exerted over long distances through empty space.

(iii) Action in a continuous medium, described by the extension of Newton's mechanics of point particles to the mechanics of continuous fluids by Euler, and extended later to the mechanics of continuous elastic solids.

This third type of interaction theory was at first of mathematical rather than physical significance, and had in fact been developed by ignoring the fundamental physical problem of the nature of the elementary parts of matter and its forces. But it provided the first mathematical models for later field theory. From the purely mathematical point of view, a *field* in physics may be defined as a region of space in which each point (with possibly isolated exceptions) is characterised by some quantity or quantities which are functions of the space coordinates and of time, the nature of the quantities depending on the physical theory in which they occur. Euler created a field theory in this sense for hydrodynamics, by characterising the field of motion of a fluid by functions of the space coordinates, such as the velocity of the fluid and its pressure at each point. In this purely mathematical sense, however, it might be said that the formulation of the theory of gravitation in terms of a potential function defined at all points of a space containing gravitating bodies, is also a field theory. But there is clearly a physical difference between a gravitational potential of this kind and the velocity-field of a fluid. The fluid field is a continuous material medium having properties other than velocity, but the gravitational "field" has apparently no properties other than the potentiality for exerting attractive force on masses introduced into it.

It was Faraday who first made explicit some of the conditions under which one can speak of physical interaction through a "real" field, as opposed to pure action at a distance. His criteria for "physical reality" are interesting in showing how the conception of "reality" is, as it were, made anew with each physical theory. It is precisely because Faraday does not accept the definitions of mechanical and chemical "matter" as the sole criteria of reality that he is able to introduce what Maxwell later called "quite a new conception of action at a distance" in which real physical processes are going on in the medium, although in the mechanical sense the space is "empty".

In a letter of 1844, published in the *Philosophical Magazine* and entitled "A speculation touching Electrical Conduction and the Nature of Matter", Faraday characterises the current view of the atomic constitution of matter: atoms have a certain volume and are endowed with powers which hold them together in groups, but chemical experiments show that they do not touch, therefore only space is continuous throughout matter considered as an aggregate of atoms. But this leads to a paradoxical consequence in regard to electric insulators and conductors, for the space between the atoms of an insulator must be an insulator, and space between the atoms of a conductor must be a conductor. This contradiction leads Faraday to the view, which he ascribes to Boscovich, that an atom is a *point* with "an atmosphere of force grouped around it". The properties of a body, such as conduction, relation to light or magnetism, solidity, hardness, specific gravity, must then belong, not to a "nucleus" abstracted from its powers (for this is in any case inconceivable), but to the forces themselves.

> But then surely the *m* [the atmosphere of force] is the *matter* of the potassium, for where is there the least ground (except in a gratuitous assumption) for imagining a difference in kind between the nature of that space midway between the centres of two contiguous atoms and any other spot between these centres? A difference in degree, or even in the nature of the power consistent with the law of continuity, I can admit, but the difference between a supposed little hard particle and the powers around it, I cannot imagine.

It follows, in contrast to the orthodox view, that "matter" is everywhere continuous and that "atoms" are highly elastic and deformable, is mutually penetrable, and that

> matter fills all space, or, at least, all space to which gravitation extends . . . for gravitation is a property of matter dependent on a certain force, and it is this force which constitutes the matter. . . . This, at first sight, seems to fall in very harmoniously with . . . the old adage, "matter cannot act where it is not".

This paper seems to mark a decisive transition from the conception of mechanical action to that of continuous action in terms of forces filling space.

Here *all* physical forces are regarded in this way, but in other papers, of 1851 and 1852, Faraday does not make the transition in the case of gravitation, but explicitly distinguishes action at a distance from the new conception of action in a real medium by means of three questions about gravity, radiation, and electric and magnetic force:

(i) Can transmission of action be affected by changes in the intervening medium, as regards, for instance, a bending of the lines of force, or polarity effects? If the answer is Yes, as in the cases of electric and magnetic induction but unlike gravitation, this is an indication of a real process going on in the medium.

(ii) Does the transmission take time? Again an affirmative answer indicates the presence of a real medium.

(iii) Does the transmission depend upon a "receiving" end? An affirmative answer here indicates an action at a distance, as in the case of gravitation, but not of radiation. This, however, is not a decisive criterion, because electric and magnetic effects, like gravitation, do seem to depend on *two* poles of the interaction, and yet Faraday wishes to regard them as actions in a medium, on the strength of the answer to (i). After investigating experimental answers to the three questions, Faraday's conclusions are that gravitation seems to be a pure action at a distance, whereas radiation is clearly action in a medium, and electric and magnetic effects are also actions in a medium, although the physical nature of this medium is still obscure.

The question of the nature of the media, both of light radiation and of electric and magnetic phenomena, brings us to the nineteenth-century theories of the ether. Early mathematical investigation of optics, following Fresnel's mathematical wave theory of diffraction, refraction and polarisation, had been content to derive Fresnel's equations from the most general principles of mechanics, namely Lagrange's equations or the principle of least action, without attempting to describe an actual etherial mechanism which would account for the phenomena of light propagation. There were, indeed, extreme difficulties in imagining any satisfactory material mechanism, since the light ether had to be at once subtle, to allow passage of bodies through it with negligible resistance, and also of great rigidity, to allow transmission of light at great speed. These and other difficulties gave rise to the invention of bizarre mechanical models of the ether, both as a medium for light and of electromagnetism, but these were generally intended merely as aids to imagination and calculation, rather than as literal descriptions of a physical reality. It has occasionally been the fashion to scoff at the workshop-full of gadgets invoked by Kelvin, and in his earlier papers by Maxwell, to provide such models, but it should be noticed that none of the mathematical physicists who put forward these models seriously intended them as descriptions of a real material ether. They were sometimes considered to be necessities of thought, but only in the sense of making the mathematical formalisms more intelligible, and sometimes as checks on their internal consistency.

In the third of his great papers on the electromagnetic theory of light, in 1864, Maxwell explicitly abandoned such models as *hypotheses* of the ether, but at the same time made clear that he did not thereby withdraw from his adherence to Faraday's conception of the real field:

> I have on a former occasion attempted to describe a particular kind of motion and a particular kind of strain, so arranged as to account for the phenomena. In the present paper I avoid any hypothesis of this kind; and in using such words as electric momentum and electric elasticity in reference to the known phenomena . . . I wish merely to direct the mind of the reader to mechanical phenomena which will assist him in understanding the electrical ones. All such phrases in the present paper are to be considered as illustrative, not as explanatory.
> In speaking of the Energy of the field, however, I wish to be understood literally.[3]

Energy had now become the fundamental physical quantity defining the field, and since it obeyed a conservation law it came to be regarded as a substance in its own right, rather than as a mere property of substantial matter. Also, Faraday's conception of a real medium had been vindicated by Maxwell's mathematical synthesis of optics and electromagnetism, in which energy was present in the intervening space as well as in material bodies. As a criterion for the reality of the field, transmission of action in a finite time now took precedence over Faraday's other criteria, for, combined with the principle of the conservation of energy, it implied that energy must be present in the medium during the time taken for action to pass from one material point to another. *How* the action passed in a mechanical sense was a question which began to lose its meaning—it was the energy of stress and motion in the field which became fundamental: the field was not to be explained in terms of matter, matter was rather a particular modification of the field. From this point of view the assimilation of mass and energy in the mass-energy of relativity theory, was only the last stage in a conceptual development initiated in classical field-theory by the treatment of energy as a fluidlike substance. Also, the abandonment of the material ether in relativity theory, as a consequence of the absence of any absolute material standard of rest, was not new, but only the culmination of the classical re-interpretation of the medium in terms of energy-flow.

---

[3] "A Dynamical Theory of the Electromagnetic Field", *Scientific Papers*, I, p. 563.

§4 *Illustrations of a Conceptual Dispute*

The field view did not, however, lack critics. Some physicists, particularly in France and Germany, refused to make the distinction Maxwell had made between imaginative mechanical models, and the conception of a real physical field. They wished, rather, to confine the scientific content of Maxwell's theory to Maxwell's equations, and to accept the field-functions such as energy and displacement-current only as implicitly defined by these equations, and not as "real" entities in a physical field regarded as more fundamental than matter. This type of "mathematical phenomenalism" has gained many adherents among physicists during the present century, since it has become progressively more difficult to give coherent and mutually consistent pictures of what Maxwell would have regarded as the real physical events underlying observable phenomena.

The issue between Maxwell and these Continental physicists was not, however, wholly one of rival philosophies of scientific theory. The Continental school had consistently maintained, throughout the nineteenth century, an action-at-a-distance conception of electric and magnetic phenomena, stemming from the attracting and repelling particle theories of the previous century. At any given point of development, it was generally possible to say that the empirical facts were explained as well by this action-at-a-distance theory as by the British fields, and even that the two theories were essentially the same, since each could be derived from the other by mere mathematical transformation. Thus Hertz asserted:

> Maxwell's theory is Maxwell's system of equations. Every theory which leads to the same system of equations, and therefore comprises the same possible phenomena, I would consider as being a form or special case of Maxwell's theory; every theory which leads to different equations, and therefore different possible phenomena, is a different theory. . . .

> If we wish to lend more colour to the theory, there is nothing to prevent us from supplementing all this and aiding our powers of imagination by concrete representations of the various conceptions as to the nature of electric polarisation, the electric current, etc. But scientific accuracy requires of us that we should in no wise confuse the simple and homely figure, as it is presented to us by nature, with the gay garment which we use to clothe it.[4]

But closer study of the development of the two types of theory shows that the question is not as simple as Hertz supposed, and it provides an excellent case-history in support of the proposition that there is a real scientific difference between what is asserted in theories employing different representa-

---

[4] *Electric Waves*, (Eng. trans. 1893), pp. 21, 28.

tions of their respective formalisms, even though they may be empirically equivalent in Hertz's sense. This question is crucial to the problem of whether there is any real difference between theories of action at a distance and those employing continuous action in a medium, and hence whether action at a distance is a real philosophical issue, or only a matter of convenience in theoretical formulation.

Comparison between the field and action-at-a-distance methods in nineteenth-century physics can be made from two points of view: that of the particulate nature of electricity, and that of the finite velocity of transmission of electromagnetic effects. The Continental methods go back to Ampère's investigation in the 1820's of the law of force between current elements, which he expressed in terms of attractions and repulsions acting instantaneously along the line joining the elements. During the century this law of force went through many *ad hoc* modifications designed to make it consistent, first, with the conservation of energy, and second with the propagation of action at a finite speed as required by Maxwell's equations. Two of the assumptions made by Ampère had to be discarded, namely the central direction of the forces (along the line joining the current elements), and the instantaneous propagation of action. Maxwell's continuous-action theory had led by natural and obvious deductive steps to the finite speed of propagation, but the acceptance of Maxwell's formal equations by the Continental school did not carry this same plausibility into their action-at-a-distance theory, where the finite velocity had to be injected *ad hoc* at the cost of some conceptual awkwardness. On the other hand, Maxwell's theory by itself was not sufficient for a solution of the problem of the law of force between current elements, towards which some progress had been made by the Continentals. In Maxwell's theory there are no independent current *elements*, since *all* currents are continuous circuits, a "material" current being completed if necessary by a non-material displacement-current in the "ether". Maxwell's theory was in fact not suited to the expression of any phenomena which depended on the discrete and atomic, as opposed to the continuous, and for this purpose the Continental theories were far more suitable. After the discovery of electron emission and transmission in high vacua by J. J. Thomson, the problem of the law of force between discrete moving electric bodies again became urgent, and was solved more easily as an extension of Ampère's formula than of Maxwell's equations.

Thus, although all the phenomena could in principle be accommodated in both types of theory, some phenomena found easier expression in one than the other, and this fact seems to show that the two theories were *not* equivalent for all scientific purposes. Theories in science are sometimes rejected, not on grounds of inconsistency with empirical facts, but on grounds of complexity, incoherence, and lack of predictive power. Now it does not *follow* from this that theories rejected on these grounds are empirically false, but if there are two theories which both lead deductively to the same empirical facts, but are inconsistent with each other, it is natural to extend the criteria of falsity to include grounds of rejection other than empirical falsification, for these are the grounds that enable us to choose between the theories. In this sense, for example, the Ptolemaic system is *false,* even though it can be made to account for the actual motions of the solar system, because it is complex, lacks predictive power, and is not coherent with the general theory of gravitation from which the Copernican (or rather Keplerian) system follows.

The nineteenth-century case-histories in electromagnetism seem to indicate that the issue between medium theories and action-at-a-distance theories was then a real scientific issue, unless we adopt a wholly positivist attitude to theories and regard all "unobservables" as mere pictorial devices denoting nothing in reality. But although the issue was a real one, its resolution was not as definite as in the case of Ptolemy versus Copernicus. In terms of nineteenth-century physics both types of theory had some justification, with a bias at the end of the century towards field theory. Twentieth-century developments have placed the issue still more in doubt, since the dualism and latent contradictions in classical physics between atomicity and continuity have become explicit and apparently insurmountable.

§5 *Interaction in Modern Physics*

Before the development of the general theory of relativity, gravitation had remained outside the synthesis of classical physics, and had been almost universally regarded as a pure action at a distance, for the sorts of reasons outlined by Faraday. Since a finite velocity of propagation was regarded as one of the chief criteria for a real field, a decisive step in favour of field theories throughout physics was taken within special relativity by the proof that neither gravitation nor any other physical process could be propagated at a speed greater than that of light.

It is important to analyse the status of this consequence of the special theory, because if it is true, it follows either that there is no action at a distance, and hence that what has been taken to be a philosophical problem has an empirical solution, or that action at a distance must be radically re-interpreted. The proof that the velocity of light in vacuum is the maximum velocity for any physical process, depends on some immediate consequences of the Lorentz transformation of space-time frameworks, of which the most important is the following: it can be shown that if any causal action were propagated with a velocity greater than light in some reference frame, then the temporal order of cause and effect would be reversed in some other frames. That is to say, if the causal action between A and B were propagated faster than light in some frames, it would always be possible to find other frames in which B precedes A; but we imply by the causal relation: "A causes B" that B never precedes A; thus causal action as generally understood cannot be propagated faster than light. This argument cannot, of course, legislate for the world. Even if the special theory of relativity is accepted as a valid description of the world, there is nothing logically inconsistent in holding that for some processes our usual notions of cause and effect are inapplicable. What the arguments do show is that we cannot retain those notions and at the same time postulate actions propagated with velocities greater than c. Common-sense ideas about causality are more fundamental to science than the notion of processes travelling faster than light, and there is nothing in the context of special relativity to induce us to abandon the former rather than the latter, so it may be concluded that if the special theory is accepted, then there is no detectable propagation of action exceeding the velocity of light.

In relativity theory, then, gravitation satisfies at least one of the classical conditions for a real physical field. In general relativity, gravitational potentials, and momentum and energy densities, are defined, and conservation laws of momentum and energy in space are shown to hold. Thus gravitational field theory satisfies another of the conditions for a real field, namely the presence of energy at space-time points remote from matter. There is no material medium in the sense of an ether, but space itself becomes the medium. This gravitational theory is indeed conceptually further from the notion of action at a distance even than the electromagnetic field theory, for here it is not the presence of the field sources which causes a field to appear in previously empty Euclidean space, but rather the geometrical properties of space itself are determined by the presence of matter, since the gravitational potentials specify the variable curvature of the non-Euclidean space. For the geometrician accustomed to thinking in terms of abstract spaces, the difference between the two models is very great: in one case geometry itself is the model, whereas in the electromagnetic case the model is still a refined version of fluid flow. The mathematical advantages of the geometrical model have led to various attempts, upon which Einstein was engaged until the end of his life, to incorporate electromagnetism and quantum theory with gravitation in a unified field theory of the geometrical type, but so far no fully satisfactory theory of this kind has been produced.

From the scientific point of view, however, relativity theory has not said the final word on action at a distance. Even if that theory is accepted, there are still two ways of avoiding the conclusion that action at a distance is a physical impossibility. One is to re-interpret action at a distance in such a way that it becomes consistent with a finite velocity of propagation: and the other is to remark that even instantaneous action at a distance might be postulated in a *theory*, so long as it is in principle impossible actually to *observe* the resulting causal anomalies. This second method depends upon the fact that causal anomalies are not self-contradictory, but merely contradictory with our normally accepted notions of causality, and these notions need not be affected in practice by unobservable violations of normal causality in a theory. The possibility of this way of escape for action at a distance was opened up by quantum theory.

The question of modes of action in quantum mechanics is closely connected with the uncertainty principle. In the first place, if this principle is expressed in terms of the complementary variables energy and time, it is clear that detection of a finite amount of energy requires a finite time-interval. Thus no action involving passage of energy from one point to another distant from it can be *observed* to be instantaneous. On the other hand, it is conceivable that instantaneous action at a distance might be postulated theoretically, as it were within the uncertainty limits, in such a way that the causal anomalies resulting from the special theory of relativity could not in principle be observed. Thus quantum theory opens a way, at least on the small scale, which appeared to have been closed by relativity theory, but so far suggestions of such instantaneous action or action faster than light have not passed beyond the speculative stage.

Continuous action on the small scale however does not fare any better. It is universally agreed that there is at present no consistent continuous-action interpretation of quantum theory analogous to that of classical theory, either in terms of particles, or waves, or fields, and it is argued by many that there cannot in principle be any such interpretation within the framework of the present theory. The most commonly accepted interpretation of the theory has been what Reichenbach calls *restrictive*, that is to say, it is held that quantum theory makes no statements at all about "real" unobservable events occurring between observations, but only about *states* which are functions of observables produced by measurements on the system. The theory speaks merely about systems passing from one state into another when measurements are made. The transitions are discontinuous, and describable neither as action at a distance nor action by contact in the classical sense. The "quantum jump" of an electron, for example, from one energy level to another, with emission or absorption of radiation, now appears as a successive occupation of energy states, with corresponding creation or annihilation of photons. There is no causal action of one state on the other; the transition is detected in a single experimental operation, that is, there is no separate description of a particle leaving one state and then entering another, so that it is meaningless, within the restrictive interpretation, to speak of the duration or any other property of the transition. This is neither a continuous action nor an action at a distance—it is not an action at all.

It does not follow, however, that continuous-action descriptions may not be possible on the macroscopic scale where the quantum of action is negligible. Indeed quantised field theories approximating to classical theories on the large scale have been developed, and these are able to some extent both to describe the interaction of particles with radiation, and to assimilate quantum theory with the special theory of relativity. Such theories have had some success in rationalizing the experimental results obtained in nuclear and other high-energy reactions, by interpreting these in terms of particle creations and annihilations. Since these creations and annihilations can occur randomly even in "vacuum", this theory in a sense re-introduces the notion of a material "ether" pervading all space, but in a relativistically-invariant way, so that this ether cannot be regarded, as the nineteenth-century ether was, as an absolute standard of rest.

The development of quantum field theory is, however, seriously hampered by the occurrence of infinite self-energy terms in its equations. Such divergence difficulties have been one of the motives for recent attempts by Wheeler and Feynman to reintroduce the conception of action at a distance on the large scale. Since it can no longer be maintained that this takes place instantaneously, action at a distance must now be understood to mean direct particle-interaction involving no independent field with its own energy and momentum. It has been assumed by most physicists that such action would violate the conservation principles, but Wheeler and Feynman have been able to show for classical electrodynamics that this is not so if the advanced as well as retarded solutions of Maxwell's equations are taken into account. This leads to the apparently paradoxical result that one event affects another at a distance r at a time r/c seconds *before* it occurs, as well as r/c seconds afterwards. It is of course contrary to all ideas of normal causality that an event should have an effect *before* it takes place, and such a notion is open to the objection that we might be able to intervene to prevent the cause occurring as soon as we had observed its prior effect, but if we did this successfully, then the prior effect which did occur should not have occurred. This paradox has in fact been urged by philosophers as a general and decisive refutation of any theory involving temporal reversal of cause and effect. Just in the case of quantum theory, however, there seems to be an escape from this paradox. If action at a distance of the Wheeler and Feynman type were incorporated into quantum field theory in such a way that the advanced effects always occurred within the uncertainty limits, it would never in principle be possible for a macromechanism (including a human agent) to detect them in time to prevent the later occurrence of the "cause".

On the whole, then, the most commonly accepted interpretation of quantum theory is agnostic about the nature of small-scale interactions, and even denies that either continuous action, or action at a distance, can be significantly postulated within the uncertainty limits. And on the large scale, quantised field theory is more analogous to classical field theory than to action-at-a-distance theory. The suggestions that have been made regarding the re-introduction of action at a distance are still speculative, but they are sufficient to show that the uncertainty principle *might* be used to confer considerable freedom upon the forms of theory, and that therefore different modes of action in future theories cannot be excluded, even within the general framework of relativistic and quantum physics.

§6 *Conclusion*

Finally, we may return to the two questions with which we began. First, the conceptual question: is it conceptually or methodologically necessary to retain an ether or medium of contact *of some kind* in theories of interaction? And secondly, the empirical question: given the fundamental concept of "matter" or "substance" in a particular physical theory, is there in fact an ether of *this* kind, or of some other kind, or does action take place across empty space?

The second question may not always be easy to answer in the case of a particular physical theory of matter, because the critical experiments may not have been performed, or their interpretations may be in doubt. But from a philosophical point of view the question seems easy to understand—we know what sort of experiences would lead to a positive or negative answer. On the other hand, fundamental theories of matter are not static, and the attempt to answer the second question often raises the first, and often involves that progressive withdrawal from empirical positions as soon as they are falsified which indicates that the question has become a metaphysical one. Thus, it is at first believed that matter acts only by contact; then faced with matter attracting at a distance with no apparent medium, subtle matter of dubious status and properties has to be postulated. All attempts to describe these precisely are refuted. So the medium is described in terms of stresses and tensions, in such a way that energy is the only material property which is located in it; then energy is itself regarded as substantial, and this is said to show that action is after all continuous. Again, although it is shown as conclusively as is possible in science that action can only be propagated with a finite velocity, this is not taken as a final refutation of action at a distance, for the antithesis turns up again within the framework of finite velocities, and presumably it will continue indefinitely to take new shapes as new theories are developed. Clearly no single or simple assertion of any kind, metaphysical or empirical, is being made when it is said that matter can, or cannot, act at a distance.

Is it, however, conceptually or methodologically necessary to postulate a medium? The preference for action-by-contact theories in physics was historically connected with the objectification and depersonalisation of nature, and the desire to eliminate from scientific explanations all "psychological" or animistic models, in favour of the model of mechanism; and it was a fact that most familiar mechanical devices acted by contact. Such consideration may help to explain the deep-seated intuition of what is possible, which forms a climate of thought favourable to continuous action. But this intuition has seldom been translated into cogent metaphysical arguments, even by those who, like Aristotle and Leibniz, held it most strongly. It has more often in fact functioned as a regulative principle, a directive to look for, or perhaps merely to postulate, a medium which is material in some sense, whenever action appears to make jumps.

The conceptual question then becomes a methodological rather than a metaphysical one. There is no doubt that apparent assertions about action at a distance and continuous action are used as regulative principles from time to time in physics, in the form "Do not postulate unobservable intermediate entities" by Newton (in the *Principia*), and by most subsequent positivists down to the restrictivists in quantum theory; and on the other hand "Always look for continuously acting causes", by the Atomists, Descartes, Newton (not in the *Principia*), Faraday, and the quantum theorists like Bohm who oppose the restrictive interpretation. At first sight it looks as though the second directive will encourage the construction of more fruitful models, because models conforming to it will have to contain descriptions of intermediate unobserved events as well as observables, and in general (in theories not governed by the uncertainty principle at least) it is eventually possible to devise further experiments to detect the intermediate events predicted. Continuous action therefore appears to be more powerful as a predictive model, and to make more claims upon the facts, and this would seem to explain why action at a distance has usually been associated with positivist views of theory, where it is desired to assert no more than is already justified by experiment. But further consideration of historical examples suggests that this is too simple a view of the matter. It does not explain why the model of attractions and repulsions at a distance was useful throughout the eighteenth century, nor why the recent theory of Wheeler and Feynman is fertile in revolutionary suggestions, even though these may turn out to be unacceptable. There is no simple equation between action at a distance and positivism, and action-at-a-distance theories do not always lack the content which makes good models.

Action at a distance and contact-action form one of those pairs of apparently contradictory principles which continually reappear in the history of science: atomicity and continuity, mechanism and vitalism, determinism and freedom, causation and teleology, body and mind. Unlike some of these, however, the dispute about action at a distance does not now raise extra-physical problems bearing on the nature of life and mind, although, as we have seen, this has not always been the case, for prior to the eighteenth century the problem was closely associated with some of these other antitheses, and was itself a dispute about the place of animistic explanations in physics. And it may be that its obvious connection with the ostensible phenomena of extra-sensory perception will in the near future raise all these problems again, but in a more extreme form, for these phenomena, if veridical, seem to involve not only "jumps" across space and time of regular and determinate amounts, but also a certain *independence* of space and time. Since existing physical theories, of continuous and distance action alike, have always depended on *regular* variation in space and time, this is one of the features of para-normal phenomena which makes their theoretical expression exceptionally difficult. But if the history of physics is any guide, it seems that if *some* theoretical explanation is available, its agreement with pre-conceived notions of matter and action is less important than its intrinsic simplicity, predictive power, and correspondence with the facts. It would therefore be rash to conclude that para-normal phenomena or any other distance actions, do not occur, *merely* on *a priori* grounds.

Meanwhile, with regard to physics, it must be concluded that continuous-action formulations while not conceptually *necessary* are at least conceptually convenient, and, with our present concept of matter and energy, seem to accord best with the empirical facts.

PART TWO

The Concept of Matter
in Modern Philosophy

KANT'S DOCTRINE OF MATTER

John E. Smith

It might be thought that a philosopher like Kant who had so much to do with the founding of an idealist tradition, would have laid little emphasis on matter. The facts do not bear out the supposition. Kant had much to say about the concept of matter and it plays a major, even if not always unambiguous, role in his thought. Matter is, for example, employed by Kant as a principle which guarantees a real, external object for knowledge and he urges it against what he takes to be the erroneous idealism of Berkeley. On the other hand, when Kant is called upon to give his own view of the status of matter we find him interpreting it as appearance (*Erscheinung*) in contrast with the thing in itself. After a century and a half of discussion we are used to the problems and difficulties surrounding all attempts to understand Kant's ideas. Achieving clarity about his doctrine of matter proves to be no exception to the general pattern. Unfortunately, many of the things Kant had to say about matter are bound up with certain other doctrines of the critical philosophy and a fully adequate account would require a full scale interpretation of the entire theory of knowledge. Kant, for example, characterizes matter rather consistently as the representation of something as *outer* and in order to grasp what this means we would need to settle certain difficult points in the doctrine of outer sense. But a full development of the critical theory of knowledge is beyond our present purpose. A working paper on Kant's doctrine of matter must be confined to more modest limits. Ideas basic to his theory must be set forth as far as possible without raising the many questions about his analysis of knowledge which long discussion has shown to be unavoidable. The chief sources of information here are the first *Critique* and the *Metaphysische Anfangsgründe der Naturwissenschaft*.[1]

These writings point to the two basic contexts in which matter figures in Kant's thought. On the one hand, Kant refers to matter and to material content repeatedly throughout the argument supporting his epistemological theory. And, on the other, the concept of matter is the most fundamental no-

---

[1] This work, the title of which I would translate as *The Basic Elements of Natural Science*, has never been satisfactorily translated. The work, moreover, has been unduly neglected, which is unfortunate because it not only throws light on Kant's own views regarding the science of nature, but it provides some clues to the interpretation of his account of empirical cognition in the *Critique*.

tion in his philosophy of nature and natural science. It belongs as much to
his account of science as to his theory of knowledge. Its dual status in the
Critique as substantial material content in space on one side and as phenomenal appearance on the other, reflects the same duality to be found in Kant's
claim that his theory is an empirical realism and a transcendental idealism
at the same time. In the *Critique* the concept of matter appears at many different points in the argument, a fact which points to ambiguities but also
helps to explain inconsistencies, real or apparent, in Kant's views as well.
The *Anfangsgründe* presents fewer problems in this regard because matter
is the key concept in the analysis, and the entire theory is built around it.
The arrangement, however, is without complication; matter is defined in
terms of motion and then considered in accordance with the four categoreal
headings, quantity, quality, relation and modality. The fourth section,
which Kant calls "Phenomenology" throws light on his view of the connection between matter and the object of knowledge.

§1 *The Doctrine of Matter in the* Critique of Pure Reason

Kant treats matter (*Materie*) in many places in the *Critique* and it is of
the utmost importance to identify the particular context within which his
statements occur. As is made clear in the Appendix on the "Concepts of Reflection" (B 316–B 324), matter may be taken in a logical sense as given concepts or material for judgment; it may be understood in an empirical or existential sense as the essential constituents of something; it also figures in a
further and difficult-to-name sense according to which it means unlimited
reality or the possibility of things in general.

Matter (*Materie*) is referred to by Kant as material content in contrast
with form in the case both of concepts and judgments. In B 322, 'matter'
means material for judgment, or the subject matter. In B 10, Kant speaks of
matter as the content of the concept in distinction from its form. The final
relation between conception and judgment need not be decided; in either
case matter means that which is thought or arranged as distinct from form
which is the manner of its arrangement.

Matter is further described as the material of knowledge, both in the
sense of something given without which empirical cognition would not be
possible and in the sense of the ultimate subject matter of the understanding. In A 223 (= B 270), sensation is equated with "matter of experience",
and in many places (B 34; B 60; B 119; B 207–8), matter is identified with
that in appearance (*Erscheinung*) which corresponds to sensation. In so far
as sensation for Kant always means the immediate presence of the sensible
content to the knowing subject (e.g. A 19 = B 34), matter must be pre-

sumed to be immediately apprehended. Kant makes the point explicitly in the First Edition version of the Paralogisms (see esp. A 371), where he maintains that matter, as constituent of appearance and as itself appearance, has "a reality which does not admit of being inferred, but is immediately perceived". As bound up with sensation, matter always belongs to the *given* and is, as Kant points out in the Anticipations of Perception, that element in the total appearance which *cannot* be anticipated (esp. A 167 = B 209). This means that matter must always be encountered; it cannot be constructed. When it is received, it is present to the perceiver.

In addition to his identification of matter with sensible content, Kant assigns to it a special status in the theory of *objective* cognition. The concept of an *object* is, of course, central to Kant's epistemology. Although there are many passages in which he writes uncritically as though objects are "given", it is clear that the central problem of the *Critique* is to isolate and justify the conditions under which representations alone may be said to possess objective validity, i.e. possess a real object. 'Object' always means "object of knowledge" for Kant and it carries with it a pejorative sense; when representations are validly said to have an object, it is because they stand under the constraint of certain conditions derivative from understanding and reason. From the Transcendental Deduction and, even more, from the Second Analogy (esp. A 197–8 = B 242–3), we learn that for representations to acquire "relation to an object" they need to be connected in accordance with necessary rules ultimately derivative from the rational faculties. Emphasis upon this side of Kant's theory has led to the supposition that his doctrine of knowledge and of truth is one of rational coherence alone. But the coherence interpretation explicitly contradicts Kant's repeated claim that empirical cognition is not possible on the basis of form alone without content. Despite the priority which Kant's theory gives to the *a priori* or necessary conditions of experience—forms of intuition, categories and principles—the fact remains that knowledge is never possible without material content which can be given only through sensation. In many passages (B 88; B 207–8; B 440; B 748; B 751), Kant makes a point of saying that matter is an essential element if we are to have an *object* as the referent of our concept. In B 88 he even identifies the two; the entire passage is instructive:

> But since it is very tempting to use these pure modes of knowledge of the understanding and these principles by themselves, and even beyond the limits of experience, which alone can yield the matter (objects) to which those pure concepts of understanding can be applied, the understanding is led to incur the risk of making, with a mere show of rationality, a material use of its pure and merely formal principles, and of passing judgments upon objects without distinction—upon objects which are not given to us, nay, perhaps, cannot in any way be given.

In B 646 (= A 618), matter is said to constitute "what is real in appearance" and in the *Anfangsgründe* Kant is even more explicit in defining matter as the *object* which is meant by or referred to in the concept. In the "Discipline of Pure Reason" (B 748), reference is made to "the matter of appearances whereby (*wodurch*) *things* are given in space and time". Regardless of Kant's tendency at the heart of his argument to place the ground of objectivity in the *a priori* conditions, i.e. on the side of form, the fact is that matter or content given in sensation also figures among the conditions necessary for having a real object. The only concept which represents the material element *a priori* is the concept of a thing or object in general. But this concept can never be more than a rule of synthesis of representations; it is not the concept of a real, i.e. determinate object. For that, matter is required.[2]

In other passages, Kant characterizes matter more directly and not merely in terms of its status in the matter-form distinction. Matter is defined further as "substance in the world" (A 627 = B 655), as "the physical" (B 751) or the constituents of things (*Gehalt* and *Inhalt*), as the substance and existence of things in distinction from their states or accidents (B 663–4), as the *substantia phenomenon* (B 333) known through the space it occupies and the effects it can produce. Matter as substance or the permanent in time bears further consideration. In several passages (e.g. A 618 = B 646) Kant cites *extension* and *impenetrability* as the distinguishing characteristics of matter in the sense of the real thing or object. The meaning of extension is given along with Kant's theory of space; the meaning of impenetrability derives from Kant's so-called dynamic conception of matter and it marks one of his distinctive ideas. It is more fully developed in the *Anfangsgründe*, but seeds of the doctrine are to be found in the *Critique*. Matter does not occupy space by "mere bulk"—the apt expression of Kemp-Smith—but rather in its force of attraction in bringing other objects to it or in repulsion which means resisting penetration by other objects. Kant seems to have thought of matter as *substantia phenomenon* (cf. B 321; B 333; B 663–4)—permanent in

---

[2] The question of ultimate consistency in Kant's position is not raised here. As against every thorough-going rationalism which would find all the necessary conditions for knowledge in understanding alone, Kant stresses the need for given material which cannot be constructed or anticipated. On the other hand that very material is frequently interpreted by him in a way which comes perilously close to reducing it to form. In B 341 (= A 285), he holds that all we can know of matter is external *relations* some of which are self-subsistent and permanent. And, he claims, it is these relations which alone give us a determinate object. As long as these relations are thought of as *spatial* and thus as determined through outer sense, the material pole is preserved. But if the spatial or intuitional element is ignored and the relations in question are taken as expressable through concepts alone, the material or given element is lost.

time, as expressed in the First Analogy (A 182 ff. = B 224 ff.)—which is known through action and the effects produced on the senses. Here as always Kant denies that we can have a wholly intelligible grasp of matter independently of its appearance to sense.

In the *Anfangsgründe,* Kant identifies matter with motion in its various aspects. The stress laid upon motion, force and permanence may seem to reduce extension to a minor role. This is only apparently the case. Kant repeatedly argues that in so far as matter means the content of a real object, it must be represented as *outer*, i.e. as being in space and sharing the characteristics of space. Thus, for example, insofar as matter has parts, these parts are outside each other as the parts of space.

The relation of matter to space is important and it brings with it questions concerning the role of outer sense in empirical knowledge. Kant, on the one hand, relies heavily on the idea that material content is given and that it affects the senses as a basis of the claim that we have a real existent object of knowledge. On the other hand, there are many passages where he claims that this same matter is "outer" only in so far as it is represented as such through outer sense. The two aspects can be combined if we recall the doctrine of empirical realism and transcendental idealism. In relation to a reality considered *in itself*, i.e. considered as the object of a purely intelligible knowledge in which the appearance of an object to the senses plays no part, all of our experience is *appearance (Erscheinung)* and thus transcendentally ideal. But in relation to a reality considered as knowable by human beings only in accordance with the conditions imposed by human capacities, experience as empirical cognition is empirically real, i.e. it is genuine knowledge of objects through the effect of material upon outer sense when the sensible material thus received is represented as in space and is judged in accordance with the necessary forms and principles for objective knowledge.

Kant was not unaware of the difficulties and he often spoke of his position as "paradoxical". His doctrine of space illustrates the point. The ultimate status of space as a form of intuition means that space is "in us" (A 370 and many other places), i.e. as a form of sensibility, and is transcendentally ideal. As such it is neither a real being in itself, nor a set of logical or purely conceptual relations. But, Kant repeatedly argued, the assigning of this ultimate status to space is not inconsistent with granting to it and to its content an empirically real status. He believed that our representations of things can be of what is external, i.e. material, not because we have a power of immediately knowing things in themselves, but because our perceptions are related to a space in which *all things are external to each other*. The distinction between what has genuine external existence in space, i.e. matter, and what is represented only as inner, is never for Kant a distinction between what is appearance and what is in itself, but *a distinction within appearance*. Matter is therefore, not the thing in itself, but the appearance as external.

The doctrine of matter is not left unaffected by Kant's doctrine of appearance as the solution to the problems raised in the antinomies. Both versions of the Paralogisms make this clear. The arguments differ considerably in the two editions, but there is a most important point of agreement between them. In the first edition (A 359 ff.), Kant places emphasis upon the idea that matter is but appearance and that nothing of it is known apart from the senses. He attempts to overcome the heterogeneity existing between soul or self and matter by interpreting the latter as a "species of representations". Soul and body are not two heterogeneous substances but rather representations of two different sorts. Matter is that representation which is external or outer and which corresponds to a something in space. Matter is *not* a kind of substance in this argument (see esp. A 385), but only the representation of something as outer. Kant goes so far as to say that matter, not being a thing in itself but only the appearance to outer sense, may very well, in itself, be simple and appear as composite only in virtue of its effect upon outer sense. The question of the nature of matter as it appears in experience is shifted from that of its inner nature as substance to that of the conditions which lead us to regard some representations as corresponding to *outer* objects. The answer is found in outer sense; matter becomes the representation of something as external made possible by the outer sense and its form of intuition.

In the second edition (B 368 ff.), Kant is concerned to show the impossibility of our knowing a substantial ego in empirical terms. Less emphasis falls upon finding a way of closing the gap between body (matter) and the self or thinking substance. Kant, however, returns (B 427–28) in the conclusion to his former point that since we do not know reality in itself, we cannot be sure that the heterogeneity of matter and mind as normally understood represents any final truth. The assumed heterogeneity, he says, between the object of inner sense (the soul) and the objects of the outer senses has always constituted a major problem. His proposal is to say that the two differ not inwardly (i.e. in their substantial character) but "only in so far as one *appears* outwardly to the other" (B 428). It may be, says Kant, that "what, as thing in itself, underlies the appearance of matter" (*was der Erscheinung der Materie, als Ding an sich selbst, zum Grunde liegt*) may not be so different from the soul. But since no final resolution of the essentially speculative problem is possible, we are forced to remain within the sphere of appearance and its distinctions. Matter, in both statements of the Paralogisms, takes on the status of a *kind of representation* with appearance or experience, i.e. the representation of a something as outer.

Kant's discussion of matter and form as concepts of transcendental reflection (B 316 ff.) provides a brief summary of the basic senses in which matter figures in the *Critique*. Concepts of reflection, for Kant, express the relations in which concepts stand to one another, depending upon the faculty in which they originate. Matter and form are said to underlie all other reflection because of their generality. Matter is understood as the *determinable* in general and form as *determination*. In the logical context, matter means "given concepts" or material for judgment; in the empirical or existential context, matter means the constituent elements (*essentialia*) of a thing; in the metaphysical context, matter means unlimited reality or the material of all possibility. Considered in relation to understanding and sensibility as separate faculties, the relation of matter to form undergoes a change. With regard to understanding, *matter is prior to form* in the sense that judgment always presupposes something as given, and this means for Kant ultimately "given to the senses". With regard to sensibility, *form is prior to matter* because sensibility is a faculty with its own structure (i.e. it is not for Kant as it was for Leibniz merely a confused form of representation or an inferior form of understanding) and no perception of material content can be had without the forms of space and time. The given element, matter, is prior to judgment, but to be given in the first instance the same material element is subject to form as a condition of perception (and, we should add, sensation as well).

As should be clear enough from the foregoing, the concept of matter figures in so many contexts in the *Critique* that no single formula will suffice to express its full meaning. Kant wanted to retain a doctrine of matter as given material of sensation which cannot be constructed but but must be encountered and he wanted this doctrine in order to have a genuine, outer object of knowledge. He shared the view of common sense to that extent; ordinary experience and refined knowledge are about "real" things. On the other hand, it is equally clear that his transcendental idealism prevented him from identifying matter with a thing in itself and led him to assign to it the status of appearance, i.e. of reality as essentially conditioned by human faculties, especially the fact of its appearance to sensibility. As appearance, however, matter is known only as outer representation (i.e., as in space or object of outer sense) and our principal access to it is through the effects which it produces. Matter takes on the character of a way of representing while the ground of the outer representation remains unknown. There is, nevertheless, no possibility of going wrong in interpreting Kant if we consistently identify matter with the *given* in any context; it represents, even if this characterization is at a formal level, what can never be anticipated as content or constructed *a priori* in determinate fashion.

§2 *The Doctrine of Matter in the* Metaphysische Anfangsgründe

The *Anfangsgründe*[3] treats the "general conception of matter as such". In accordance with his view that a general science of nature cannot be built up in disregard of the general principles of understanding and experience, Kant proposes to analyze the concept of matter under the four headings of his table of categories—quantity, quality, relation and modality. The work thus divides into four sections according to an architectonic principle. Each section is to consider the concept of matter in one of its aspects and each is said to add "a new determination" (366) to it. Kant's description of the fundamental characteristics of matter represents a return to the *Critique* and shows at the same time that "phenomenological" considerations, although they appear to be confined to but one section of the *Anfangsgründe* (Section IV is entitled, "*Phenomenologie*"), actually determine the whole discussion. In defining matter in its basic sense, Kant says: "the fundamental characteristic of a Something as an object of outer senses (*äusserer Sinne*) must be motion, because only through motion can these senses be affected" (366). Motion is fundamental and all other predicates of matter are said to find their ground in it; the doctrine of matter is said to be a *Bewegungslehre*. The subject matter of the four sections is determined accordingly. In Section I (*Phoronomie*), matter is considered as pure quantum of motion, or more accurately, motion is considered as pure quantum; in Section II (*Dynamik*), matter is considered as having the quality of an original power of motion; in Section III (*Mechanik*), matter is considered as having the quality treated in the previous section in relation to another; in Section IV (*Phenomenologie*), matter is seen as having motion or rest relative to a mode of representation, i.e. as appearance of outer sense.

It is important to notice Kant's identification of matter with the outer or with the object as falling beyond the concept. In the Preface to the work (362), he refers to matter as that which makes it possible for a concept to have "its own object" and he refers to the subsidiary concepts of motion, inertia and the filling of space as conditions for applying concepts to "outer experience" (*äussere Erfahrung*). Matter is here in its familiar role as the guarantor of an existing object in space. The curious fact is that Kant speaks of "outer senses" in the plural—not his usual practice—rather than the representation of something as outer through "outer sense" as form of intuition. Although this issue cannot be discussed here, it is noteworthy that mat-

---

[3] Citations are to Hartenstein's edition of the *Sammtliche Werke*, 1867, *4*; all translated material is my own.

ter appears as having to do with the external and with the securing of an object for a concept.

The most elemental definition of matter in the *Anfangsgründe* is as follows: "Matter is that which is movable in space" (369). The force of the term *"Das Bewegliche"* is that of a something which is capable of moving or of being moved; it embraces both the fact of the motion and a "that which" moves. In relation to the cognitive faculty, matter means an object of outer senses (the plural persists in this work) and in contrast to form it means the "object of sensation" (*Gegenstand der Empfindung*). Once again, matter is said to constitute what is "real" in experience, i.e. that it is experience of an existing object. Rest belongs as much to matter as motion does; Kant understood rest (*Ruhe*) as the existence of something in the same place throughout a time. Rest means duration (374).

Under the heading of "Dynamics", matter is analyzed as pure quality of motion or power and in this connection Kant introduces the concept of filling a space. This is one of the distinctive features of his theory. To fill a space means the power (*das Vermögen*, 387) to resist penetration. The initial proposition of the section is most explicit on the point: "Matter occupies a space, not through its bare existence, but through a specific moving force" (388). Kant's argument is that the occupancy of space by a body means the capacity of that body to alter—obstruct or redirect—the motion of another body. Matter is thus a cause of motion in the sense of being able to alter motion by occupying space with intensity. Kant calls the cause of motion "moving force" (*bewegende Kraft*) and holds that this property and not mere occupancy of space defines what we mean by matter filling a space.

The doctrine of power or force is put forward by Kant in explicit opposition to the view that the occupancy of space means primarily the "solidity" of matter. He is arguing against certain of his contemporaries who held that, according to the principle of contradiction, it is inconsistent for the same space to be occupied by two different things at once. In accord with his doctrine that the principle of contradiction is not sufficient to determine questions of material content, Kant says: "All by itself the principle of contradiction does not exclude any material which approaches and seeks to penetrate a space in which something is already to be found" (389). In so far as there is any contradiction involved, it would have to find its ground or explanation in the more basic fact that for something to assume a position in space means that it has the power to exclude all movable things external to it. It is contradictory to say that something is in a space which is at the same time penetrated by something else. Kant goes on to analyze other aspects of the occupancy of space by matter—repulsion, elasticity etc.—in terms of his dynamic conception. One of the principal features of his theory is that impenetrability, and indeed force in general, is subject to degrees. Occupancy of space, though it becomes a matter of degree, is not definable in mathematical terms alone but has a physical basis.

Kant assigns to impenetrability the status of the fundamental property of matter whereby it "reveals itself to our outer senses as something real in space" (400). But by itself this property is insufficient to account for an object. A second, fundamental force is required; the force of attraction. Kant argues that in so far as matter consists in or maintains itself through repelling —via the force which is at the root of impenetrability—other forces, in the absence of such forces it would have no cohesion; it would be infinitely distributed and we could not explain the consistency of matter occupying a given space. If repulsion were the only force, space would be empty of matter; it follows that in addition we need an intensive (*zusammendrück-end*) force which is opposed to extension. All matter (Kant speaks here of the "possibility of matter") must have, as an original property, a force of attraction or cohesion.

In a most revealing comment about the force of attraction (401–2), Kant asks why impenetrability or the principle of repulsion is made fundamental or a matter of immediate apprehension while attraction is assigned the status of an inference. Although the question put is clearer than the answer, Kant's chief point is that, since matter must appear to sense, the property of filling a space is the most immediate we apprehend whereas for attraction we have no immediate sensation and we do not readily refer it to an object in experience.[4]

In the third section, "Mechanics", Kant characterizes matter as having its force in relation to other bodies. In the previous section, matter was considered only as occupying a space, i.e. as possessing force in its original constitution without regard to anything else. Mechanics, however, is concerned with the imparting of motion and with the formulation of laws which are based on the *relation between many bodies* or parts of matter. Such concepts as quantity of motion, mass, velocity, quantity of matter etc., are defined and determined through the relations between bodies in space.

In the final section, "Phenomenology", matter is defined in its modal characteristic; it is described as that which has motion (*das Bewegliche*) in so far as it can appear as an object of experience (450). The main theme is the relation of matter to the human faculties. The opening passages are most explicit in identifying matter with the *object* of experience (451). This sec-

---

[4] The distinction in this passage between a predicate connected directly to a subject concept and one which is connected only through some other predicate is at the heart of Kant's distinction between analytic and synthetic judgments. The passage reflects exactly the same position Kant adopted in his reply to Eberhard in the *Streitschrift*.

tion is of the utmost importance for Kant's conception of space and especially his contention that motion and rest as properties of matter can never be thought absolutely but only relatively. Matter in space can be related to matter in space, but not to space itself. Space cannot be perceived, but must be intuited. What can be perceived is matter in space and such perception is always a necessary condition for empirical cognition purporting to be true of a real object.

§3 *Epilogue*

A most satisfactory way of understanding Kant's view of matter and the material world is to view it against the background of positions held by his contemporaries. On the one hand there were the adherents of the Cartesian tradition according to which the material world is characterized primarily by extension understood as a mathematical property capable of being expressed in a purely conceptual way. It is important to notice that on this view, extension means not so much the *visible* spread or bulk of something appearing to sense, as *mathematical* property which can be directly grasped by the mind and expressed through geometry. In addition to extension the Cartesian tradition stressed the purely mechanical nature of matter, especially its complete otherness from the world of mind and feeling.

Against this position Kant had two principal objections. He refused to accept a view of knowledge which excluded the contribution of the senses. Human beings are incapable of attaining purely conceptual knowledge of the world; appearance to sense is always a necessary condition. Consequently, he rejected the Cartesian mathematical extension as the essence of material substance. In addition, he was opposed to the purely "materialist" conception of matter according to which all qualitative changes and characteristics are functions of quantity of matter. In this regard, Kant was closer to Leibniz and the Leibnizians. Like them, he wanted a place for power and what he called intensive magnitude.

The Leibnizian view, on the other hand, presented other problems; Leibniz, following in the Aristotelian tradition, would not acquiesce in the banishing of the forms of things from the real world. He wanted, moreover, to retain the concepts of potentiality and possibility, something which seemed impossible on the Cartesian view. Kant was on the side of Leibniz in his "dynamic" view of matter and in his refusal to allow that all of its characteristics can be deduced from extension. On the other hand, Kant placed Leibniz in the same camp as Descartes on the issue of purely intelligible knowledge. Kant repeatedly criticized Leibniz for trying to reduce the sensible component in knowledge to a mere form of confused conception.

To Kant this meant the attempt to construe the world in exclusively mathematical and conceptual terms. As against this view, Kant was the empiricist in every respect. Knowledge demands that the world appear to the senses; we have no power of grasping it directly by means of the understanding and the reason. Mathematics is important, but in addition to the fact that (on Kant's view, of course) it is not a purely conceptual instrument, we can never construe the world completely *a priori* since it contains a material element which cannot be anticipated.

In addition, Kant was uneasy about Leibniz' ontological idealism, especially his attempt to identify the ultimate nature of things with monads or points of force. Apart from Kant's general objections to a naive (because not preceded by criticism) ontological theory, he regarded the Leibnizian view as not sufficiently empirical. For Kant matter is more brute in character than can be described in terms of monads. And indeed Kant stressed this point the more it became necessary for him to have a genuine "outer" object as a contrast to inner sense or self consciousness.

The ultimate tension in Kant's theory of material existence stems from his attempt to maintain his empirical realism with a genuine outer material object and also to solve the antinomies of reason through the doctrine of appearance at the same time. As against some forms of idealism, he urged the reality of a material pole. As against the materialist interpretation of the physical world he urged that matter is appearance and is known, in so far as it is known at all, only as object of outer sense.

There is also a something-less-than-perfect integration in Kant's view between his doctrine of matter as it figures in the theory of knowledge and as it occurs in his theory of science and cosmology. In the former, it appears as a limiting concept the meaning of which is determined by the nature of our knowing equipment. Matter, that is to say, is understood only in so far as it figures among the necessary conditions of experience. In the *Anfangsgründe,* on the other hand, *matter* is treated more directly as an essential concept in the interpretation of the natural world. We discover what matter is more directly in its own right and not only through its role in the cognitive process. Even when we say this, however, and further note that in the four-fold treatment of matter in the *Anfangsgründe,* considerations of Phenomenology are confined to but one part, the fact remains that at the outset of the discussion Kant defined matter in terms of its having to affect the senses. His primary characterization of matter as motion is said to be dependent upon the fact that it must exert an influence upon us in order to be known. The long arm of epistemology was difficult to avoid.

In the subsequent history of the philosophy of nature and of science, Kant's influence is seen chiefly in his having established the *reflective ap-*

proach. The older cosmology and the philosophy of nature gave way, under the impact of positions influenced by him, to the study of the structure of our knowledge of nature. We speak not so much of matter as of the concept of matter and not so much of levels or forms in nature as of the role played by principles in theory construction. In this regard the "critical" philosophy is still very much alive.

APPENDIX

A check list of important passages in *The Critique of Pure Reason* upon which an account of Kant's theory of matter must be based.

A 6 = B 10	B 88	A 618 = B 646
A 20 = B 34	B 321	A 627 = B 655
A 86 = B 119	B 322–23	B 663–4
A 185	B 333	B 748
A 223	B 341 = A 285	B 751
B 207–8		

In addition to the above, sections such as the First Analogy; the Anticipations of Perception; the Paralogisms; the Refutation of Idealism and the Second Analogy should be consulted as a whole.

MATERIALISM AND MATTER
IN MARXISM-LENINISM

N. Lobkowicz

卐 According to the recently published Soviet *Philosophical Encyclopedia*,[1] the Marxist-Leninist world-outlook considers the problem of the relation of consciousness to matter as:

> the central question of philosophy, i.e. as the starting-point for the solution of all other questions of philosophy in general and of Dialectical Materialism in particular. This problem has always been and still is at the very core of philosophical thought. All philosophical trends, of the past as well as of the present, fall into two fundamental and irreconcilable groups, according to the way in which they solve this central question of philosophy: *materialism* and *idealism*.[2]

And in the (quasi-official) textbook, *The Principles of Marxist Philosophy*,[3] first published in 1958, it is said:

> The question of the relation of thought to being, of Spirit to Nature, is the central question of world-outlook. Which is primary and primordial: Nature (being, matter) or Spirit (reason, consciousness, idea)? In other words: does matter precede consciousness or, on the contrary, does consciousness precede matter? Do being and matter determine thinking and consciousness or is it the opposite that is true? . . . According to the solution they offer, philosophical doctrines divide into two fundamental trends: materialism and idealism. Those philosophers who acknowledge matter as primary, are called materialists; they hold that nobody created the world surrounding us, that nature exists eternally . . . Idealists, on the contrary, consider thinking or "Spirit" to be fundamental. They maintain that Spirit existed before Nature and independently of it.[4]

---

[1] *Filosofskaya Entsiklopediya*, ed. by F. V. Konstantinov, Izd. "Sovetskaya Entsiklopediya", Moskva, 1960, *1*. Abbreviation: *FE*. This first volume of the *Philosophical Encyclopaedia* was signed for print on August 15, 1960; three other volumes are to be expected in the course of 1962/63.

[2] *FE*, p. 483/b ff.

[3] *Osnovy marksistskoy filosofii*, Moskva 1959. This is the 2nd edn. of a work, first published in 1958; see the summary by J. M. Bocheński, *Die dogmatischen Grundlagen der sowjetischen Philosophie*, Dordrecht-Holland, 1959. It is, however, not expressly characterized as 2nd edn., and slight changes have been introduced after the XXI Congress of the C.P.S.U. Signed for print on October 19, 1959 (1st edn.: August 7, 1958). Abbr.: *OMF*.

[4] *OMF*, p. 11.

It is well known that the interpretation, the concepts of *materialism* and *idealism*, as indicated in the texts quoted above, originates from F. Engels. In his pamphlet *Ludwig Feuerbach and the End of Classical German Philosophy*, published in 1888, Engels points out that "the great basic question of all philosophy is that concerning the relation of thinking and being":

> The question of the relation of thinking to being, the relation of Spirit to Nature—the paramount question of all philosophy . . . which is primary, Spirit or Nature?—that question, in relation to the church, was sharpened into this: did God create the world or has the world been in existence eternally? According as this question was answered this way or that, philosophers divided into two great camps. Those who asserted the primacy of Spirit to Nature and, therefore, in the last instance, assumed world creation in some form or other . . . comprised the camp of idealism. The others, who regarded Nature as primary, belong to the various schools of materialism. This and no other is the original meaning of the two expressions: idealism and materialism.[5]

This strange identification of two quite heterogeneous problems, that of the relation of thinking and being and that of the relation of Spirit and Matter, has had a decisive effect on all later Soviet handling of the two concepts, *materialism* and *idealism*. Thus Lenin, who explicitly refers to the passage in *Ludwig Feuerbach*,[6] declares:

> The existence of something reflected that is independent of the reflectants (i.e. the independence of the outer world of consciousness) is the fundamental premise of materialism;[7] matter is a philosophical category for the denotation of objective reality which is given to man in his perceptions, and which is copied, photographed and reflected by our perceptions, while existing independently of them;[8] the *sole* "property" of matter with whose recognition philosophical materialism is bound up is the property of *being objective reality*, of existing outside our consciousness.[9]

---

[5] F. Engels, *Ludwig Feuerbach und der Ausgang der klassischen deutschen Philosophie*, Stuttgart, 1888, p. 15 ff.

[6] V. I. Lenin, *Materializm i empiriokrititsizm*, Moskva, 1909, in V. I. Lenin, *Sochineniya*, 4th edn., Moskva, 1941 ff., *14*, p. 87 ff. Abbr.: *ME*.

[7] *ME*, p. 111.

[8] *ME*, p. 117.

[9] *ME*, p. 247. It is interesting to note that as Stalin did not refer to this famous Leninist definition of matter, it is reported neither in the *Short Philosophical Dictionary* of M. M. Rozental' and P. Yudin (1939; 4th edn. 1955) nor in G. F. Aleksandrov's textbook, *Dialectical Materialism* (1953). It is, however, literally quoted in the first edn. of the *Bol'shaya Sovetskaya Entsiklopediya*, Moskva, *22*, 1935, p. 133, and it reappears again in the 2nd edn., Moskva, *36*, 1954, p. 522/a. We find it in *OMF* (1st edn., p. 119; 2nd edn., p. 115), in *FE*, p. 482, etc.

In Stalin too, we find the same identification; whereas the first of his famous "principal features" of Marxist philosophical materialism is that "the world is by its very nature material", the second feature is that:

> contrary to idealism . . . Marxist philosophical materialism supposes that matter, Nature, being, is an objective reality existing outside and independent of our consciousness, that matter is primary, since it is the source of perceptions, representations, consciousness, and that consciousness is secondary, derivative, since it is a reflection of matter, a reflection of being.[10]

In other words: whereas in the Western usage 'materialism' and 'idealism' denote an ontological and an epistemological theory respectively, i.e., two philosophical theories of entirely different orders, Marxism-Leninism uses 'materialism' as connoting realism and similarly 'idealism' as connoting any negation of materialism.[11] This idiomatic peculiarity seems to originate from the following misunderstanding. It seems at least probable that, as ontological materialism entails epistemological realism and epistemological idealism entails some kind of spiritualism, ontological materialism and epistemological idealism are incompatible. But Marxism-Leninism mistakes this entailment for an equivalence and argues, moreover, that any epistemological realism entails ontological materialism[12] and that any ontological position other than materialism entails epistemological idealism. Finally this alleged equivalence is claimed to be a simple identity: a non-materialistic realism is but a hidden idealism, therefore there is no difference between materialism and realism—both are "materialism"; anybody who asserts a primacy of Spirit to Nature is a disguised idealist, therefore ideal-

---

[10] I. Stalin, *O dialekticheskom i istoricheskom materializme*, contained both in *Istoriya Vsesoyuznoy Kommunisticheskoy Partii (bol'shevikov)*. *Kratky kurs,* Moskva, 1953, pp. 99–127; here p. 106 ff., and in I. Stalin, *Voprosy Leninizma*, 11th edn., Moskva, 1952, pp. 574–602; here p. 580 ff. Abbr.: *DHM*, with references to both editions.

[11] Some examples: according to Lenin's *Philosophical Notebooks*, Aristotle wavered between idealism and materialism, which is repeated by *FE*, p. 91/b. According to *FE*, 448/b, "Descartes' idealism is complicated by the religious presuppositions of his system". Compared with older works, however, *FE* is unusually gentle in this respect; thus to the authors of the *Short Philosophical Dictionary*, philosophers like Plato, Hume, Comte, Spencer and Dewey are all idealists, whereas F. Bacon, Spinoza, Hobbes, Locke and even Darwin are materialists. The decision as to whether a philosopher is to be classified as materialist or idealist often depends upon some rather incidental text in the "classics".

[12] Most recent Soviet sources have begun to realize that there is something wrong. Thus, e.g. *FE*, p. 210/a, admits that the acceptance of the "objectivity of being" does not by itself characterize sufficiently a materialist position. "It can get along together with objective idealism, with the religious philosophy of Neo-Thomism, such as, for example, in the work of Bocheński, Sciacca, and others."

ism and any negation of materialism are one and the same—they are both "idealism".

But whether this Marxist-Leninist idiom be based upon a fatal misunderstanding[13] or rather, as all communist philosophers would claim, upon a profound philosophical insight,[14] in any case it decidedly calls for an examination as to what Marxism-Leninism does claim by characterizing itself as "materialism". As we cannot, however, discuss all aspects of Marxist-Leninist materialism, we may restrict ourselves to an examination of the following problem: what exactly does Marxism-Leninism claim by saying that the world is by its very nature material or, to use another formula of Stalin, that all the "multifold phenomena of the world constitute different forms of matter in motion"?[15]

Before we, however, approach this problem, the following three introductory remarks may be useful:

(1) Contrary to most other contemporary philosophies, Marxism-Leninism is an ideology, i.e., a system of thought thoroughly ordered towards political action. Most of its ontological theories have to be understood as extrapolations of a social theory which, in its turn, is used as an "ideological weapon" by the Communist Party, whose history, then, must be regarded as "the visible incarnation of the ideas of Marxism-Leninism".[16] It is, therefore, always to some extent questionable to discuss Marxist-Leninist ontological or epistemological theories in themselves, leaving out of account their ideological and political background; for at a purely theoretical level the

---

[13] See G. A. Wetter, *Der dialektische Materialismus*, 4th edn., Freiburg i. Br., 1958, p. 330 ff.; English translation *Dialectical Materialism*, transl. by P. Heath, New York, 1958, p. 281 ff.; J. M. Bocheński, *Der sowjetrussische dialektische Materialismus,* Bern, 1950, p. 89 ff.; and especially J. de Vries, *Die Erkenntnistheorie des dialektischen Materialismus,* München, 1958, p. 92 ff.

[14] This "insight" has, most obviously, a Hegelian background. If one presupposes with Hegel that Nature and the absolute Spirit are related to one another like a thinking and its immanent *Setzung*, and if one identifies, moreover, the human mind with the Absolute Spirit, the identification of idealism with any recognition of a "creating" Spirit seems to be a logical consequence.

[15] *DHM*, p.106/p.580.

[16] "Za tvorcheskoe izuchenie marksistsko-leninskoy teorii", *Kommunist, 14,* 1954, p. 6, quoted by Wetter, *op. cit.*, p. 319. A good example of the dependence of ontological on social and political theories is the recent theory of "gradualness" (*postepennost'*). In 1950, as a help in arguing for the non-revolutionary character of the passage from socialism to communism, Stalin introduced the concept of a gradual revolutionary leap, see I. Stalin, *Marksizm i voprosy yazykoznaniya*, Moskva, 1950, p. 58 ff.; some years later, B. M. Kedrov wrote an article that described "gradualness" as an ontological category, see *Voprosy Filosofii* (Abbr.: *VF*), 2, 1954, p. 50 ff. See also my article "Sowjetideologie und Volksdemokratie", *Wort und Wahrheit* (Wien), *11,* 1960, p. 709 ff.

rigidity and the dogmatism of Marxism-Leninism (which may seem
acceptable as long as one considers it mainly as an instruction for political
action) easily comes to appear not only intolerable, but entirely unintelligible.

(2) This does not mean, however, that there exist no true Marxist-
Leninist philosophers. The widely accepted opinion that no true philosophy
is possible as long as there is no absolute freedom of thought does not seem
to be altogether correct; the Christian and the Moslem philosophies of the
Middle Ages are probably the best counter-instances. The basic tenets of
such thinkers may be grounded in dogma; but within such a frame, no
matter how inhuman and narrow, one may always find serious attempts to
philosophize. One must, however, avoid applying the high critical standards
of contemporary Western philosophy to Marxism-Leninism; up to now at
least, communist philosophy is of interest less to the philosopher properly
speaking than to the historian of philosophy.

(3) Although it would be hardly correct to speak of a decisive liberalization of Soviet philosophy since Stalin's death[17], in the last four or five years
so many original contributions have been published that it would be simply
inexcusable to discuss a problem like ours mainly on the strength of texts
written before 1953. For this reason, much more weight will be given in
this essay to more recent Soviet works, on the assumption that these are, on
the whole, more competent and more revelatory than those written under
Stalin's rule.

§1 *Materialism*

At first sight it might seem that materialism is a highly unequivocal
position. Any materialist philosophy will deny the existence of supernatural
forces as well as of spiritual beings and try to reduce everything knowable
to matter. The question might, of course, arise as to how such a position can
possibly be proved; as to the meaning of the basic tenets, however, there
seems to be little to ask.

And yet it is easy to show that this first impression is treacherous. There
are, indeed, several ways in which the materialist contention that everything is material and that nothing but matter exists is fundamentally
ambiguous.

(a) If someone claims that nothing but matter exists, we can under-

---

[17] It is often overlooked that the decisive liberalization of Soviet philosophy was
stimulated by Stalin himself. His authoritative decision on linguistics of June/July
1950 that declared language not to be class-conditioned, made possible all later discussions on logic, on philosophical questions of physics, etc. None of the later changes,
e.g. the reintroduction of certain "laws of dialectics" omitted by Stalin, is as important
as the earlier one.

stand him as saying either that *matter* is the most general category, or that there are no other than material existents (substances, individuals). As can easily be seen, these two assertions are not strictly equivalent; for if I say that there are no other than material existents, I can still maintain that some of their properties are not to be described as material (provided that 'material' does not simply mean *property of material existents*) and thus avoid at least an intensional identification of the two concepts: *matter* and *being*. One could argue, for example, that *matter* and *material* can be predicated only of existing individuals and not of their properties; or again, one could admit that all properties of material individuals are themselves material, without necessarily allowing that properties of such properties could be said to be material. In other words: any materialist will maintain that *matter* is a transcendental concept; yet it might be only an extensionally transcendental concept that cannot, without further ceremony, be identified with the intensionally transcendental concept of being.[18] On the other hand, someone might want to claim that there are entities which are neither existing individuals nor their properties, things which, therefore, cannot be called material. Thus e.g. the Stoics who denied the existence of non-material individuals and are, therefore, usually considered as materialists, nevertheless held that the true (αληθές) is incorporeal, since it is neither an individual nor one of its properties, only an ideal being, a λεκτόν[19]. We may, therefore, distinguish between *matter* as a radically extensionally transcendental concept (in a materialist system that denies any non-material entities at all, although it still may admit non-material properties of material entities) and an extensionally transcendental concept of matter which embraces only real being, even though ideal entities like meaning, propositions or even Platonic ideas are admitted in the system. (Though there might be some doubt as to whether such a system would still be truly materialistic.)

Thus we get a whole scale of possible materialist positions. We might speak of a "maximalist" materialism that takes *matter* to be an intensionally transcendental concept and thus identifies it with *being;* of a "minimalist" materialism that denies spiritual individuals and, of course, supernatural forces, but admits ideal entities and non-material properties of material existents; and of several intermediary positions.

(b) This is, however, by no means the only ambiguity of the basic

---

[18] 'Extensional' and 'intensional' are used here in a special sense. An "extensionally" transcendental concept is one which applies to all individuals, but not to the properties of these individuals. An "intensionally" transcendental concept is one which would apply to both.

[19] See Sextus Empiricus, *Adv. Mathem.*, 7, 38 ff.; Sexti Empirici *Opera*, Lipsiae, 1912 ff., B 81.

materialist tenet. For a materialist not only denies the existence of a certain kind of entities, individuals or properties, he maintains moreover that the actually existing things are such and such, namely, material. Once again, this might seem a thoroughly unequivocal assertion: whatever exists, is material, i.e. it is of the same kind as the palpable objects we usually call "material"—stones, water, fire and the like. Such statements might help us to know what exists and what does not exist; they yield, however, very little information as to how and what the actually existing things are. In other words: if whatever exists, is material, what does this tell us of what exists? If nothing but matter exists, what are, then, the properties of whatever exists? It is easy to see that these questions can be answered in several different ways. Someone might want to claim that matter is to be defined as that which really exists and whose existence can be "scientifically" acknowledged; this is, of course, the least satisfactory of all possible answers. Another might want to say that matter is what can be known through the senses; or: what is in space and time; or: what is changeable; etc.

(c) Finally, there is a third ambiguity which is probably the most important of all and, unfortunately, at the same time, the least easy to describe. If we inquire into the exact meaning of assertions like "the world is by its very nature material" or "nothing but matter exists", we are interested not only in the common attributes of material things (like *knowable through senses, existing in space and time,* etc.), but, moreover, in the semantics of the expression 'matter', especially insofar as it reveals, so to speak, the "ontological function" of matter.

Someone might want to claim that 'matter' is a generic term for material things like stones, trees and dogs; another might want to go so far as to add that properties of such things are matter, too. In this case, the semantics of 'matter' will be similar to that of 'a (subsistent) being' and 'being' respectively.[20] This is, however, neither the only possible nor even the usual interpretation. Matter, conceivably, might also be understood as something things are made or consist of. Thus one could suggest, for example, that 'matter' designates anything of which some other thing consists: flesh and bones are the "matter" of animals, chemical compounds the "matter" of flesh and bones, atoms the "matter" of chemical compounds, etc. Or again, one could lay down the convention that only those things *everything* consists of, like elementary particles, are to be called "matter". Finally, someone might want to say that 'matter' designates not a thing at all, but a metaphysical principle of all material things; and in that case it might be something things consist of, for example, or a "radical subject", a "substratum", a "substance" or whatever else might be denoted by such ambiguous expressions.

[20] The German distinction between *'Seiendes'* and *'Sein'* and the French terminology introduced by Sartre (*'étant'*, *'être'*) would be more appropriate.

These preliminary reflections indicate that the basic materialist tenet of the thorough materiality of the world is susceptive of many different interpretations. The concept of matter can be transcendental in several different ways; material things can be characterized this way or that; and the ontological function of "matter" will differ from system to system. There are probably still other such sets of ambiguities; for the purpose of our inquiry, however, the three forementioned will be sufficient. As it is somewhat difficult to keep asunder the second and the third group of problems, we shall treat them jointly; and as the transcendentality of the concept of matter will largely depend upon the definition of matter itself, we shall invert the order followed up to now and dispose our essay according to the following two questions:

a) By what attributes does Marxism-Leninism define matter and which ontological function does it impute to it?

b) Are there, according to Marxism-Leninism, any entities or properties not to be qualified as "material"?

As to both questions, we shall try to point out by what kind of proofs Marxism-Leninism usually substantiates the respective answers; we shall, however, not discuss such proofs in detail, since their hidden logical structure cannot be made intelligible without entering into the highly complex matter of Soviet philosophical methodology.[21]

§2 *The Marxist-Leninist Definition of 'Matter'*

In order to understand how the Marxist-Leninist concept of matter arose, one has to remember the complications and difficulties in which classical materialism was entangled by the developments taking place in physics during the closing years of the last century. Because of the discovery of radioactivity, the previously simple concept of matter suddenly became obscure. Many of the properties, previously ascribed to matter, like compactness, impenetrability or even extension, seemed no longer acceptable; the new discoveries showed that atoms are neither unchangeable nor indivisible; the disappearance of matter in radioactive decay, in particular, came as a considerable shock. In a word, it seemed no longer possible to give an exact and authoritative definition of matter; many philosophizing scientists, like L. Houllevigue, H. Poincaré and A. Righi, simply spoke of an abandonment of the traditional concept of matter.

In order to evade the difficulties arising from such discussions, Lenin decided to define matter in purely epistemological terms as "objective reality given to us in sensation".[22] There is, of course, the famous passage in which

---

21 See T. J. Blakeley, *Soviet Scholasticism*, Dordrecht-Holland, 1961.
22 E.g. *ME*, p. 254.

Lenin maintains that "the *sole* "property" of matter with whose recognition philosophical materialism is bound up is the property of *being objective reality,* of existing outside our consciousness",[23] and no Soviet philosophical textbook has ever omitted to quote or at least to paraphrase this "classical definition of matter".[24] It is, however, obvious that the contention that matter is objective reality, though it may inform us that matter exists, tells us neither what it is nor, consequently, what actually does exist; and, in particular, it does not exclude spiritual beings, since everybody who maintains that such beings exist, obviously takes them to be objective reality. It seems, therefore, beyond question that Lenin had in mind objective reality which is knowable through the senses; the severe criticisms of the Leninist definition of matter as objective reality without qualification, which have by now become usual in the West, hit the letter rather than the spirit of Marxism-Leninism. One might add that many Soviet textbooks quote the definition of matter as "objective reality given to us in sensation" as well; and at least one of them, published in 1958, says explicitly:

> According to the Leninist definition, matter is the philosophic category for the denotation of objective reality which exists outside of man; matter is that which, acting on our organs of sense, produces sensation. This definition of matter makes reference to its most essential traits: objective existence and ability to act on the sense organ.[25]

It seems, moreover, that Lenin was quite aware of the sensualism entailed by his definition of matter: he accuses E. Mach, B. Petzold as well as the Russian V. M. Tshernov of trusting too little in the testimony of our sense-organs, of being "inconsistent in the carrying through of sensualism",[26] and some pages later he explicitly approves the following definition in A. Franck's *Dictionnaire des sciences philosophiques,* published in 1875:

> Objective sensualism is materialism, for matter and bodies are, according to materialists, the only objects that can reach our senses.[27]

Since Marxism-Leninism is, however, by no means a consistent empiricism—it recognizes the importance, for example, of abstractions and abstract theories[28]—it takes everything for material that can be known either directly

---

[23] *ME*, p. 247.

[24] I. D. Pantskhava, *Dialektichesky materializm*, Moskva, 1958, p. 133.

[25] *Ibid.*

[26] *ME*, p. 116.

[27] *ME*, p. 118.

[28] Thus e.g. *FE* criticizes the positivist "principle of verification", and tries to show that all basic questions of philosophy would become meaningless if one were to presuppose it; see p. 242/a.

by our senses or through abstraction and inferences from our sense-experience, i.e. again simply everything that can be acknowledged as "objectively real". It might seem, therefore, in spite of what we just said, that it is Lenin's definition of matter as objective reality without qualification that Marxism-Leninism is mainly relying upon.

It is, indeed, true that Marxist-Leninists, when trying to show that something is material, often simply prove that it really does exist. Thus, in 1953, the philosophizing physicist S. G. Suvorov advanced the following "proof" of the materiality of the electromagnetic field: it undergoes influences of compact matter, therefore it is not simply a formal device for calculation, but reality—and no other proof for its materiality is needed.[29] On the other hand, however, no Marxist-Leninist would admit a proof of the objective reality, for example, of God; he would argue that God cannot be objective reality, since he is not material. But one cannot, obviously, consistently prove the materiality of A by proving its existence and, at the same time, reject a proof of the existence of B on the grounds that B is not material—especially, if the contention that only material things exist (which may perhaps be presupposed as a sort of axiom) amounts to saying that only objective reality exists.

Thus only the following interpretation seems possible. According to Marxism-Leninism, matter is to be defined as objective reality which is either knowable through the senses or at least can be conceived in analogy to things knowable through the senses. Thus the magnetic field, for example, although it can be neither seen nor heard, is an objective reality detected by methods which can be understood as a sort of enlargement of direct sense-knowledge (or, at least, its "being objective reality" is detected by such methods); it can, therefore, be conceived as analogous to customary material things. Vague as such a definition may sound, in any case it seems to exclude spiritual as well as ideal entities. There still remain, of course, serious difficulties, especially concerning the matters discussed by Marxist-Leninist political economy. Thus, according to G. F. Aleksandrov, the value of goods (Marx' "*Warenwert*") is something real and, consequently, material—a "social relation that arises from the production process and is revealed during the process of exchange";[30] yet it seems difficult to admit that phenomena like values are objects of sense-knowledge, no matter how the latter be "enlarged". It might be, however, possible to understand such phenomena as properties of material things, an interpretation which would attenuate the difficulty considerably.

---

[29] S. G. Suvorov in *Uspekhi Fizicheskikh Nauk*, *1*, 1953, p. 135, quoted by S. Müller-Markus, *Einstein und die Sowjetphilosophie*, Dordrecht-Holland, 1960, p. 372.

[30] *Dialektichesky materializm*, ed. by G. F. Aleksandrov, Moskva, 1953, p. 390.

The epistemological definition of matter just described strongly suggests a distinction between two different concepts of matter: a philosophical and a scientific. It was, indeed, Lenin himself who distinguished sharply between the "matter" concerning which materialism and idealism differ in their respective answers and "the question of the structure of matter, of atoms and electrons" which, according to him, "concerns only this "physical world" ".[31] Regarding the latter, Lenin had recognized that no physical interpretation of the structure of matter may be considered as definitive:

> The "essence" of things, or "substance", is also relative; it expresses only the deepening of man's knowledge of objects; and if yesterday the profundity of this knowledge did not go beyond the atom, and today does not go beyond the electron and ether, Dialectical Materialism insists on the temporal, relative, approximate character of all these *milestones* in the knowledge of Nature gained by the progressing science of man. The electron is as inexhaustible as the atom; Nature is infinite. . . .[32]

Yet this distinction was not without its drawbacks. First, it encouraged an emancipation of the sciences from their ideological tutelage; and secondly, there was the disturbing possibility that, by using a more technical and therefore narrower concept of matter, scientists might discover that besides or even beyond matter there is something different from matter, e.g. energy or the field. Therefore, in 1951, at a meeting of the Institute of Philosophy of the Soviet Academy of Sciences, this very distinction between a philosophical and a scientific concept of matter was rejected; while hitherto it had been described as "a necessary presupposition for the consistent maintaining of a materialistic line in philosophy",[33] now suddenly it became "totally opposed to scientific philosophy, to Dialectical Materialism".[34] From now on, only the philosophical concept of matter was to be used, by philosophers as well as by scientists; according to F. T. Arkhiptsev, this philosophical concept of matter:

> embraces the whole objective world, everything that existed, exists or will in the future be produced by Nature. The concept of matter embraces all known and unknown kinds of Nature; it is the all-embracing, absolutely universal concept. And as it was elaborated by philosophy, it is called a philosophical concept.[35]

This condemnation[36] of a probably quite useful distinction throws an

---

[31] *ME*, p. 246 ff.

[32] *ME*, p. 249.

[33] *Bol'shaya Sovetskaya Entsiklopediya*, 22, 1935, p. 134.

[34] S. G. Suvorov, *loc. cit.*, quoted by S. Müller-Markus, *op. cit.*, p. 370.

[35] F. T. Arkhiptsev in *VF*, 6, 1951, p. 52.

[36] More recent Soviet textbooks seem to tend to revive the old distinction between

interesting light on Soviet philosophic method. On the one hand, Marxism-Leninism claims that materialism is daily confirmed by all the sciences; as one of the authors puts it: "the entire path of the development of science in the last half-century is a triumphant confirmation of Dialectical Materialism".[37] Yet on the other hand its own exceedingly broad definition of matter is said to be the only scientific one,[38] and it is declared obligatory for all sciences. In other words: since all conceivable accesses to scientific falsification are barricaded by a concept of matter broad enough to permit the recognition of *any* scientifically discoverable reality as material, by definition sciences cannot falsify the basic Marxist-Leninist tenet. Yet if a falsification is strictly inconceivable, the statement about the "triumphant confirmation" through science is not very meaningful either; for the possibility of proof seems to depend upon, at least, the conceptual possibility of disproof.

One might be tempted to add that if *matter* is an all-embracing, absolutely universal concept, then the statement "everything is material" cannot be proven or meaningfully confirmed anyway, since it is analytic or tautologous. There is, however, at least one instance wherein this observation would not hold true. Suppose that A's, B's and C's be the only conceivable beings; and suppose, moreover, that it be proved that neither B's nor C's, although being conceivable, actually can exist. In this case, then, A will be, in a sense at least, a transcendental concept (one might call it a *"quoad se"* transcendental concept, since, in that case, the statement 'only A's exist' will be, so to speak, analytic or tautologous only *quoad se,* not *quoad nos*); and yet it will be correct to say that the statement 'only A's exist' has been proven. This point is of some relevance, since Soviet thinkers seem constantly to presuppose that the existence of entities other than material ones has been, once for all, disproved by their "classics". It is on this assumption that the concept of matter is said to be all-embracing and absolutely universal; and the constant recurrence to sciences is a self-satisfied applause rather than a verification. To put it more bluntly: like any other believer, a Marxist-Leninist enjoys the "corroborations" and, as regards negative cases, he will try to exclude them by an appropriate re-interpretation. There is, of course, a difference between a Marxist-Leninist and any other believer: whereas religious believers usually claim that God told them what to believe,

---

a philosophical and a scientific concept of matter. See for example *Osnovy marksistskoy filosofii. Popularny uchebnik,* Moskva, 1960 (not to be confounded with *OMF*): "The philosophical concept of matter has to be distinguished from the way in which natural sciences describe the world (*ot estestvennonauchnoy kartiny mira*), i.e. from those ideas on the structure . . . of concrete forms of matter which arise in the development of science." p. 51.

[37] N. M. Sisakyan in *VF,* 2, 1959, p. 89.

[38] See e.g. F. T. Arkhiptsev, *loc. cit.*

Marxists-Leninists maintain that their basic tenets, the so-called "principles of Marxism-Leninism", have been scientifically proven by the "classics". In the last resort, however, this difference is rather verbal: for these "principles" are being accepted not so much on the ground of their respective proofs as because they have been promulgated by the "classics"; as has been recently shown by T. Blakeley, a reference to a "classical" text is usually considered to be a sufficient proof.[39] The question as to whether the "classics" actually *did* prove their point is of little relevance; Soviet thinkers are satisfied simply with a claim in this respect. It is highly significant of the Soviet philosophic atmosphere that, since 1931, there are only three cases in which a proposition of a "classic" has been called in question (one of the three critics being, by the way, a "classic" himself).[40]

Up to now, we have heard about only two attributes of matter: it is objective reality, existing outside our consciousness, and it is, usually at least, knowable through the senses. Both are characteristic of Lenin's conception of matter in purely epistemological terms: objective reality as opposed to pure impressions, Kantian "phenomena" and the like, and knowableness through senses as opposed to any kind of intellectual intuition or mystical experience. There are, however, according to Marxism-Leninism, several other attributes of matter, although Soviet thinkers seem not to be inclined to introduce them into a definition of matter; since Lenin maintained that the *sole* property of matter with whose recognition philosophical materialism is bound up is the property of being objective reality, all other properties of matter (except, usually, the sensuous perceptibility which is often only indirectly alluded to) are treated independently of the definition. And yet it was Lenin himself who asserted (only about ten pages later, by the way) that:

> a Dialectical Materialist considers motion as an *inseparable property of matter*.[41]

Once again, therefore, it would seem that there is not just one property of matter with whose recognition Dialectical Materialism is bound up;[42] be-

---

[39] T. Blakeley, *op. cit.*, especially p. 13 ff.

[40] *Ibid.*, p. 18 ff.; also my work *Das Widerspruchsprinzip in der neueren sowjetischen Philosophie*, Dordrecht-Holland, 1960, p. 5 ff. The "classic" in question is Stalin who, sometime around 1947, criticized a small passage in Engels' *External Politics of Russian Tsarism;* the two other criticisms, by E. A. Asratyan (1955) and A. Kol'man (1958), concern passages in Engels, too.

[41] *ME*, p. 257. Our italics.

[42] This statement, contrary to what H. Dahm, "Der Streit um die Materie des Diamat", *Ost-Probleme* (Bonn), 27, 1956, pp. 290 ff., thinks, must not be understood to contradict Lenin's statement about the "*sole* property". For in the latter passage, Lenin speaks of "philosophical materialism" in general, whereas in the former he is concerned with Dialectical Materialism.

sides being objective reality given to us in sensation, matter is always and necessarily in motion. Although emphasizing (against energeticists like W. Ostwald) that there never can be motion without matter, Lenin does not shrink from asserting that:

> whether we say the world is moving matter, or that the world is material motion, makes no difference whatever.[43] This assertion does, however, not entail any denial of bodies at rest; rest is said to be a limiting case of motion, a "relative equilibrium" that "only has meaning in relation to one or other definite form of motion".[44]

Matter without motion, therefore, is "just as unthinkable as motion without matter"; according to Engels, motion is the "mode of existence of matter (*Daseinsweise der Materie*)".[45] More recent Soviet sources, however, usually reserve this latter definition (*sposob ili forma sushchestvovaniya materii*) to space and time, and speak of motion as of the "form of being of matter (*forma bytiya materii*)";[46] according to the *Philosophical Encyclopaedia* (which, incidentally, uses the older definition of Engels), motion is "an inseparable property (of matter), an attribute belonging to matter intrinsically (*vnutrenne*)".[47]

But as matter "cannot move otherwise than in space and time",[48] there are still two more properties of matter, both of which are said to be "basic forms of the existence of matter".[49] "A being outside of time makes as little sense as a being outside of space".[50] As the essence of space and time consists in motion,[51] matter, motion, space and time are absolutely inseparable.[52] We are told that real space is three-dimensional, and time one-dimensional

---

[43] *Ibid.*

[44] F. Engels, *Herrn Eugen Dührings Umwälzung der Wissenschaft ("Anti-Dühring")*, 11th edn., Berlin, 1958, p. 70.

[45] *Ibid.*, also F. Engels, *Dialektik der Natur*, 3rd edn., Berlin, 1958, p. 61.

[46] This was first observed by H. Dahm, *art. cit.*, p. 921 ff. The difference between these two formulae is often overlooked, thus e.g. by G. A. Wetter, *op. cit.*, p. 346 ff.; and by Bocheński, *op. cit.*, p. 8. It would seem, however, that even Soviet philosophers do not stress this difference very much. Thus, contrary to *BSE, 14*, 1952, p. 291, or to *OMF*, p. 122, more recent works use the classical formula of Engels: the popular version of *OMF*, p. 54 ff.; I. D. Andreev, *Dialektichesky Materializm*, Moskva, 1960, p. 118 ff.; etc.

[47] *FE*, p. 433.

[48] *ME*, p. 162.

[49] *OMF*, p. 132; *FE*, p. 298/b; I. D. Andreev, *op. cit.*, p. 125 ff., etc. H. Dahm points out that the *Great Soviet Encyclopedia* (Abbr.: *BSE*) defines space as a *universal* and time as *one of the basic* forms of existence of matter; see *BSE, 9*, 1951, p. 272, and *35*, 1955, p. 105.

[50] F. Engels, *Anti-Dühring*, p. 61; also *Dialektik der Natur*, p. 25.

[51] This queer contention goes back to Lenin's *Philosophical Notebooks* (V. I. Lenin, *Filosofskie Tetradi*, Moskva, 1947, p. 241).

[52] See e.g. *BSE, 9*, 1951, p. 273.

and irreversible,[53] but their exact relationship to matter remains somewhat in the dark, as no Soviet source risks discussing the crucial question of the exact kind of distinction that holds between matter and its "attributes". It is even difficult to say whether, according to Marxism-Leninism, space and time are really properties of matter or rather aspects of its motion;[54] the *Philosophical Encyclopaedia* evades this problem simply by saying that space and time are "forms of the existence of matter in motion".[55]

Last but not least, matter is said to be infinite; some texts prefer the pleonasm "eternal, infinite and boundless".[56] Infinity is said to be "the most universal qualitative characteristic of matter in motion"; accordingly, the world is:

> infinite as to space, infinite as to time and infinitely multiform as to its properties, as to those concrete forms in which matter is in motion. Whereas infinity in space and time concerns the world as a whole, the infinity of properties characterizes not only the world as a whole, but each separate material object, too. From the point of view of Logic, the infinity of the material world might, in all three of its aspects, be understood as a consequence of the substantial character of matter. As matter is a unique substance which is *causa sui*, its evolution and motion cannot be limited by anything whatsoever.[57]

These last two sentences and, indeed, the whole passage, call our attention to the difficult problem as to what "ontological function" Marxism-Leninism imputes to matter. Strictly speaking, there are at least two different, though closely connected questions involved: in what sense is matter said to be unique? and: is matter a thing or rather some metaphysical principle of things?

We may begin by pointing to a certain disparity between the attributes which Marxism-Leninism ascribes to matter. As we have seen, there are at least five such attributes: matter is "objective reality", it is knowable through the senses, it is in motion, it is spatio-temporal, and it is infinite; as regards infinity, it is infinite as to space and time, and, moreover, infinitely multiform as to its properties. The disparity that we have in mind is simply the following. Each material thing, taken separately, is a sensuously perceptible objective reality which moves in space and time and is infinitely multiform as to its properties. It would, however, be absurd to maintain that each material thing is infinite as to space and time; these two attributes "con-

---

[53] G. F. Aleksandrov, *op. cit.*, p. 307 ff.; *OMF*, p. 136 ff.; *FE*, p. 298/b; etc.
[54] For more details, see G. A. Wetter, *op. cit.*, p. 357 ff.
[55] *FE*, p. 437/b.
[56] *OMF*, p. 121.
[57] *FE*, p. 154/a.

cern the world as a whole", i.e. the spatio-temporal *totality* of material things. The *Principles of Marxist Philosophy* is definite on this point:

> No material thing is unchangeable, all material things are finite. Yet wherever one thing disappears, another thing comes to be and replaces it, and this in such a way that not a single particle of matter disappears without trace or turns into nothing . . . Where the limits of one material object come to an end, the limits of another object begin, and there is no end as to this boundless interchange and interaction of material objects. Matter . . . is eternal, endless and boundless.[58]

There is, of course, nothing to object to in a distinction between properties of "separate material objects" and properties of the "material world as a whole"; yet it is somewhat difficult to see how *both* kinds of properties could be attributes of "one and the same" matter. It would be of no great help, either, to say that, in the last resort, as it were, *all* attributes of matter are properties of the "world as a whole". For though we might disregard the oddity of such an assertion, it seems impossible to consider the knowableness through senses, at least, as a property of the totality of all material things (nobody ever saw all material things). Moreover, it seems beyond question that Soviet thinkers consider all attributes of matter, except, precisely, infinity as to space and time, as properties of material things, taken separately.

This ambiguity, if an ambiguity it is, seems to originate from Engels. According to this "classic" of Marxism-Leninism, "matter as such is a creation of thought and an abstraction":

> We disregard qualitative differences of things by comprehending them under the concept of matter (*indem wir sie unter dem Begriff Materie zusammenfassen*). Unlike definite existing matters, matter as such is, therefore, nothing sensuously perceptible and existing (*nichts Sinnlich-Existierendes*). When natural science goes in search of a uniform (*einheitliche*) matter as such . . . it acts as if, instead of asking for cherries, pears or apples, it would like to see the fruit as such; or, instead of cats, dogs, sheep, and the like, the mammal as such; or the gas as such, the metal as such, the stone as such, the chemical compound as such, motion as such.[59]

Except for the somewhat queer terminology (and in spite of the fact that Engels paraphrases Hegel here),[60] this passage is unequivocal. There exists no matter as such, since it is nothing more than a generic concept which embraces all kinds of "definite existing matters". There exists no matter as such, only this quantity of a definite chemical compound or that piece of stone. Another passage seems to point in the same direction:

---

[58] *OMF*, p. 121.
[59] F. Engels, *Dialektik der Natur*, p. 271 ff.
[60] See G. W. F. Hegel, *Sämtliche Werke*, Stuttgart, 1932 ff., *8*, p. 59.

Nobody has ever seen or experienced matter as such or motion as such, only the different actually existing stuffs and forms of motion. Stuff, matter is nothing but the totality of stuffs (*die Gesamtheit der Stoffe*), from which this concept is abstracted, and motion as such is nothing but the totality of all sensuously perceptible motions; words like 'matter' and 'motion' are nothing but *abbreviations* in which we comprehend many different sensuously perceptible things according to their common properties (*nach ihren gemeinsamen Eigenschaften*). Matter and motion *cannot*, therefore, be known otherwise than by studying the single stuffs and forms of motion; through knowing the latter, we know *pro tanto* matter and motion *as such*, too.[61]

Again, except for the expression 'totality', it seems clear what Engels has in mind: 'matter' is a sort of generic term for all the different materials like water, wood or chemical compounds. There is, however, one major ambiguity indicated by the expression '*totality* of material things'. It is, to begin with, a highly unusual contention that matter should be the totality, i.e. the sum of all material things, since terms of this semantic type rarely if ever are used in a collective sense. Secondly, it seems that 'matter' cannot, at one and the same time, be a generic term for all kinds of stuff, i.e. be predicable of *any* quantity of *any* stuff, *and* designate *all* quantity of *all* stuffs, *taken together*. In the latter case, namely, 'matter' seems to be less a generic term than the name of a sort of vague, but necessarily unique individual. In other words: as unequivocal as the passage just quoted might seem, there are at least two ways in which it can be understood. We might disregard the expression 'totality', and then we may say that *anything* endowed with certain properties is matter; 'matter', then, is an abbreviation for *any* objective reality that is knowable through the senses and moves in space and time, whether it consists of gold, water or flesh and bones. We might, however, stress the expression 'totality', and then we may say that, strictly speaking, only the totality of *all* material things and stuffs deserves to be called 'matter'; 'matter', then, would be a sort of collective term, or else the name of a sort of unity of which the individual things that consist of definite stuffs would be sections or "appearances".

Although it seems highly probable that Engels had in mind the first rather than the second alternative, Soviet philosophers do not appear to be able to decide for one of them. Very often, they prefer to speak of matter as if it were a sort of totality, for example, when they argue that "the moving and eternally developing matter is the substance, i.e. essence, the ground of the world".[62] Taken literally, this statement would seem to suggest that there is but one Matter with a capital 'M', a sort of materialist *élan vital* which eternally develops and of which concrete things are but appearances.

[61] F. Engels, *Dialektik der Natur*, p. 251.

[62] M. Rozental', P. Yudin, *Kratky filosofsky slovar*, 4th edn., Moskva, 1954, p. 467.

As regards the expression 'substance', one has to remember that Lenin himself disliked it and maintained that only pompous professors would use it instead of the "clear and exact expression, matter";[63] if today Soviet philosophers sometimes do use it, they have in mind not so much the Aristotelian οὐσία as the infinite and thoroughly independent *substantia* of Spinoza. Most of all, however, Marxist-Leninist matter is reminiscent of Hegel's eternally evolving and self-differentiating "Spirit" translated into materialistic terms. Thus, occasionally, we meet statements like the following:

> The world is the motion of matter according to law (*zakonomernoe dvizhenie materii*);[64] matter is cause of itself, and hence nothing can act upon it, since there is nothing in the universe apart from self-moving matter and its forms of appearance.[65]

On the other hand, however, Marxism-Leninism is much more conscious of the existence of distinct individual things than either Spinoza or Hegel ever was; we might even say that it contains a considerable dash of "Aristotelian" (as opposed to "Hegelian") mentality.[66] Hegel and all Hegelians start from the "whole" which is itself conceived as a sort of individual, and reduce individual things to parts, aspects or appearances of this very "whole"; only the latter really does exist, whereas individuals are subordinated to it, they exist only in virtue of the totality. Consequently, Hegelians are used to thinking in terms of a universal drift, of motion and of relations, but not of things. On the contrary, Aristotle and, by the same token, all materialists not influenced by Hegel, start from individual things, ask what properties they have in common and subsume them under the appropriate concepts; according to them, what really does exist are things like individual cows and definite pieces of stone in comparison to which any commonness of properties and any being related to one another is a highly secondary affair. It is well enough known that Marxism-Leninism has inherited quite a lot from the Hegelian totality-thinking; yet there are, as we have just said, many "Aristotelian" elements, too. As regards the problem of matter, such an "Aristotelian" mentality turns up whenever Marxist-Leninists, instead of speaking of matter in general (of the "unique substance of the world"), begin to discuss the structure of concrete material

---

[63] *ME*, 157.

[64] *ME*, 156.

[65] I. B. Novik in *VF*, 3, 1954, p. 142.

[66] The distinction between a "Hegelian" and an "Aristotelian" trend among Soviet philosophers has been introduced by J. M. Bocheński, "Einführung in die sowjetische Philosophie der Gegenwart", *Aus Politik und Zeitgeschichte. Beilage zu "Das Parlament"*, 45, 1959, p. 608 ff. By "Aristotelian trend", here, is meant a naturalist Aristotelianism, like that of Alexander of Aphrodisias or, more recently, of Randall.

things. Thus they will distinguish between different types of matter, for example, between compact matter (*veshchestvo*) and fields, both of which are said to possess mass as well as energy[67]—and treat them as though they were some kind of individual substances. According to the *Principles of Marxist Philosophy*:

> matter exists in the form of an infinite variety of bodies and objects which, quantitatively and qualitatively, are different from each another . . . The different types of matter differ by a greater or smaller complexity and are objects of investigation of different sciences: physics, chemistry, biology, etc. The "elementary" particles of matter are relatively simple: photons, electrons, positrons, mesons, protons, anti-protons, neutrons, anti-neutrons, etc. More complicated forms are atoms and molecules. The next degree of complexity is represented by gases, liquids and solid bodies . . . and by the heavenly bodies too: planets, stars and the starry systems. Considerably more complicated are the bodies of organic Nature and, especially, its highest product, man. A special material object is human society . . .[68]

Occasionally, they will even quote Marx' statement that matter is the subject (*sub'ekt*) of all changes,[69] and they do not seem to have in mind a unique universal subject (like Spinoza's *substantia*), but rather *any material thing*. Here, once again, 'matter' seems to become a generic term that designates individual material things, i.e. any individual that exists outside our consciousness, is knowable through the senses and moves in space and time. As, however, this individual is neither infinite as to space and time, nor *causa sui,* nor the unique substance of the world, nor anything of that kind, we can, it seems, conclude that Marxism-Leninism tries to combine two thoroughly different conceptions of matter and that, consequently, its materialism is anything but uniform and consistent.

Yet before drawing such a conclusion we should consider whether we have not overlooked an interpretation which might remove at least some of the above-mentioned ambiguities. We have been presupposing that, except for the expression 'totality', passages in Engels indicate that he understood 'matter' as a generic term for material things. Actually, however, Engels does not speak of material things alone, but of "stuffs" as well. We must, therefore, consider the following possibility: do not Marxist-Leninists consider matter as a sort of "stuff" of which all material things consist? It is obvious that there are many arguments which seem to support this interpretation. First of all, it would seem much more natural to take 'matter' as a stuff-term (like 'gold' or 'wood'), instead of considering it as

---

[67] See e.g. G. F. Aleksandrov, *op. cit.*, p. 293; the popular version of *OMF*, p. 52; I. D. Andreev, *op. cit.*, p. 110. Not to be found in *OMF* itself.

[68] *OMF*, p. 121 ff.

[69] G. F. Aleksandrov, *op. cit.*, p. 282.

a generic term for material things (like 'body' or 'tree'), not to speak of the queer contention that it is a name for the totality of material things. In this event, the disparity of the properties of matter mentioned above would seem to disappear: there is no difficulty in saying that one and the same stuff is knowable through the senses as well as infinite in space and time; it seems easy to admit that a stuff might exist everywhere and always (though there are some difficulties as to what exactly this *does* mean). Besides, a stuff-term has as referent neither individual things, taken separately, nor all things, taken together, but rather a component of any of them; and as, by "adding" stuff to stuff, we never get anything other than again the same stuff (for example: water plus water is still water, whereas a body plus another body is not again a body), it would seem that the difference between 'taken separately' and 'taken together' does not matter. Last but not least, we have ourselves admitted that Engels, when speaking of matter, has in mind stuffs rather than material things.

In spite of all these arguments which probably might still be multiplied, the interpretation of Marxist-Leninist "matter" as some kind of "stuff" is untenable. First of all, so far as we know, Marxist-Leninists never actually do say that things *consist* of matter; occasionally, they seem to have a stuff in mind, but they will never use other expressions than 'things and nature are material', 'the phenomena of the world are different forms of moving matter', and the like. Secondly, when speaking about matter, they never use stuff-terms. When giving examples of "types of matter", they do not mention chemical elements or compounds, but "things" like elementary particles, atoms, living cells, organisms, animals and society. Thirdly, a materialism for which matter is a stuff of which things consist, cannot consistently say that there is nothing in the universe apart from self moving matter; for the definition of matter as of a stuff of which things consist entails that there exist, at least, the things that consist of it. Besides, if things really consist of matter, they cannot, in the last resort, consist of matter alone, since, in that case, they would *be* matter and not *consist* of it; yet the only co-principle that comes into question is motion—and, once again, Marxist-Leninists never say that man, for example, consists of matter and motion. He *is* "matter in motion", in the same way as this *is* an animal and that *is* a tree, and everything *is* a being. Finally, it seems somewhat difficult to see how motion and, indeed, *all* motion should be an "intrinsic" property of a stuff; for it would seem that things themselves, rather than anything of which they consist, change. It would be odd to maintain that if an animal dies, it is the stuff of which it consists that undergoes the change in question; though the stuff may change, too, it is most obvious that it is the animal, not the stuff it consists of, that dies. If an animal *is* a stuff, or if it *is* matter (it would be more appropriate to say: if it is a quantity of a stuff), then, of course, this stuff, this matter (this quantity of a stuff) "dies"; yet even in this case, it is not the stuff of which the animal *consists* that dies, but, at best, the stuff which it *is*.

This last remark shows that there is, after all, a sense, in which Marxist-Leninist matter might be understood as a sort of "stuff". Provided that "stuff" be not defined as something another thing consists of (as a bed, for example, consists of the "stuff", wood), but rather as something which "another" thing *is* (as wood might be said to "consist" of the "stuff" wood, taking 'consist' as equivalent to 'is identical with') there is no reason why Marxist-Leninist matter should not be described as a sort of "stuff". Yet it is easy to see that if "stuff" is to be defined as something that material things *are*, 'stuff' is a generic term for material things. 'Matter', therefore, designates here not something things *consist of* (in the ordinary sense of the phrase) but things themselves. Thus we may return to our original contention that, although Marxist-Leninist matter is a sort of materialist *élan vital,* material things are called "matter", too. The inconsistency of this conception of matter might, however, be somewhat attenuated by saying that matter, though being a sort of universal drift, actually never exists otherwise than in the form of a number of distinct, though closely interconnected, individual things; though it seems difficult to say what exactly such a statement might mean, it does not seem to be inconsistent. Another way, and probably a more appropriate one, to describe the situation in question would be as follows. It is well-known that some metaphysicians and, in particular, some Neo-Scholastics, when speaking of "being", have in mind individual beings as well as a sort of all-embracing Being with a capital 'B', the latter being neither God nor a pure concept, but rather a sort of universal bond. The similarity of such a "being" to Marxist-Leninist matter is patent; and as earlier Soviet thinkers often treated 'matter' exactly like Scholastics treat 'being' (we shall see this later), one might even say that the ambiguity of the Marxist-Leninist concept of matter gives form to a sort of materialist "problem of being". Yet we have to hasten to add that such reflections are far beyond the intellectual frame of contemporary Marxism-Leninism; the heights of metaphysics are—for the moment at least—out of reach for most Soviet thinkers.

There still remains a minor question to be considered, namely, whether Marxist-Leninist matter is truly a "subject of change"; the reader will remember that it was Marx himself who has used this expression ('subject of all changes'). This question may be considered in order to compensate, as it were, for the fact that we cannot anymore meaningfully ask whether matter is to be taken as a thing or as a "metaphysical principle". To a

Hegelian the problem is not very meaningful anyway; and as far as the materialist component of Marxism-Leninism (which we have called "Aristotelian") is concerned, 'matter' designates things.

It would go far beyond the scope of this essay to analyse in detail the Marxist-Leninist theory of change; it is as interesting as it is inconsistent, since most Soviet thinkers are prepared to allow, in this context at least, a violation of the principle of non-contradiction. We should like only to point out that Marxism-Leninism is very definite in saying that matter "is nothing inert to which motion has to be added . . . no unaffected "subject" (*podlezhashchee*[70]) for the predicate 'it moves ";[71] any "unchanging substance of things" is most emphatically denied.[72] Consequently, matter has to be considered as a "subject" that undergoes change not so much as to its accidents or "forms" as in itself. The definition of motion as of a "form of *being* of matter" seems to point in the same direction: materiality and motion are, so to speak, only two aspects of one and the same reality.

This denial of any unchanging subject of change has its counterpart in the thesis that matter is "throughout infinite in depth".[73]

> Just as the relation between things, and their modes of change, are endless, so too is the number of stages leading *into the depth* of things, *into the depth* of their being . . . However simple and elementary a given particle of matter may appear to us, in reality it can never be absolutely simple, absolutely elementary, there can never be any sort of ultimate building-stone, any sort of mythical *materia prima*, from which the whole world might seem to be built up.[74]

It is, therefore, impossible to say that, even if matter and change are not really distinct, there might still be some element of matter which is the last subject of change. When going "into the depth" of matter, we shall never find anything other than some kind of material things which are intrinsically in motion.

It would, however, be too simple to say that, in this case, matter would not be much more than a sort of subsistent motion; in any case, Marxist-

---

[70] '*Podlezhashchee*' is the technical term for 'grammatical subject'; although it uses the present participle instead of the perfect participle, the expression is a literal translation of the Latin '*subiectum*'. FE, p. 92/a, describes the Aristotelian material cause as "*materiya i podlezhashchee (substrat)*", as "matter and subject (substratum)".

[71] FE, 435/b. The passage, however, admits that matter is the "universal bearer" ('*nositel*') of motion; yet it is not clear what 'bearer' might mean, especially as opposed to 'subject'. BSE, 30, 1954, p. 191/a, treats under '*nositel*' only 'bearers for catalysers" in chemistry.

[72] See e.g. M. Rozental', P. Yudin, *op. cit.*, p. 467.

[73] V. I. Lenin, *Filosofskie tetradi*, p. 86.

[74] B. M. Kedrov in *Bol'shevik*, 2, 1948, p. 45, quoted by G. A. Wetter, *op. cit.*, p. 342.

Leninists are not Hegelian enough to say anything of that kind. Similarly, it would not seem appropriate to say that matter and motion are related to one another, say, like prime matter and substantial form.[75] Such interpretations seem inevitable only as long as one adheres to the schema "subject-form" which our everyday language suggests. Though Marxist-Leninists never brought forward any other schema, it is highly probable that they never permit the correlating of matter and motion in the same way that Scholastics and, in a sense, everybody except Hegelians, would do, namely, as a subject to its properties, "forms" or the changes that it undergoes. If they were forced to produce a schema, they probably would grasp at a picture and say that matter is "soaked through" with motion, or that motion is at the very "heart" of matter. "Whether we say the world is moving matter, or that the world is material motion, makes no difference whatever".[76]

§3 *Matter, Being and Consciousness*

We have seen that, except for matter as "universal drift" which we shall disregard from now on, 'matter' designates material things, i.e. any subsistent entity[77] which exists outside our consciousness, is knowable through senses and moves in space and time. There remains, however, the question as to the materiality of *properties*. It is, of course, obvious that if all things are material, all properties are material, too, in the sense that they precisely are properties of material things. Yet there is still another meaning of 'material' whose recognition might permit one to say that there exist non-material properties of material things. Suppose that we define material properties by saying that they exist outside our consciousness, are knowable through senses and change in space and time; in that case, if there are any properties of material things which, for example, are neither sensuously perceptible nor spatio-temporal, such properties will be "non-material".

This remark is of some interest, since many Soviet thinkers deny that consciousness, though being the property of a type of matter, is material. Before, however, treating the special case of consciousness, we should like to say some words on the type of transcendentality that Marxist-Leninists ascribe to the concept of matter.

Earlier Soviet thinkers unquestionably tend simply to identify *matter* with *being*. Thus e.g. the Deborinist B. Bykhovsky in his *Outlines of a Philosophy of Dialectical Materialism*, from 1930, speaks about matter in terms that a Thomist would use when speaking about being:

[75] This has been suggested by H. Dahm, *art. cit.*, p. 924 ff.

[76] *ME*, p. 257.

[77] Not 'any individual', since society is a form of matter, too.

Matter is everything that is, the most general concept, the genus of all genera. Whatever is, is a kind of matter; but matter itself cannot possibly be defined as a special case of some genus. If whatever exists is matter, then it is preposterous to consider matter as a characteristic differentiating anything from anything else, for such a something else cannot be but non-existent, i.e. it cannot be.[78]

A similar mode of expression is to be found in a textbook of Dialectical Materialism, from 1933, edited by M. B. Mitin:

Matter is the entire world existing independently of us. The concept of matter is the most general of all concepts. Everything that exists is matter in one form or other, though matter itself cannot be defined as falling under any other class.[79] (According to Mitin, too, it is simply impossible to assign a specific difference to the concept of matter.)

Although neither of these texts (which could easily be multiplied) mentions anything about properties, they seem to suggest that *matter* is an intensionally transcendental concept. As, however, philosophers of the early Stalinist period, especially some like M. B. Mitin, rarely used to think, it is possible and, indeed, highly probable, that they simply forget to think about properties. Mitin knew as well as any contemporary Soviet philosopher that, according to the official teaching, there exists a "subjective reality" too; if, therefore, only what exists "independently of us" is matter, *matter* cannot be the "most general of all concepts".

In any case, more recent texts hardly ever use such terminology. It is, however, very difficult to discover exactly which kind of transcendentality they ascribe to *matter*. The only texts which touch upon this problem are those which discuss the relation of *matter* to *being;* but although it would seem that today Soviet philosophers are inclined to count *being* among the so-called "philosophical categories",[80] there are very few texts which treat this kind of problem. In 1931, the otherwise little known Deborinist Gonikman proposed a "closed system" of categories which was to begin as

---

[78] B. Bykhovsky, *Ocherk filosofii dialektiches̆kago materializma*, Moskva, 1930, p. 78, quoted by N. Lossky, *Dialektichesky materializm v SSSR*, Paris, 1934, p. 27. Lossky quotes also a passage from A. M. Deborin's introduction to Hegel's complete works, 2nd edn., *1*, 1929, p. xvi: "being is by its very essence a material category".

[79] M. B. Mitin-I. Razumovsky, ed., *Dialektichesky i istorichesky materializm*, part I, *Dialektichesky materializm*, ed. by M. B. Mitin, Moskva, 1933, p. 107, quoted by G. A. Wetter, *op. cit.*, p. 343.

[80] The "philosophical categories" are generally defined as concepts designating the "most general objects, properties, aspects and connections in reality"; see e.g. V. P. Tugarinov, *Sootnoshenie kategory dialekticheskago materializma*, Leningrad, 1956, p. 4; also H. Fleischer: "On Categories in Soviet Philosophy—A Survey", *Studies in Soviet Thought, 1*, Dordrecht-Holland 1961, p. 64 ff. *FE*, p. 207, describes *being* as such a "philosophical category".

well as to end in the category of being;[81] yet this point of view has been explicitly rejected. The notion of *being* has always, in fact, been relegated entirely to the background in comparison with the central category of *matter;* thus the second edition of the *Great Soviet Encyclopaedia* treats *being* in half a page, for example, mentioning first the opposition between materialism and idealism and then proceeding immediately to a discussion of "social being".[82]

The recent *Philosophical Encyclopaedia,* however, devotes nearly seven pages to this subject, treating successively the concept of being in Buddhism, in ancient Chinese thought, in Greek, medieval and modern philosophy, and ending with a section on the Marxist-Leninist interpretation. Yet although the "historical" section distinguishes at the end no less than eleven fundamental meanings of 'being' (among them *matter, existence, essence, substance* and *idea*),[83] from the "systematical" section we learn little about the properly Marxist-Leninist concept of being. Being is said to be the "infinite subsistence (*sushchestvovanie*) of matter that is independent of consciousness", whereas non-being is said to be neither emptiness nor absolute disappearance, but rather the "conversion of matter from one form of being to another, i.e. the formation of a new form of being of matter".[84] As for the relation of *being* to *matter,* we are told that "there is no "being as such" without substance, without matter":

> As it denies the infinity of matter and immediately leads to idealism,[85] the very assumption that being precedes matter is a failure ... Only matter subsists and its *being* is proved by the law of conservation and transformation of energy.[86]

Yet this does not mean that being and matter are simply one and the same, though it would seem that 'being' designates only the "epiphenomenon" of the existence of matter (an epiphenomenon, since matter exists by definition, as it were). We are referred to a passage in Engel's *Anti-Dühring,* according to which:

> the unity of the world does not consist in its being, although its being is a prerequisite of its unity, since it first must *be* in order to be *one* ... The real

---

[81] See G. A. Wetter, *op. cit.*, p. 428.

[82] *BSE, 6,* 1951, p. 432/b ff.

[83] *FE,* p. 209/b.

[84] *Ibid.* This strange definition of *non-being* is confirmed by what is said on 'annihilation', p. 69/a: "an interaction of particles and anti-particles of such a kind that they disappear and change into other particles. The expression 'annihilation' is inexact. In the phenomenon in question there happens ... exclusively a transformation of one form of matter into another."

[85] The idea seems to be as follows: if being is prior to matter, then matter proceeds from something non-material.

[86] *FE,* p. 209/b.

unity of the world consists in its materiality, and this is proved . . . by a long and protracted development of philosophy and natural science.[87]

From this text which in itself is queer enough, the *Philosophical Encyclopaedia* concludes to the still odder result that the "coincidence and identity" of being and matter "presuppose a distinction and are relative":

> The relativity of the coincidence of the notions of being and matter is easy to discover, since matter, not being, is the substance of the world, and the unity of the world does not consist in its being, but in its materiality.[88]

It is difficult to decide what such expressions could possibly mean; and, in any case, it does not help us much to answer the question as to the type of transcendentality of *matter*. The only conclusion that we might draw is that Marxism-Leninism probably would not be prepared to admit the existence of ideal entities, since they hardly are a being "with substance, with matter". Yet as we shall see in a moment, not all Soviet philosophers agree on this point, though most of them tend to reduce meanings, propositions, and the like to "subjective reality", i.e. to activities of human mind.

The only relatively well-reasoned essay on our problem that we know of was published by the Slovak philosophical quarterly in 1958.[89] F. Cízek, the author, starts from the observation that the category of *reality* which simply states the fact of being, is wider than the categories of *objective reality* and of *matter*:

> A phenomenon can be objective as well as subjective, ideal as well as material, . . . but if it exists, it is reality. The category of reality implies, indeed, not only objective, but also subjective phenomena, and not only material, but also ideal events.[90]

Thus, taken purely as such, the category of *reality* is in a sense neutral and "can equally well be used by a materialist as abused by an idealist". Marxism-Leninism does not, according to Cízek, deny that there are subjective and ideal realities; yet contrary to all forms of idealism it does not reduce objective and material reality to ideal products of any consciousness (which, according to *all* Marxist-Leninist authors, would be the case, if matter were to have been created by God). Marxism-Leninism tries rather to distinguish as clearly as possible "the problem of the objectivity of the existence of reality and the problem of the objectivity of our knowledge of

---

[87] F. Engels, *Anti-Dühring*, p. 51.

[88] *FE*, p. 210/a.

[89] F. Cízek, "Poznámky k materialistickému pojetí kategorie reality", *Slovensky Filozoficky Casopis* (Bratislava), 4, 1958, pp. 401–408. See also my book *Marxismus-Leninismus in der CSR. Die tschechoslowakische Philosophie seit 1945*, Dordrecht-Holland 1961, p. 134 f.

[90] *Ibid.*, p. 405.

it".[91] All things that exist independently of our mind and, consequently, independently of any mind at all, are objective reality, i. e. "certain specific forms of existence of matter in motion"; but psychic events are subjective, not objective realities, at least for the subject that experiences them. Finally, though the results of such psychic events, i.e. our ideas, are neither objective nor material, they are still realities.

According to this analysis, then, there would be at least one kind of reality that is not material: our ideas and thoughts on matter, i.e. material realities as reflected by our mind; as Cízek points out, the most Utopian ideas are, philosophically speaking, as "real"[92] as the ideas of Marx and Lenin, for example. As to the "subjective reality", i.e. the psychic events, they are neither explicitly classified as material, nor is their materiality explicitly denied. Some passages, however, indicate that they may be understood to differ from properly material realities:

> Subjective reality has its own relative independence. That the objective is reflected by the subjective does not entail any identity of the objective and the subjective . . . If, therefore, we emphasize the unity and the "identity" of objective and subjective reality as to their content, we may not forget their relative difference and independence as to their form.[93]

This analysis is remarkable for several reasons. First of all, it is the only essay of which we know that discusses the relation of *matter* and *being* (or, to use Cízek's terminology, of *objective reality* and of *reality* without qualification) in terms of their transcendentality. Secondly, it is unusually unequivocal as to the distinction between *subjective* and *ideal* reality, i.e. between psychic events and their intramental object-correlates; as we shall see, Soviet thinkers often confuse them. Finally, it seems to suggest that Marxist-Leninists, some of them at least, might be prepared to admit that their concept of matter is only partially extensionally transcendental (see §1, (a)). As regards this last point, however, we have to be careful. For the recognition of truly ideal entities seems to be bound up with the acceptance of entities which are neither material things nor, in particular, their properties; yet Cízek does not explicitly say that ideal reality is not a sort of qualitative determination of "thinking matter". Nevertheless, it is now clear that some Marxist-Leninist thinkers would deny that *matter* is an intensionally transcendental concept like *being* or *reality*.

The last paragraphs may have indicated that the crucial point of the

---

[91] Notice the ambiguity of the expression 'objectivity' which is common in Marxism-Leninism. "Objectivity of existence" is given, when something actually does *exist;* "objectivity of knowledge", when we *recognize* the former.

[92] 'Real', here, embraces also ideal phenomena.

[93] *Ibid.*, p. 406.

Marxist-Leninist conception of material properties is its interpretation of consciousness. It is, however, interesting to note that the discussion about whether consciousness is material or not, never even touches the problem as to the exact meaning of 'material'. In other words: one would expect that the decision as to whether consciousness is material or not would depend upon whether 'material property' means *property of a material thing* or, for example, *spatio-temporal and sensuously perceptible property;* and if it means the latter, we would expect that Marxist-Leninists would discuss whether consciousness is spatio-temporal and sensuously perceptible or not. This is not, however, what happens. Instead, all Marxist-Leninists refer to a passage in Lenin, according to which it is incorrect to call thought 'material', and then they try to adjust the meaning of 'non-material property' to the requirements of materialism and, in particular, of Dialectical Materialism as opposed to "vulgar materialism" of nineteenth-century philosophers like L. Büchner, J. Moleschott and K. Vogt.

The passage in Lenin in question criticizes J. Dietzgen for having said that:

> the non-sensual idea (*die unsinnliche Vorstellung*) is not only sensual, but also material, i.e. real (*wirklich*) ... Spirit is not more different from a table, the light or a sound than those things are different each from another.[94]

Whereon Lenin:

> This is, obviously, incorrect. Rather it is correct that both thought and matter are "real", i.e. that they exist. But to call thought material is to make a wrong step towards a confusion of materialism and idealism.[95]

Yet besides the authority of Lenin, there is another and, indeed, equally important reason why Marxism-Leninism does not dare simply to identify consciousness with matter. This reason might be best described by reporting the following event. In 1954, the editors of the Soviet philosophical journal *Voprosy Filosofii* received from seven students of the University of Leningrad a letter submitting the following question: "Is it correct to assert that consciousness is material?". In the name of the whole editorial board, V. N. Kol'banovsky answered decidedly in the negative:

> No, it is not correct. Those who maintain it, confuse the material base of consciousness with the very essence of consciousness and thinking which are not material.[96]

---

[94] J. Dietzgen, *Das Wesen der menschlichen Kopfarbeit, dargestellt von einem Handarbeiter. Eine abermalige Kritik der reinen und praktischen Vernunft*, Hamburg, 1869; reedn. Berlin, 1955, p. 82 ff.

[95] *ME*, p. 231 ff.

[96] V. N. Kol'banovsky, "Pravilno li utverzhdat', chto soznanie material'no?" *VF*, 4, 1954, p. 236.

Some lines later the reason we alluded to crops up:

> Those comrades who argue for the materiality of consciousness have, obviously, forgotten that the conflict of materialism and idealism on the question of the relation of matter and consciousness is the very basis of all history of philosophy. What meaning would all this struggle have if consciousness were to be material? If, from the point of view of Logic, it were necessary to argue (*esli logicheski posledovatel'no*) for the materiality of consciousness, then it would be equally necessary to arrive at a denial of the fundamental question of philosophy, at the conclusion that all philosophy is needless . . . It is, therefore, wrong to include thought in the concept of matter.[97]

It is easy to see the point: if the question of the relation of thought to matter, of Spirit to Nature, is the "central question of world-outlook", then matter and thought simply cannot be one and the same, for otherwise the whole question would be meaningless. One might even suggest that this was what Lenin had in mind when he said that to call thought "material" is "to make a wrong step towards a confusion of materialism and idealism"; for all Marxist-Leninists agree that to deny the importance of the "central question of world-outlook" is a typical feature of idealism.

Yet on the other hand it was Lenin himself who agreed with Dietzgen's assertion that thought is "a function of the brain, just as to write is a function of the hand",[98] and who says in another passage:

> To be sure, the opposition of matter and consciousness has absolute significance only within a very limited field, in our case exclusively within the limits of the fundamental epistemological question, namely, what is to be acknowledged as primary and what as secondary. Outside such limits the relativity of this opposition is beyond all question.[99]

Accordingly, most Soviet philosophers maintain that the opposition of matter and consciousness is only "relative". Thus, e.g. M. B. Mitin who, as we know, described *matter* as nearly an intensionally transcendental concept, speaks of a purely epistemological opposition:

> We distinguish matter from consciousness, oppose them one to another; but this opposition is nonetheless relative, and is meaningful only in relation to the "epistemological" problem . . . The contrast between knowledge and being is a contrast between knowing matter and known matter, and nothing more. The wholly proper and legitimate contrast between subject and object loses its meaning outside the theory of knowledge. If we were to begin contrasting matter, from the scientific point of view, with Spirit, this would amount to a betrayal of materialistic monism.[100]

---

[97] *Ibid.*
[98] J. Dietzgen, *op. cit.*, p. 81.
[99] *ME*, p. 231.
[100] M. B. Mitin, *op. cit.*, p. 107, quoted by G. A. Wetter, *op. cit.*, p. 343.

Similarly, B. Bykhovsky tries to describe the relation of matter to consciousness in a way that reminds one of G. T. Fechner's *"Zweiseitentheorie"*, according to which the mental and the physical would be related one to another as the concave and convex aspects of the circumference of a circle:

> Physical and mental are one and the same process, only seen from two different sides . . . That which is seen from the outer, objective side as a physical process is equally perceived from within, by the material being itself, as a phenomenon of will or sensation, as mental in character.[101]

This amounts to saying that it is but a "being experienced from within" that gives to an otherwise thoroughly material process its "non-material" character. The expression 'subjective reality' which is constantly being used by all Marxist-Leninists, points in the same direction: the "non-materiality" of a psychical act is but subjective, i.e. it appears only from within. This interpretation (which, incidentally, was adopted by Stalin)[102] is followed by many Soviet philosophers up to this day. It contains, however, a serious inconvenience. No Marxist-Leninist will maintain that the "non-materiality" of mental events is subjective in the sense that, strictly speaking, we are wrong in maintaining it; though it appears only "from within", it is, in a sense at least, "objective". This being presupposed, one might admit that if matter experiences itself, the object of this experience is material; yet though the "spirituality" of a mental act might consist only in its being experienced, the experiencing itself has to be explained, too. Either it is again material "objective reality", and then there seems nowhere to be a "subjective reality"; or it is not material, and then this non-materiality cannot be purely "subjective" and "epistemological", but must have some objectivity, too.

Such considerations compel Soviet thinkers to look out for new interpretations. Thus Kol'banovsky, for example, distinguishes between a "material base" of consciousness and the very essence of consciouness itself:

> The material base of thinking and consciousness are the nervous processes that happen in man's brain; but the very essence of understanding and awareness that gives to man the possibility of abstracting from concrete objects and phenomena, the possibility to universalize them, to analyse and to synthesize, to discover complex connections of the multiform phenomena of reality . . . this essence of thinking and consciousness consists in an ideal reflection of the objective world.[103]

To the decisive question of how Marxism-Leninism understands the "essence of the ideal reflection of reality", Kol'banovsky has, however, no

---

[101] B. Bykhovsky, *op. cit.*, p. 83 ff., quoted by N. Lossky, *op. cit.*, p. 44.
[102] See J. V. Stalin, *Sochineniya*, Moskva, 1946 ff., *1* (1906/07), p. 312.
[103] V. N. Kol'banovsky, *loc. cit.*, p. 237/b.

other answer to offer than the classical formula of Marx, namely, that the ideal is nothing more than the material, transplanted in man's head and transformed in it. One finds the same lack of clarity in A. G. Spirkin's article on *Consciousness* in the *Great Soviet Encyclopaedia* (1956). Spirkin explicitly stresses that consciousness does not, in essence, consist of the material events that occur in the brain; yet he too has no other explanation to offer than the queer "epistemological distinction":

> Dialectical materialism proceeds from the assertion that material objects, phenomena and events . . . that are reflected by our consciousness, exist in the latter as pictures, as thoughts on bodies and phenomena, not in a material, but in an ideal form . . . The nervous stimulations and physiological events make up the necessary basis of the subjective picture, but not the picture itself . . . Consciousness is an ideal, i.e. subjective picture of the objective world. From the point of view of epistemology, consciousness is an ideal phenomenon and as such opposed to matter; but this opposition is relative.[104]

This text, besides explaining very little, suffers from a fundamental obscurity which is common in Marxism-Leninism. It confuses, namely, the "subjective" and the "ideal", i.e. mental activities and their intra-mental object-correlates, the "pictures". This is apparent, too, from Spirkin's definition of consciousness:

> a singular and unique process of reflection of reality that includes all forms of man's psychic activities: sensations, representations, thinking, awareness, emotion and will.[105]

It is difficult to see in what sense psychic events, especially those of emotion and will, could reflect anything at all; and, in any case, such mental activities are most obviously not the "things transposed in our head", although it may be possible to qualify ideas in such a way. Yet this confusion in a way explains why Marxism-Leninism maintains that the opposition of matter and consciousness is "purely epistemological"; for it seems, at least, conceivable that mental acts and their "ideal" object-correlates, ontologically speaking, are one and the same reality, though from the point of view of subjective experience and of epistemology they have to be distinguished. Yet what, in that case, are opposed are by no means consciousness and its material basis, but rather the "content" of consciousness and consciousness itself. In other words: if someone can show that conscious matter reflects not in a material, but in an ideal way, he has not proved that consciousness is not material; he has only pointed out that the "content" of consciousness is not the material object itself, but rather its ideal copy. If, therefore,

---

[104] *BSE, 39,* 1956, p. 658/b.
[105] *Ibid.,* p. 656/b.

Marxism-Leninism wants to maintain that it differs from all forms of "vulgar materialism", it has to prove that the mental "operation" of thinking itself, and not only the material thing as thought (the "picture" or "copy") is different from material (physiological) events.

We must, therefore, focus our attention on the question as to whether Marxist-Leninists are prepared to admit that consciousness is not only epistemologically, but also in some fundamental ontological sense different from matter, and, consequently, in some sense non-material. So far as they admit it, we shall then be able to say that the Marxist-Leninist concept of matter is extensionally rather than intensionally transcendental; if, however, they do not admit it, we shall have to say that only an "epistemological attitude" (whatever that may be) allows them to keep asunder the two concepts of *matter* and *being,* and, consequently, to distinguish Dialectical Materialism from a form of "vulgar materialism".

If we restrict ourselves to more recent contributions, it seems that we can distinguish at least three different groups among Soviet philosophers:

(1) Those who defend the materiality of consciousness, although admitting that, "epistemologically speaking", thought is ideal, not material. This, in a sense classical, point of view has recently been again advocated by N. V. Medvedev who emphatically denies that his position could be qualified as mechanism or vulgar materialism. According to Medvedev, vulgar materialism simply identifies consciousness with matter, whereas Dialectical Materialism calls consciousness only "material" insofar as it is a motion and thus a mode of existence of matter:

> Any motion whatsoever is inseparable from matter, has a material subject—and is in this sense material. Yet motion is a mode of existence of matter and not matter itself.[106]

It is easy to see that this position evades the problem. Nobody questions the fact that if man is matter, consciousness is a material property in the sense that it is the property, or motion, or "mode of existence", of a material thing. The real problem, however, concerns the materiality of the property itself.

(2) Those who hold that consciousness cannot be considered as material, either from an epistemological or from a properly ontological point of view. This interpretation was defended by F. I. Georgiev in the beginning of 1959:

> Psychic events appear to be ideal, not material, both considered in their relation to the brain and in their relation to objective reality . . . How can psychic events be at one and the same time, though from different points of view, both ideal and material? . . . To say that the epistemological character-

---

[106] N. V. Medvedev, "K voprosu ob otrazhatel'noy rabote mozga", *VF, 6,* 1960, p. 109.

istic determines the "ontological" characteristic of psychic events is to turn the whole problem head over heels and to be overreached by idealism.[107]

(3) Finally those who try to overcome the dilemma by maintaining that consciousness is neither material nor non-material, since neither 'material' and 'non-material' can be properly predicated of consciousness. As a matter of fact, we do not know of any Soviet philosopher who has maintained such an "intermediary" position; it was, however, explicitly proposed by G. Klaus, professor at the Karl Marx University in East Berlin, in his polemics against G. A. Wetter. Whereas in the first edition of his book, *Jesuits, God and Matter,* he had defended a position similar to that of Medvedev,[108] the second edition offers the following quite original analysis. According to Klaus, it is not altogether correct to say that consciousness is a property of matter; rather it is a property of the movement (*Bewegungsvorgänge*) of the brain and thus a property of a property of matter. But a property of a property of a thing is, "contrary to what Aristotle believed", not necessarily a property of the thing itself; thus e.g. the pendulum of a wall-clock has the property of being in motion, and this motion again the property of being periodical—and yet we cannot say that the pendulum is periodical. In a similar way, according to Klaus, neither of the two expressions 'material' and 'non-material' can be predicated of a property of a property of matter, although matter has still to be considered as the last subject of the property, consciousness.[109] In spite of its originality, however, this analysis does not seem to lead out of the dilemma. For if someone maintains that there be properties that cannot be described either as f or as non-f (in a qualified sense), he has, it seems, to admit that there exist properties of such a kind that they cannot be qualified as f, i.e. those properties which are described as non-f (in an unqualified sense); if, therefore, somebody maintains that there are things which are neither material nor spiritual (or neither material nor ideal), he has to admit that not everything is material.

It would seem, therefore, that only two alternatives are open for Marxism-Leninism: consciousness is to be described ontologically either as material or as non-material. Both alternatives lead to serious difficulties. Those who maintain that consciousness is material probably cannot escape some kind of "vulgar materialism" which reduces everything to matter and to which the question as to the priority of Matter to Spirit is hardly meaningful. It is

---

[107] F. I. Georgiev, "V. I. Lenin o vzaimootnoshenii psikhicheskogo i fiziologicheskogo", *Nauchnye doklady vysshey shkoly. Filosofskie Nauki* (Moskva), *1,* 1959, p. 22 ff.

[108] G. Klaus, *Jesuiten, Gott, Materie. Des Jesuitenpaters Wetter Revolte wider Vernunft und Wissenschaft,* Berlin, 1957; 2nd edn., Berlin, 1958, p. 138 ff.

[109] *Ibid.,* p. 356 ff. (added in the 2nd edn.).

of little help to maintain that this alternative still permits an "epistemological" distinction between matter and consciousness; for in this case "vulgar materialism" will be only "epistemologically" and "subjectively" mistaken, whereas, ontologically speaking, it would be correct to say with K. Vogt that thought stands in the same relation to the brain as the bile to the liver —which all Soviet philosophers violently deny. Yet the second position has not a much easier stand; for how can matter have truly non-material properties? It is not by chance that, beginning with Aristotle, the Occidental philosophical tradition always maintained that from non-material activities or properties of a seemingly material object we have to conclude to a relative non-materiality of this object itself, to a relative "independence of its form from matter", or the like. Even admitting the basic assumption of Marxism-Leninism that consciousness genetically arises from matter as a late product of its historical development, we have to ask whether such a production of something non-material does not require in matter itself a certain trans-materiality, or else a radical change that has to be described as a sort of "dematerialisation". When F. I. Khaskhachikh, the late specialist for epistemological questions of the Stalinist period, says that:

> inorganic matter contains the possibility of sensitive and later on of thinking beings; but this possibility becomes actual only at a definite stage in the historical development of matter[110]

we have, it seems, to conclude to a basic homogeneity of matter, sensation and consciousness, or else to assume that at a definite stage of its development "matter" by a "dialectical leap" became sufficiently non-material[111] to produce non-material results like sensation or consciousness. Thus the second of the two positions would seem to break through the fundamental framework of Marxist-Leninist materialism. We cannot, therefore, but agree with G. A. Wetter who points out that:

> the view of Soviet philosophers would be bound, logically, to lead either to acceptance of the soul, or to vulgar materialism, if the problems involved were to be rationally thought out to a conclusion . . . The very conditions imposed upon the enterprise of Soviet philosophers, namely, the construction of a psychology devoid of a psyche . . . , without at the same time reducing the mental to the physiological, compel them to enter a blind alley from which there is no escape.[112]

We conclude this essay by pointing out that the dilemma that we have just been describing, is but a counter-part to the inconsistencies of the con-

---

[110] F. I. Khaskhachikh, *Materiya i soznanie*, Moskva, 1951, p. 47 ff.

[111] It would not seem sufficient to say "sufficiently complex", or the like. For if the "matter", man, for example, is only more *complex* than other matters, its properties may be more complex, too, yet they never will be *less material*.

[112] G. A. Wetter, *op. cit.*, pp. 552, 556.

cept of matter mentioned above. All such inconsistencies and dilemmas are, in the last resort, due to the basic paradox of Marxism-Leninism, namely, that it wants to be a materialism without leaving the heights of Occidental metaphysics which, to Soviet philosophers, is exemplified in Hegel. As has been often pointed out, such syntheses are hardly possible; and the synthesis of an extreme idealism with materialism, this most threadbare of all philosophical positions, can amount, to use a favourite expression of Lenin's, only to "*galimat'ya*" (rubbish).

PHENOMENALISM WITHOUT PARADOX

Kenneth Sayre

§1 *Obscurities in Formulation*

Refuting Phenomenalism has become a fashionable exercise among philosophers in recent years, one result of which has been to point up a rather wide variety of theses which the Phenomenalist ought not to hold if in fact he were ever tempted to do so. Another result has been to obscure those aspects of Phenomenalism which make the effort to refute it worthwhile. Thus, however unfashionable it may be to take the Phenomenalist's part, it is not inopportune at this time to attempt to clarify some of the issues involved in this controversy, and to show that the Phenomenalist need not always come out second-best. My purpose in this paper is to formulate Phenomenalism in a way which avoids the usual traps and snares, and which at the same time preserves what I take to be the heart of the Phenomenalist's traditional contribution to the conceptual analysis of matter.

Not all the blame for obscurity in the Phenomenalistic thesis lies with the critics, for Phenomenalists themselves often have been content with a paradoxical manner of expressing their insights. Berkeley's Philonous, for example, confronts the bewildered Hylas with the assertion that "a cherry is nothing but a congeries of sensible impressions".[1] Mill speaks of that which we call by the names 'ice' and 'log' as "Permanent Possibilities of Sensation".[2] And Mach writes that "thing, body, matter, are nothing apart from their complexes of color, sounds, and so forth . . ."[3] More recently, Russell has suggested that "the thing" of common sense may in fact be identical with the whole class of its appearances";[4] the sun, for instance, is "a whole assemblage of particulars, existing at different times, spread out from the centre with the velocity of light, and containing among their number all those visual data which are seen by people who are now looking

---

[1] *Three Dialogues between Hylas and Philonous*, George Berkeley, LaSalle, Illinois, 1947, p. 81.

[2] *An Examination of Sir William Hamilton's Philosophy*, John Stuart Mill, 1872, p. 257.

[3] *The Analysis of Sensations*, Ernst Mach, Chicago, 1914, p. 6.

[4] *Mysticism and Logic*, Bertrand Russell, London, 1953, p. 147.

at the sun".[5] And Price, although himself not a Phenomenalist, understands that theory to assert that "a *system* of sense-data is a material thing".[6]

The main drawback of these traditional formulations, apart from their
hyperbole, is that they tend to suggest that Phenomenalism has implications
in areas where the Phenomenalist would do well to leave himself un-
committed. The correct and often repeated objection to the doctrine that
material objects are composed of sense-data is that it makes material objects
out to be much too flimsy. If sense-data, or sensible impressions, are entities
of any sort at all, then surely they are not entities substantial enough to
compete with electrons or quanta for the role of basic building block of the
material world.

Wishing to avoid any apparent conflict with physical theory, Phenomenalists recently have proposed a variety of alternative formulations. Russell
makes it clear that he considers material objects to be "logically constructed"
and not physically constructed out of sense-data. In Russell's sense, for a
thing of type A to be logically constructed out of things of type B is for any
statement about things of type A to be replaceable without change in truth
value by statements about things of type B.[7] As Ayer remarks, "those who
assert that material things are "logical constructions" out of sense-data must
be understood to claim . . . that a proposition which is expressed by a sentence
referring to a material thing can equally well be expressed by an entirely
different set of sentences, which refer only to sense-data."[8] It is now customary among Phenomenalists to present their thesis as an assertion about the
meaning of statements about material objects. But this formulation has its
own weaknesses. Not only is it unclear what sense of 'meaning' is involved
in this assertion, but moreover no Phenomenalist has been able to produce
a set of sense-datum statements which comes close to being able to do the
job of the material-object statement it might be considered to replace. The
Phenomenalist seems to have committed himself to a thesis in semantics
which he is either unable or unwilling to pursue in any practical detail.

The formulation of Phenomenalism which I wish to present is intended
to avoid these apparant entanglements with physics and semantics. I believe
that it retains nonetheless the central insight which Phenomenalists of the
traditional stripe have attempted to convey. It was at least part of the claim
of the traditional Phenomenalists that we have no empirical evidence that

---

[5] *Mysticism and Logic*, p. 130–31. Russell in this work defends the Selective Theory,
which differs from Phenomenalism primarily in sanctioning unsensed sense-data.

[6] *Perception*, H. H. Price, London, 1932, p. 282.

[7] *Mysticism and Logic*, p. 149.

[8] *The Foundations of Empirical Knowledge*, A. J. Ayer, London, 1953, p. 233.

there is anything in what we call a material object beyond that which possibly could be presented to us in ordinary experience. Thus any assertion that there is more to an object than can possibly appear to an observer is a speculative hypothesis for which we have no empirical evidence. As Berkeley suggests, what we mean in non-theoretical contexts by saying "that an orange tree exists is that we perceive it by our senses".[9] Mill writes that all changes of the sort involved in the combustion of a log and in the melting of ice are "intelligible without supposing the wood, the ice, or the water, to be anything underneath or beyond the Permanent Possibilities of Sensation".[10]

Insofar as we have empirical evidence for an assertion about a material object, that assertion cannot entail that there is more to the object than can be disclosed in sensory observation. The emphasis is on the evidential status of material-object statements in Russell's assertion that if we can state the law by which appearances of a thing vary, "we can state all that is empirically verifiable; the assumption that there is a constant entity (which "has" these appearances) is a piece of gratuitous metaphysics".[11] And Ayer remarks that the purpose of the Phenomenalistic analysis is to elucidate "the meaning of statements about material objects by showing what is the kind of evidence by which they may be verified".[12]

The thesis I shall defend is that any assertion about material objects which can be completely warranted on the basis of ordinary sensory experience can be expressed with reference exclusively to sense-data. This is weaker than the claim that all statements about material objects can be expressed in sense-datum statements, for it allows that there may be material object statements, for which we do not have complete empirical justification, which may not be warrantable on the basis of sense-data alone, and which therefore cannot be expressed in sense-datum statements.

A corollary of this thesis which is of particular relevance to the present topic of discussion is that no statement which asserts a distinction between matter and form as fundamental components of material objects can be wholly warranted on the basis of ordinary sensory experience. On this basis it may be maintained that in analyzing what we can say about material objects without exceeding the warrant of sensory observation no need arises for introducing the concept of matter as a principle apart from the sensible appearances of material things.

---

[9] *Three Dialogues*, p. 66.
[10] *Hamilton*, p. 251.
[11] *Our Knowledge of the External World*, London, 1914, p. 111–12.
[12] *Foundations*, p. 235.

Before expanding this thesis, it will be helpful to examine certain objections to the more traditional formulations of Phenomenalism, and to attempt in the process to secure a defensible definition of the term 'sense-datum'.

§2 *Phenomenalism Is Not Descriptive*

Persons whose acquaintance with Phenomenalism is second-hand seem to protest most against the notion that material objects might be made up of sensations. But it is clear that material objects are not composed of sensations, in any ordinary sense of these terms, and insofar as the traditional Phenomenalists have suggested that this is the case they have exhibited a bad sense of public relations. Our best information is that material objects are made up of atoms, which in turn are made up of elementary particles which can be described only in terms of scientific theory. It is not the part of the Phenomenalist to suggest anything to the contrary, and it is to the discredit of the traditional authors that their formulation did not make this clear. It is all too easy to find reason in Berkeley, Mill, and Mach to think that they conceived sense-data to be much too concrete. And the price of this manner of speaking is still being paid in misunderstanding and lack of sympathy. Even in recent literature we find critics professing to be unable to find sense-data with characteristics which would enable them literally to be parts of material objects, and pronouncing the downfall of Phenomenalism (or "The Sense-Datum Theory") on that basis.[13]

Another understandable misreading which, like this one, construes talk about sense-data as talk about entities in some sense present in the external world, is the notion that Phenomenalism is a theory about the phenomenological character of perceptual experience. It is then objected that Phenomenalism carries with it a false phenomenology and consequently is incorrect. Critics who take this tack cite with glee Lewis' remark in *Mind and the World Order:*

> It is indeed the thick experience of the world of things, not the thin given of immediacy, which constitutes the datum for philosophical reflection. We do not see patches of color, but trees and houses; we hear, not indescribable sound, but voices and violins. What we most certainly know are objects and full-bodied facts about them which could be stated in propositions.[14]

[13] For an example, see "The Myth of Sense-Data", W. H. F. Barnes, in *Proceedings of the Aristotelian Society*, 45, 1944–45, pp. 89–118.

[14] *Mind and the World Order*, C. I. Lewis, New York, 1929, p. 54.

Wild, for example, approves of this statement, but criticizes Lewis for going on to abandon "the aim of classic philosophy to describe the *thick* experience of the world of things *as it is given*" in favor of "the procedure of modern empiricism [which] . . . singles out a certain portion of the given as peculiarly accessible or *given* in some special sense".[15] Stout, for another example, maintains that the question what it is that we perceive "is one of fact and cannot . . . be answered on *a priori* grounds. It can only be answered by an analysis of the process of sense perception as it actually occurs in particular instances."[16] Stout's analysis concludes that there is an element in perception which draws our attention to actual physical objects as distinct merely from their sensible appearances. It would seem to follow that this element cannot be described in terms of sense-data alone.

Criticism of this sort may emphasize noteworthy aspects of perceptual experience. Its effect on the Phenomenalist, however, has never been what the critics hope for. The Phenomenalist in fact may explicitly subscribe to the proposition that material objects and not sense-data are presented in perceptual experience. No Phenomenalist in this century, as far as I know, has proposed an analysis of the phenomenological character of perception which would conflict with any of the analyses mentioned above. The Phenomenalist is concerned not with describing how things appear but rather with the problem how what is given in sense-perception (however described) can furnish evidence for statements about the material world. Phenomenalism entails no specific phenomenology, and hence cannot be criticized legitimately for a false phenomenology.

Objections of this sort have had the worthwhile effect of showing the Phenomenalist what he should not commit himself to, and of encouraging him to seek formulations of his position which make his commitments more explicit. Thus in recent years Phenomenalism has been presented almost exclusively in what, following Carnap, is called the "formal mode" of speech. In this formulation, Phenomenalism is made out to be a thesis about the meaning of material object statements instead of about the constitution of the material world. In order to give a just hearing to this formulation it will be helpful to make the concept of sense-datum more precise. This is appropriate for other reasons as well, for some critics seem to think that a difficulty common to *all* forms of Phenomenalism is its reliance upon an indefensible notion of sense-data, and I would like to exempt in advance

---

15 "The Concept of *The Given* in Contemporary Philosophy", *Philosophy and Phenomenological Research*, *1*, 1940, pp. 70–71.

16 "Phenomenalism", G. F. Stout, *Proceedings of the Aristotelian Society*, *39*, 1938–39, p. 7.

from this difficulty the formulation of Phenomenalism which I propose below.

§3 *In Defense of 'Sense-datum'*

It has been suggested at one time or another that sense-data are entities characterized 1) by being unaltered by mental attitudes,[17] 2) by being indubitable,[18] or 3) by being in no sense inferred.[19] Philosophers who have maintained that sense-data possess one or more of these characteristics have maintained moreover that they belong to sense-data uniquely, and hence that sense-data can be defined by the possession of these characteristics. Another approach to the problem of defining 'sense-datum' has been by use of technical verbs of perception, such as 'to be aware of' or 'to see' in a special sense of the word 'see'. Broad once suggested that 'sense-data' be used to refer to objects of which we are "directly aware".[20] And Ayer has used 'sense-datum' to designate what we see (or feel, etc.) in a sense of these verbs which entails that what is seen (or felt, etc.) really has the properties it seems to have.[21] As an example of a third approach, we have Russell's attempt to define 'sense-datum' in terms of a special sort of knowledge: "Let us give the name of 'sense-datum' to the things that are immediately known in sensation: such things as colours, sounds, smells, hardnesses, roughnesses, and so on."[22]
Each of these approaches has been roundly criticized in recent years. Barnes has pointed out that if sense-data are entities at all they are entities of a very queer sort, such that they fail to possess determinate characteristics which can be re-examined to settle questions about their nature.[23] Ryle has contended that the identification of sense-data with sensations results from the "logical howler" of assimilating the concept of sensation to the concept of observation.[24] The mistake allegedly is in thinking of sense-datum terms as objects of verbs which properly take as objects only terms which refer to phyiscal objects. And the attempt to identify sense-data as objects of a special sort of knowledge has been criticized in detail by Prichard, who claims that the mistake in this procedure stems from the erroneous notion

---

[17] Lewis, in *Mind and the World Order*, p. 66.
[18] Price, in *Perception*, p. 3.
[19] Russell, in *Our Knowledge of the External World*, pp. 75–78.
[20] *Scientific Thought*, C. D. Broad, London, 1923, pp. 234–40.
[21] *Foundations*, p. 24.
[22] *The Problems of Philosophy*, Bertrand Russell, London, 1946, p. 12.
[23] "The Myth of Sense-Data".
[24] *The Concept of Mind*, Gilbert Ryle, New York, 1949, p. 213.

that perception is a form of knowing.[25] These are samples of a large body of literature critical of sense-data.

As well taken as these points may be, I cannot see that they in any way affect the Phenomenalist's thesis about the relation between sense-datum statements and statements about material objects. The Phenomenalist is responsible for making clear his use of the expression 'sense-datum statement', a task to which we return shortly. But if he can do so without assuming that sense-data are *entities* of any sort, it is not germane to his thesis to point out that sensing is not knowing, that sense-data are not objects, or that they do not possess determinate characteristics. What is germane, and what the Phenomenalist cannot afford to ignore, are questions about the evidential status of sense-datum statements in respect to statements about material objects.

It is part of the traditional concept of sense-data that they are in some sense indubitable and in no sense inferential. Let us consider some of the difficulties involved in these notions, and see whether they still can be put to good use. To avoid issues connected with the charge of inventing entities, however, let us speak of sense-datum statements instead of sense-data. Thus we shall ask: "Is there any sense of 'dubitable' in which sense-datum statements are less dubitable than all material object statements?"; and "What is meant by the claim that sense-datum statements are non-inferential in a sense not characteristic of other statements?"

Russell once wrote that "a man possessed of intellectual prudence will avoid such rash credulity as is involved in saying "there's a dog!" "[26] The grounds for the prudent man's avoidance would be the fact that artificial stimulation of the optic nerves, hypnosis, a clever technicolor projector, or perhaps drugs, could induce the sort of experience a person has when confident he is seeing a dog. The possibility that his experience originates in one of these anomalous ways renders the belief of the prudent man less than certain even when he has every other reason to believe he actually is seeing a dog. What is certain is what can be expressed in the proposition: "There is a canoid [sic] patch of colour".[27] There is also allegedly a sense in which the belief that this is a dog is inferred from other beliefs. If one had never seen a dog before (or had never learned about dogs from other people), his experience on this occasion would not lead him to state that this is a dog. Moreover, at best one does not see a *whole* dog, but only part (one side) of one. Yet the belief that this is a dog includes the belief that there is more to this dog than the side which happens to be seen. Belief that this is a dog,

---

[25] *Knowledge and Perception*, H. A. Prichard, Oxford, 1950, pp. 45–46.
[26] *An Inquiry into Meaning and Truth*, London, 1940, p. 151.
[27] *Inquiry*, p. 139.

then, seems to depend at least partially on prior information, and in some sense may be said to be inferred from this information. Belief that there is a dog-like patch of color, however, is dependent for its veracity only upon the experience of the moment (disregarding the verbal facility needed to express it), and hence is in no sense inferential.

Now we may well agree that it is possible that the experience in question could be induced by hypnosis, drugs, or the like: and certainly it is true that we commonly believe there is more to dogs than what we actually see of them at a given moment. But the conclusions Russell draws from these facts are not equally obvious. First, in what sense am I called upon to doubt the statement: "There is a dog", in circumstances under which normally I would say confidently and deliberately that I am seeing a dog? It would be irrational flatly to *disbelieve* this proposition, for this would be to believe the proposition "There is no dog", and there is far more evidence for the former than for the latter. Nor would it be rational to suspend belief in the matter, especially if the dog shows signs of unfriendliness. However, the sense in which I am to doubt there is a dog is not a practical, but supposedly a philosophical or logical sense of 'doubt'. To doubt in this case involves admission of the logical possibility that my belief that there is a dog might be mistaken. If this is what Russell meant by doubting that there is a dog, then certainly he is right in maintaining the dubiety of this proposition. We do doubt in this sense, for this is no more than to admit that the denial of the proposition "There is a dog" is not contradictory.

But plainly speaking, our admission that the denial of a belief is not contradictory is not to doubt that belief at all. If Russell means by 'doubt' only admission of the logical possibility that one *might* be wrong, then no philosophical clarity results from giving it that name. For to call this admission 'philosophical doubt' only leads one to seek pointless explanations of how he can doubt when actually he does not doubt at all.

Nor is it easy to find a sense of 'inference' in which my belief that this is a dog is inferred from other beliefs. It may be true that this belief would not have arisen had I no previous experience with dogs. And certainly it is true that one does not see *all* there is to see of a dog at any one glance. But there is nothing in these facts to require our admission that the proposition that there is a dog is inferred, in any clear sense of the term 'infer', from other beliefs or propositions. It is not *deductively* inferred, for the denial that there is a dog is in no way inconsistent with any proposition about dogs which I entertain currently or have entertained. Nor is the proposition that this is a dog *inductively* inferred from previous information about dogs, for the denial of any part of my previous information about dogs seems in no way to weaken my present claim that this is a dog. There is no proof in these remarks that beliefs like the belief that this is a dog are in no sense inferential, for some unusual sense of the term 'infer' might be tailored to turn the trick. But we see that the sense in which such beliefs might be inferential does not come clear in the ordinary use of the term 'infer', and it is good counsel at this time that the defender of sense-data speak plain language.

Considerations regarding the inferential status or the dubiety of material object statements do not rescue the concept of sense-data from philosophic limbo. The type of perceptual misinformation against which Russell cautions us to be on our guard, however, can be interpreted in a less misleading way, and it will be helpful to do so in an attempt to preserve what is worthwhile in the concept of sense-data.

Imagine Jones seated at his desk, anticipating the visit of a friend. Presently he hears the sound of footsteps, and arises to let his visitor into the room. But if Jones should open the door onto an empty hallway, he would be no more than mildly surprised. For he has only tentative reasons for believing his friend has actually arrived. It is quite possible that he heard only the floor squeaking from an unknown structural strain, while the familiarity of the sound was contributed by his anticipation. Less probably, but possible nonetheless, Jones might have imagined the sound, or have been the victim of a hoax. The point of such consideration is not to advise Jones and others like him to be more cautious. Intelligent persons form habits early in life to guard them against being misled by their senses in ordinary circumstances. The point rather is that the sound of footsteps alone is not sufficient warrant for a firm belief that the friend has arrived.

But evidence from audition has a low confidence rating. Let us consider another example involving visual perception. Imagine Jones walking along a wooded path in the twilight, when suddenly he notices an ominous form a short way down the path. It looks very much as if a man is lying in ambush, partially concealed by a tree. But from where Jones stands it is hard to tell; perhaps it is only a large shrub which gives the appearance of a man in the twilight. Since there have been reports of robberies in this area, Jones is concerned to determine whether in fact it is a man or something else. There are several means by which to settle the problem; Jones could issue a challenge, throw a rock, or intrepidly continue along the path. Regardless of how the matter is settled, more evidence is needed than Jones has at the moment to tell what he is seeing. The point again is that Jones' *present* awareness does not furnish sufficient warrant for believing either that another man is present or that instead what he sees is only a bush or some other object.

Admittedly these examples are cases in which the average rational man would be skeptical about his evidence if he had compelling reasons to avoid mis-judgment. Consider, however, the case of a nearly complete illusion. Not long ago a popular magazine told of the return of a group of Arctic explorers to a bay they had visited before, but had explored only in part. The reason they had not previously explored further was that access to the inner part of the bay apparently was blocked by an enormous wall of ice. For some reason (perhaps air reconnaissance) they believed the obstructed part of the inlet has been cleared since their previous visit. Upon their return they were surprised to see the inlet completely blocked as before. They approached for a closer look at the blockade; and as they approached the wall of ice it began to change radically in appearance. Suddenly it was gone and the inlet opened clear before them. They had been victims of a rather costly illusion. Atmospheric conditions had caused the appearance of a huge block of ice, an appearance which has been convincing enough to cause them to turn back once before. According to their testimony, and photo graphs they took of the illusion, the mistaken appearance could not be distinguished from an appearance of the real thing.

Even when we are convinced without reservation of the evidence furnished by our senses, we *may* be mistaken. The moral is not that we should be generally distrustful of our senses. What is to be learned from these, and other more imaginative exercises in "philosophic skepticism", is that more evidence is needed for the adequate warrant of any belief about the material world than can be provided in a moment's sensory awareness. Regardless of the conviction which a moment's awareness might inspire in us, the test of veracity of any belief we might base on that awareness is in the accord of that belief with past and future experience. No sensory awareness of momentary duration, however clear, carries with it proof that it is not illusory. It is worth insisting again that this is not a call to general skepticism. Sensory awareness obviously is the basis of most of our beliefs about material objects, and some of these beliefs are highly warranted. I would want to argue, indeed, that some of these beliefs are warranted as completely as possible. But no momentary instance of sensory awareness in itself completely warrants any belief about a material object.

Awareness of this momentary character, however, does offer complete warrant for beliefs of a more limited sort. Any given sensory awareness, however described and under whatever circumstances, furnishes complete warrant for the belief that this awareness is occurring. The proposition that such an awareness is occurring is informative, since it might have been false, and is empirical since it is not decidable apart from the occurrence of the awareness which it reports.

The expression 'complete warrant' is central in these remarks, and may be explained with reference to a statement A which is claimed to introduce complete warrant for a statement B. To say that A introduces complete warrant for B is to say that it is sufficient for the justification of B that A be true. More formally, it is to say that both A and B are true, and that there is no other statement C compatible with A such that if C were true then B would be false. If there were such a statement C and if B is true, then the negation of C would combine with A in the introduction of complete warrant for B. The character of the warrant introduced by A may be either empirical or theoretical, since this sense of 'warrant' includes not only empirical evidence but also evidence in the form of theoretical propositions which bear upon the truth of B.

The importance of this notion of complete warrant in theory of knowledge can be indicated by formulating a tautology:

> All assertions about material objects can be divided into two classes, 1) those which can be completely warranted by momentary sensory awareness, and 2) those which cannot.

This statement itself surely is unexceptionable. Yet one of the central problems of epistemology is how beliefs which fall into class 2) are warranted; and disagreements among epistemologists arise immediately with the attempt to determine what beliefs fall into class 1). Defenders of sense-data have often maintained that only statements about sense-data are properly classified under 1), and have given the impression that any disagreement regarding the propriety of this classification could be settled by careful observation. Russell, as we have seen, defined sense-data as those aspects of our awareness which are impervious to skeptical doubt, and then attempted to show that statements about these are the most highly warranted of empirical statements and hence alone belong in class 1). This procedure is open to two types of objection. First, it could be claimed that there are no elements of awareness which properly go by the name 'sense-data' thus defined, and hence that sense-data are non-empirical. Second, it could be objected that statements about sense-data, if such exist, are no more completely warranted than other statements which clearly are not about sense-data. Both objections have been ably developed, and I do not wish to quarrel with them.

Instead of arguing whether there are assertions which merit the name 'sense-datum statement' and which fall into class 1), I wish to leave the membership of class 1) an open question. This of course is not to leave the *character* of assertions in class 1) an open question, for by definition this class contains only assertions which are warrantable in a particular way. Further, I would like to provide a way of expressing assertions which are reasonable candidates for membership in class 1) without using the term 'sense-datum' or equivalent terminology. This should enable us to discuss the membership of class 1) without arousing suspicion of our having introduced an ontology of reified appearances. The expression 'sense-datum statement' then can be explicitly re-introduced as an optional but convenient generic name referring to all assertions which may in fact fall into class 1).

The expression 'sense-datum statement' will be used in this way during the remainder of this discussion. Philosophers who do not wish to use this expression may follow my argument by reading 'assertion of class 1)' wherever I write 'sense-datum statement'.

This procedure has the considerable advantage of avoiding the interminable problem of "whether sense-data exist". If there is any call to use the term 'sense-data' (as against 'sense-datum statement'), it can be understood as referring to whatever in sensory awareness warrants statements of class 1). Questions regarding "the nature of sense-data" may then be settled by the individual on the basis of his own experience, and no question regarding the degree of warrant of sense-datum statements will arise, for such statements will have been defined as those which are completely warranted by momentary sensory awareness.

How may beliefs, the statements of which fall into class 1), be expressed without use of 'sense-data', 'appearance', or other nouns which purport to have similar reference? Beliefs of this sort are those which are completely warranted on the basis of momentary awareness. Thus we may consider first what sort of belief can be correctly held without exceeding the warrant of momentary sensory awareness. The individual epistemologist has no more ultimate authority than his own experience in deciding this issue. In the next few paragraphs I describe relevant experience of my own, and assume that it is not radically different in important respects than that of other individuals interested in these problems.

Consider my awareness of the magnolia tree outside the window. I am firmly convinced that there is a magnolia tree there, and my belief that there is a magnolia tree there of which I am now aware, is based not only on my present awareness, but also (in a way difficult to explicate) upon memory and upon my general confidence that objects like trees do not suffer radical alterations from one moment to the next in the normal course of events. Although I have no reasonable doubt concerning the existence of that tree, I am not fully warranted in my belief by my current awareness alone. The statement that there is a magnolia tree outside the window falls into class 2).

What *is* warranted by my momentary awareness of the magnolia tree? I am warranted in believing, first, that this awareness is occurring *now* and, second, that *I* am having it. Third, I am warranted in believing that I do at least *seem* to see a magnolia tree. This belief is warranted simply because I *do* seem to see a magnolia tree, which is a matter admitting the relevance of no judgment other than my own. Even if I am the victim of some sort of illusion, which is perhaps remotely possible, I nonetheless do seem to see a magnolia tree. Moreover, no memory of past experience, nor any eventuality of future experience, can in any way influence the fact that now I seem to see such a tree. Since there is no possible additional information which would affect my present belief, and since I *do* seem to see a magnolia tree, my belief that this is so could not be more fully warranted. A statement of belief, then, which is completely warranted by my present awareness, can be expressed in these terms: 'It seems to me now as if I am seeing a magnolia tree'.[28] In the statement of this belief there is no occurrence of 'sense-data' nor of equivalent nouns of appearance.

It may be pointed out that it goes against the grain of language to say, in these circumstances: "It seems to me now as if I am seeing a magnolia tree". The reason this assertion seems out of place is that in fact a much stronger statement could be made. Counting on evidence accumulated from past experience, and from a closer inspection of the tree, if needed, I would indeed be justified in saying flatly: "I see a magnolia tree". As a convention of common use, the locution 'it seems as if. . .' is reserved for occasions on which evidence is obviously incomplete. When we wish to call attention to the fact that a belief we hold is mere opinion, or is held with admittedly incomplete evidence, a common way is to use 'it seems' or 'it appears'. Two points constitute my rejoinder. First, although I am indeed warranted in making a stronger statement than "It seems to me now as if I am seeing a magnolia tree", my warrant exceeds my awareness of the present moment. Second, whenever a statement of the form 'I see *x*' is warranted, a statement of the form 'It seems to me as if I am seeing *x*' also is warranted. Evidence for the former includes warrant for the latter. It may seem odd to use the latter locution when the former is justified, but at least one is not exceeding his warrant in doing so.

Another objection might be raised to the effect that even the statement 'It seems to me now as if I am seeing a magnolia tree' is not completely warranted, for it is not inconsistent to deny this statement in the face of the awareness which I have said furnishes its warrant. This is not a serious objection, for the relationships of consistency hold among statements, not between statements and sensory events. Moreover, although no sense-datum statement is *entailed* by any sensory occurrence, there is a sense in which a

---

[28] This is similar to expressions used for similar purposes by Ayer and Warnock. See *The Problem of Knowledge*, A. J. Ayer, London, 1953, p. 111; and *Berkeley*, G. J. Warnock, London, 1953, p. 169.

sense-datum statement is rendered logically incorrigible by the appropriate sensory awareness. By saying that a statement is logically incorrigible I mean that its function is such that no evidence which would tend to decrease the warrant of the statement is admissible. What I intend by the statement: "It seems to me now as if I am seeing a magnolia tree" is such that no added information (added now or at any other time) could possibly discredit that statement. The statement is meant to preclude from relevancy to its warrant all considerations other than the occurrence of the appropriate sensory awareness.

With respect to language usage, of course, the statement is not incorrigible. The language I use may be corrected; I might be informed for example, that such things are usually called "bushes" and not "trees". This correction would be admissible. But no proffered correction regarding my grounds for the sense-datum statement is acceptable. Not even if I were to glance away for a moment and look back to find, for some unknown reason, a large oak in place of my magnolia tree, would I admit any weakening of my previous statement about my seeming to see a magnolia tree. For at that time it did seem to me as if I were seeing a magnolia tree (although later I might not recall this with complete confidence, which is not to the point). Now, at this later moment, I would be warranted in saying that it seems to me as if I am seeing an oak tree, but this has no bearing on what I seemed to see before. Further, even though I were to have the most trustworthy information that there is no magnolia tree outside my window, this information would not bear upon the fact that I now seem to see one there. It would only tend to make me disbelieve that I would continue to seem to see a magnolia tree if I splashed cold water on my face, or advanced for a closer look.

Russell's criteria for the identification of sense data may be re-interpreted. A sense-datum statement may be called incorrigible in the sense that no added information could attest to its falsity, and non-inferential in the sense that no added information could bear upon its truth. But this is so only at the moment signified by the term 'now' occurring in the sense-datum statement when uttered. "It seems to me at t_0 that I am seeing a magnolia tree" is fully warranted and incorrigible only at t_0. At t_n, later than t_0, the statement is warranted primarily by my memory, and hence is neither incorrigible nor certain. And even if at t_n there were evidence against the statement "It seemed to me at t_0 as if I am seeing a magnolia tree", and if I had some sort of advance notice at t_0 that this evidence would be available at t_n, still this foreknowledge would have no bearing upon my warrant at t_0 for saying then "it seems to me now as if I am seeing a magnolia tree". Sensory warrant of a sense-datum statement might be thought of as an "all or nothing" affair. If such a statement has any warrant at all, it has complete warrant.

§4 *Difficulties with the Translatability Thesis*

The remarks of the previous section may not persuade everyone of the usefulness of the expression 'sense-datum statement', but they should suffice to preserve that expression in currency for those who do find it useful. The present section reviews the attempt of contemporary Phenomenalists to explicate the use of material object statements in terms of sense-datum statements.

Contemporary writers generally formulate Phenomenalism in a way which no longer suggests that it has anything to say about the actual composition of material objects. It is now customary instead to present Phenomenalism as a thesis about the meaning of material-object statements. Stout, for example, has defined Phenomenalism as the theory that any proposition about material objects can be *translated* into statements about actual and possible sense-data.[29] Ayer claims that every empirical statement about a material object is *reducible* to a statement or set of statements which refer exclusively to sense-data.[30] Firth defines Phenomenalism as the thesis that the meaning of any statement about the material world can be expressed, at least in theory, by a combination of hypothetical statements which refer only to sense-data.[31] A way of putting it which would perhaps be acceptable to each of these writers is that whatever one might wish to say of an empirical nature about material objects can be said in sense-datum statements. If this thesis is correct then all reference to material objects theoretically could be eliminated from our vocabulary and replaced by expressions which refer only to sense-data. In that case, *the concept of matter as distinct from our concept of the appearances of material objects would have been shown to be expendable.*

The best way for the Phenomenalist to support this thesis would be to produce a few examples and to give reasons for our thinking that these examples are typical. But in fact no sample translation has been given, and as far as I know no serious attempt has ever been made to produce one. The reason usually offered for this omission is the shortage of sense-datum terms in our working vocabulary. We do have a limited supply of terms for de-

---

[29] "Phenomenalism", p. 6.

[30] *Hume's Theory of the External World*, H. H. Price, Oxford, 1950, p. 177.

[31] "Phenomenalism", Roderick Firth, in *American Philosophical Association Eastern Division*, *1*, 1952, p. 5.

scribing colors, shapes, sounds, and the like, but it is not adequate to provide for the complete translation of even the simplest statement about material objects. Hence the Phenomenalist is constrained to use in his illustrations expressions like 'the sort of experience one gets when he is looking at a chair', or 'kinesthetic data typical of moving into the living room'. Translations which involve this sort of expression obviously have not been purified of all reference to material objects, but the fault is claimed to lie not with Phenomenalism but in the limitations of available language. If the critic is content to allow that the *de facto* failure of any attempt to produce an actual translation of any material object statement is attributable to a deficiency in language, then this failure cannot be construed as a disproof of Phenomenalism.

It is quite another thing, however, to claim that no Phenomenalistic translation could be completed *even* if adequate sense-datum terminology were available. This claim has been raised recently by a number of critics, and is intended to challenge the theoretical possibility of a sense-datum translation of material object statements.[32] The argument runs along the following lines: If the meaning of a material object statement can be expressed by a set of sense-datum statements, then those statements must be entailed by the object statement whose meaning they express. There are always circumstances, however, under which a given object statement would be true, but any sense-datum statement which might be taken to provide part of its meaning would be false. For example, part of the meaning of 'This is a telephone' might be expressed in terms of the tactual data one would expect to experience in picking up the receiver. But if one's fingers were numb, these data would not be obtained; nonetheless, 'This is a telephone' might convey a true statement. Similar conditions could be conceived which would prevent the sensing of any sense-datum one might refer to in the translation of any material object statement. From this it appears that no specific set of sense-datum statements can be entailed by any material object statement, and thus that Phenomenalism is incorrect.

This difficulty has been raised most forcibly in an argument by Chisholm, which may be paraphrased in our terminology.[33] Chisholm begins his argument by pointing out that before a material object statement could be translated into sense-datum statements, something would have to be known not

---

[32] Ayer in *The Problem of Knowledge*, pp. 138–39; Roderick Chisholm in "The Problem of Empiricism", *The Journal of Philosophy*, *45*, 1948, pp. 513–16; Paul Marhenke in "Phenomenalism", *Philosophical Analysis*, Max Black ed., Ithaca, 1950, p. 316; R. B. Braithwaite in "Propositions about Material Objects", *Proc. Aristotelian Society*, *38*, 1937–38, p. 275.

[33] "The Problem of Empiricism".

only about the object itself but also about the conditions under which it is perceived. Consider the material object statement P and the sense-datum statement R:

(*P*) This apple is red.
(*R*) It seems to me now as if I am seeing a red apple.

Since there is no contradiction in affirming P and denying R, P does not by itself entail R. In addition, a statement like Q is needed:

(*Q*) I am observing this apple under normal conditions; and if this apple is red and is observed under normal conditions, it seems to me now as if I am seeing a red apple.

The compound statement PQ does entail R. But another statement can be found which, in conjunction with P, entails not-R. Consider:

(*S*) I am observing this apple under conditions which are normal except for the presence of blue lights; and if this apple is observed by me now under conditions which are normal except for the presence of blue lights, it does not seem to me now as if I am seeing a red apple.

But if PS entails not-R, and if P and S are consistent, then it follows deductively that P does not entail R. Chisholm asserts generally that for every statement P' and every condition statement Q' there can be formulated another condition statement S' such that, if $P'Q'$ entails R', then $P'S'$ entails not-R'. It follows that there is no sense-datum statement R' which is entailed by any material object statement P', and hence that no material object statement can be translated into an equivalent set of sense-datum statements.

This argument is intended to constitute a disproof of Phenomenalism. In fact it does not have that force. The Phenomenalist may point out that a typical sense-datum statement entailed by P would be in conditional rather than categorical form. He could maintain that P entails, not R alone, but rather a conditional statement of the form: 'Q only if R', when Q is a statement like "It seems to me now as if my eyesight, and other conditions of observation, are normal, and that I am glancing at an object with the characteristics of an apple". The statement of entailment pertinent to his analysis, then, would be of the form 'P entails (Q only if R)'; and *this* statement contrary to what is the case with 'P entails R', *is* consistent with 'PS entails not-R (since 'PQ entails R' is consistent with 'PS entails not-R', and 'PQ entails R' entails 'P entails (Q only if R)'). Moreover, he could reject Chisholm's claim that there is a further statement S' which in conjunction with P entails that 'Q only if R' is false. Such a statement would either (i) be materially equivalent to Q, or else (ii) fail to be materially equivalent to Q. If (i), then the assertion that PS' entails not-(Q only if R) would be equivalent to the assertion that PS' entails $S'\overline{R}$, and obviously PS does not entail $S'\overline{R}$. If (ii), then Q could be false when S' is true, in which case PS' could be true when $Q\overline{R}$ is false. But if PS' could be true when $Q\overline{R}$ is false, then PS' could entail neither $Q\overline{R}$ nor its equivalent: not-(Q only if R). In neither case, then, does PS' entail the negation of (Q only if R).

The Phenomenalist who takes this line of reply would encounter difficulties of admitted severity in providing a statement like Q which would be entirely satisfactory for the purpose above. But his burden in this at least is not one of logical inconsistency. Thus Chisholm's argument is not conclusive against Phenomenalism in this form. It does, however, augment our reasons for being dissatisfied with the ordinary formulation of the Phenomenalistic thesis. But I think that most Phenomenalists would feel no more compunction in giving up this formulation of their thesis, if it seems to commit them to providing actual translations of particular material object statements, than they felt in giving up the earlier formulation which seemed to commit them to a thesis regarding the composition of material objects. In the remaining pages I wish to suggest a reformulation which allows the Phenomenalist to avoid problems of the nature of material objects and the actual translation of material object statements without giving the appearance of shifting his ground.

§5 *Phenomenalism and Empirical Warrant*

It is necessary first to secure a sense of the term 'observation' which can be used to refer to our ordinary sensory awareness of objects in the material world without theoretical overtones. In this sense of the term, for someone to say that he observes a chair or a magnolia tree is not to take a stand on the composition of a material object, on the relation of sensing to perceiving, or on any of the traditional problems of mind and matter. To observe a material object is something anyone can do whose senses are functioning normally and whose mind is not systematically disordered, and which he can do without benefit of the special training of the tea-taster or the special instruments and conceptual formation of the experimental scientist. It is on the basis of this non-technical sort of observation that we cross streets without mishap, recognize friends, and learn to apply the term 'material object'. It is also this sort of observation which furnishes whatever warrant we have for making statements of the ordinary variety about objects in the material world.

The main insight of the Phenomenalist tradition, I believe, regards the sort of statement for which warrant is provided by ordinary sensory observation. In order to bring this out more clearly, the Phenomenalist might explain the point of his thesis as follows: By asserting that material object statements theoretically can be expressed in sense-datum statements, he intends to say something not about their "meaning" but about the sort of evidence we normally have for the things we say about material objects. He intends to say that the only warrant we have in ordinary observation is the sort of warrant which backs up sense-datum statements, and that any material object statement which asserts more than can be expressed in sense-datum statements cannot be warranted by ordinary sensory obervations alone. It is apparent that Phenomenalists traditionally have meant to say this much at least, and perhaps this is the main point of what they have tried to say.[34]

The Phenomenalist might continue to explain that the customary formulation of his thesis, which seems to concern the meaning of material object statements generally, reflects the influence of the verifiability theory of meaning. If the only warrant we have in ordinary sensory observation is the sort of warrant which backs up sense-datum statements, then that warrant can be reported in sense-datum statements. And if it is correct, as suggested by the verifiability theory, that the meaning of any statement can be specified by specifying the sensory occurrences which tend to verify that statement, then it follows that the meaning of any material object statement can be given in sense-datum statements. A variety of advantages are to be gained by the Phenomenalist, however, if he explicitly dissociates his thesis from the verifiability theory of meaning. He no longer gives the appearance of having something of general significance to say in the area of semantics; he is not bound to seek solutions for the nest of problems attached to the verifiability principle; and he need be embarrassed no longer by pointed requests to provide actual translations of material object statements.

Thus, instead of trying to say something about the meaning of material object statements generally, the Phenomenalist may reformulate his thesis to assert that the only evidence available from ordinary observation is evidence which would count as warrant only for sense-datum statements, and that any material object statement which can be wholly warranted on that evidence is one which says no more than can be expressed in sense-datum statements. It is compatible with this formulation of Phenomenalism that some material object statements are not identical in meaning with any set of sense-datum statements. It is compatible even that some material object statements which are completely warranted are not warranted on the basis of evidence expressible in sense-datum statements. In such cases, however, the warrant could not come from ordinary sensory observation alone. The thesis which the Phenomenalist has now to defend is weaker than the tra-

---

[34] Ayer agrees; see *The Problem of Knowledge*, London, 1956, pp. 147–48.

ditional formulation which makes an assertion about the meaning of all material object statements. He has only to maintain that all material object
statements which are completely warrantable by sensory observation can be
expressed in sense-datum statements.

Although it would be fruitless to attempt to establish this claim in a way
which leaves no logical room for disagreement, the Phenomenalist can give
cogent reasons for accepting rather than rejecting his thesis. Consider Jones
observing a candle, which is situated on the mantel and burning normally.
Jones is a normal observer in normal conditions, and no questions cross his
mind regarding the actuality of the candle or its state of slow combustion.
We assume that Jones has been observing the candle for some time, and has
the best warrant imaginable for asserting that he is seeing a burning candle.
Consider now that Jones is asked the philosophic question, "What do you
now see?", and that he replies "It seems to me now as if I am seeing a candle". Jones would have been warranted in saying simply, "I am seeing a candle". But his warrant for that assertion would have exceeded the warrant
provided by his observation the moment the question was asked. His response was calculated not to go beyond the warrant provided by that momentary observation. If Jones had only now entered the room, or had just
noticed the candle for the first time, he would not have the warrant to claim
that he actually is seeing a burning candle, although of course he would
have complete warrant for claiming that it seemed to him then as if he were
seeing a burning candle. Before he could assert confidently that there is a
burning candle on the mantel, Jones would have to observe the candle from
several points of view, and perhaps to check the space above the candle for
convection currents and to touch the candle to make sure it is not a card-
board imitation. These further observations which would warrant Jones' assertion that there is a candle could be specified, along with the conditions
under which the observations would have to be made to be relevant. The
number of conditions necessary for further observation would be small, and
certainly not infinite as some Phenomenalists have suggested. The point is
that each of these additional conditions itself yields an observation which
warrants completely only a sense-datum statement, for the expression 'sense-
datum statement' has been taken to refer to all assertions completely warrantable on the basis of momentary sensory observation.

We have assumed that Jones has complete warrant for making the assertion that he is seeing a burning candle, and that his warrant comes entirely
from these corroborating conditions. Since his observation in each of these
conditions provides complete warrant only for what can be expressed in
sense-datum statements, it follows that Jones' assertion that he sees a burn-

ing candle, which is completely warranted, can be expressed in sense-datum statements.

Similar remarks could be made for any statement about material objects which might be claimed to be warrantable strictly on the basis of sensory observation. Consider any material object statement P concerning which this claim is made, and designate the observations which contribute to its warrant by the series of symbols '$C_1 \ldots C_n$'. No one C_i taken alone offers complete warrant for P, but we assume that C_1 through C_n collectively offer complete warrant for P. As argued previously, each C_i offers warrant for a sense-datum statement of the form 'It seems to me now as if . . . .', and for that statement alone offers complete warrant. But since each C_i offers complete warrant only for sense-datum statements, the series of observations $C_1 \ldots C_n$ offers complete warrant only for a set of sense-datum statements Now if P asserts anything which cannot be expressed in sense-datum state ments, then contrary to assumption it is not warranted completely by C_1 through C_n. It follows that P makes no assertion which cannot be expressed in sense-datum statements.

It might be objected that this argument involves what logic books call "the fallacy of composition". I have argued that since individual observations warrant only sense-datum statements, any statement completely warranted by a group of observations can be expressed in sense-datum statements. Although the argument does not involve that mistake, an explanation why it does not will help clarify the modest character of the Phenomenalist's claim. From the fact that the warrant provided for material object statements by each observation C_i of the observations C_1 through C_n warrants only sense-datum statements, it does not follow that *all* the evidence we might have for an assertion about material objects can be expressed in sense-datum statements. It follows only that assertions which cannot be expressed in sense-datum statements cannot be wholly warranted on the basis of ordinary sensory observation alone. The observations C_i and C_j might support each other in such a way that together they offer warrant for a material object statement which far exceeds the warrant offered by the set of observations C_i and C_j considered independently. But this would be the case only in the context of some *theory* which relates C_i and C_j in a way which makes them mutually corroborative. The added element which increases the warrant of C_i and C_j collectively beyond the set of sense-datum statements warranted by the observations separately considered is not another observation but rather a *theory*. It might be a theory of science or conceivably a philosophic theory. In either case, the statement P which would be warranted by C_i and C_j together in that context is not completely warranted by the observations C_i and C_j alone, but by the observations and that theory according to which these observations corroborate each other.

Phenomenalism as I have interpreted it thus does not deny that there are some assertions about material objects which cannot be expressed in sense-datum statements, for there may be statements about material objects which require warrant beyond that available in ordinary sensory observation. It is compatible with my thesis even that no assertions about material objects can be expressed in sense-datum statements. I think it would be incorrect to maintain this, however, for there are assertions which seem to me to be wholly warrantable on the basis on ordinary observation and which therefore I should say can be expressed in sense-datum statements. Among these are assertions about the way things appear in ordinary circumstances. According to the way I construe the assertions that my tea tastes bitter or that my toast feels hot, for example, what is asserted could be expressed by statements like "It seems to me now as if I am tasting bitter tea", and "It seems to me now as if I am touching hot toast". I would not insist upon these examples, however, for some people might not construe such assertions in the way I construe them. It might be even that some people never make assertions which, according to the way they construe them, can be expressed in sense-datum statements; and whether a person has made such an assertion on a given occasion is for him alone to decide. It should be clear from these remarks that my thesis is not strictly incompatible with any philosophic position in the context of which it is consistent to deny that any statement about material objects can be completely warranted in sensory observation.

This thesis has important consequences nonetheless for a variety of philosophic theories about the composition of material things. For if this thesis is correct, then any statement intended by its author to make an assertion which relies upon a fundamental distinction between that in a material object which can be given in appearance, and an underlying principle which cannot be given in appearance, is a statement which cannot be completely warranted on the basis of sensory observation.

A conclusion may be drawn in particular about any assertion to the effect that material objects are composed of two fundamentally distinct principles, matter and form. In asserting that matter and form are fundamentally distinct, one would be asserting, I think, that there are at least some statements about material objects which cannot be expressed solely in terms of the properties of these objects. And this assertion I take to entail that some things we say about material objects cannot be expressed in terms of the appearances we associate with material objects. In the terminology of this paper, the assertion that matter and form are fundamentally distinct principles in the composition of material things entails that some statements about material objects cannot be expressed in sense-datum statements. Any statement about the matter of an object apart from its form would serve as an example.

The conclusion of my paper which is directly relevant to the topic of the symposium is this: *no statement about matter which makes matter out to be a principle in material objects fundamentally distinct from appearances can be completely warranted by sensory observation*. If this conclusion is correct, it becomes the burden of a philosopher who would maintain such a distinction to explain how his statements about matter can be or could be warranted. It is sufficient here to point out that it is not enough for him to say that these statements can be warranted on the basis of a philosophic theory of matter and form, since statements in the theory of matter and form themselves are statements about matter, and surely these statements cannot be claimed to be self-authenticating.

My general conclusion, in behalf of what I take to be the only currently defensible form of Phenomenalism, is a conclusion about our evidence for material object statements. The term 'ordinary observation' has played a central role in this discussion, and in a sense my conclusion may be construed as a remark about the mode of observation by which we gain information about material objects. Concerning the ordinary sense of 'observation', I have argued that sensory observation provides complete warrant only for sense-datum statements. The central thesis of Phenomenalism follows directly from this: no statement which cannot be expressed in sense-datum statements can be completely warranted on the basis of our ordinary experience of material objects.

COMMENT

MY CRITICISM OF MR. SAYRE'S PAPER IS THAT IF PHENOMENALISM IS WHAT HE says it is, then it seems non-controversial and trivial. Mr. Sayre wants to separate out the pure phenomenalist insight from the irrelevant metaphysical, semantical, and phenomenological theses which have been grafted onto it. I should hold that if the phenomenalist does *not* intend to make either a metaphysical, semantic, or phenomenological claim, then what he has to say is so obviously true as to be uninteresting.

Mr. Sayre says that "the central insight" of phenomenalism is that "the only warrant we have in ordinary observation is the sort of warrant which backs up a sense-datum statement". Now by 'sense-datum statement' here he presumably means what I shall call "seems"-statements—that is, statements of the form 'I now seem to be sensing . . .'. (I cannot accept Mr. Sayre's proposal that 'sense-datum statement' in his argument can be treated as synonymous with 'statement about the material world which can be completely warranted by momentary sensory awareness'. If this synonymy is granted, then his thesis: "any assertion about material objects which can be completely warranted on the basis of ordinary sensory experience can be expressed with reference exclusively to sense-data"—seems to become sheerly tautologous. The insight of phenomenalism, then, is that "seems"-statements are the only sort of statement which can be completely warranted by ordinary observation.)

Now if this insight were taken as a phenomenological thesis about what "ordinary experience" is and isn't like, then perhaps it might be non-trivial. But Mr. Sayre doesn't want us to take it this way. He wants us to take it as a remark which is directed toward the problem of "how what is given in sense-perception (however described) can furnish evidence for statements about the material world". But what sort of a problem is this?

Mr. Sayre does not, I take it, conceive it to be the traditional problem of how to get around Descartes' demon. It is not, in other words, the problem of whether the confidence we repose in momentary sensory awareness, if we do repose confidence, is *justified*. Apparently the problem is simply one of describing accurately how we do behave. So the phenomenalist's insight is presumably an answer to the question: "Under what circumstances do we dismiss objections to our statements on the ground that they are adequately backed up by our ordinary sensory experience?" Taken as an answer to this question, the phenomenalist's insight is unquestionably sound. The answer is that the only circumstances under which we dismiss such objections to our statements is when our statements are "seems"-statements. "Seems"-statements (suitably restricted

212

to the specious present, as Mr. Sayre restricts them) are, indeed, the only variety of statements which, when uttered by other people, are not controverted by us except by an accusation of lying. Taken the other way, they are the only sort of statements we make which are not such as to require us to attend to possible correction by other people. This is because a "seems"-statement is, as Mr. Sayre says, "meant to preclude from relevancy to its warrant all considerations other than the occurrence of the appropriate sensory awareness".

Now if this is all phenomenalism comes to, it is hard to see why anybody should want to refute it. In order to find something more controversial, we may turn to the corollary which Mr. Sayre draws from phenomenalism at the end of his essay: "any statement intended by its author to make an assertion which relies upon a fundamental distinction between that in a material object which can be given in appearance, and an underlying principle which cannot be given in appearance, is a statement which cannot be completely warranted on the basis of sensory observation". Now suppose an opponent saying: "On the contrary, it seems to me that I am now seeing the difference between the appearance of that magnolia tree and an underlying principle of that magnolia tree which cannot be given in appearance". The opponent, having been careful to phrase his remark as a "seems"-statement, may now insist that "no added information could possibly discredit" it. Presumably Mr. Sayre would reply to such an opponent that it was a contradiction in terms for him to have claimed that the difference between what appears and what doesn't itself *appear*. But on what basis can Mr. Sayre make good the assertion that this is a contradiction? It seems to me that it can only be a phenomenological basis. Granted that although only "seems"-statements are warranted by momentary sensory experience, not *all* "seems"-statements are so warranted, it is hard to see how one can tell which such statements get warranted and which don't, except in terms of a phenomenological thesis about how momentary sensory awarenesses should be described.

This point may be put in another way by looking at a particular example of a statement which, according to Mr. Sayre, cannot be "completely" warranted by sensory observation: the assertion that "material objects are composed of two fundamentally distinct principles, matter and form". Mr. Sayre says that he takes such an assertion to entail "that some things we say about material objects cannot be expressed in terms of the appearances we associate with material objects"—because one of the things we say in this assertion is that it has matter as well as form. But why cannot the referent of the term 'matter' be expressed in terms of the appearances we associate with material objects? (If "matter" is leather, for example, it can be so described.) Presumably only because "matter", as Mr. Sayre is using the term, is by definition something that *cannot* be so expressed. But if this is the meaning that he gives to 'matter', his conclusion again seems trivial. If we define "matter" as that which we can't be aware of in momentary sensory awareness, then it is not surprising to be told that we can't warrant our remarks about matter by momentary sensory awarenesses.

It seems to me that in his thesis about statements referring to matter, as well as in his statement of the central insight of phenomenalism, Mr. Sayre is faced with a choice between phenomenology and triviality. In other words, I do not see how what he says in his paper, if it is not to belabor the obvious, can avoid commitment to a certain way of describing the immediately given.

In particular, I think that Mr. Sayre is implicitly committed to a way of describing momentary sensory awareness which precludes the occurrence of such predicates as 'form', 'matter', 'appears', 'awareness', and similar higher-level theoretic terms in such descriptions. I think his claims depend upon the assumption that such "seems"-statement as "It now seems to me that the matter and the form of the magnolia are distinct" can be dismissed as attempts to disguise theoretical speculation as unbiased reporting. I agree with his commitment, in that I find it pragmatically useless to attempt to defend a metaphysical thesis, such as the distinction between form and matter, by reporting what "seems to me". But I think that his argument in his paper depends upon this commitment, rather than providing reasons for it, and that in this sense his paper begs the question. I entirely agree with Mr. Sayre that it is "the burden of a philosopher who would maintain such a distinction to explain how his statements about matter can be or could be warranted". Shouldering this burden seems to me the most difficult meta-philosophical task which confronts anyone who employs the distinction between matter and form, or any similar distinction. But if a philosopher is unwilling to shoulder it, and prefers to just stand there pointing at magnolias and demanding that we divide them in the way in which he does, then I don't see that Mr. Sayre's version of phenomenalism can stop him.

Richard Rorty

DISCUSSION

Mc Mullin: I want to ask a question about the notion of warrant, principally in order to orient this paper towards the doctrine of matter. It seems to me that the paper is of relevance to the concept of matter primarily in raising the question of what sort of warrant we have for talking about matter. In other wards, the emphasis of the paper I take to be on the epistemological notion of warrant, rather than on the concept of matter itself. Now the two are obviously very closely connected, but I think perhaps they are not quite the same.

Sayre: I think the paper is both about the concept of matter and about warrant, in this respect. The theory of perception, of which Phenomenalism is an example, is primarily involved, I believe, in the analysis of the concept of matter, but in a particular context with a particular set of problems in focus. One problem is to determine how matter must be *conceived* in order that statements about material objects can be warranted in the way we usually think them to be warranted—on the basis of sensory observation. The Phenomenalist traditionally has responded by maintaining that the concept of matter must be so explicated as to enable statements about matter and material objects to be expressed in the sense-datum language. I have tried to expound this thesis in a way which throws the emphasis more on the question of warrant than upon the notion of sense-data or the particular way in which we chose to talk about our sensory experience.

Mc Mullin: This is where my difficulties begin. In effect, you drop the notion of sense-datum entirely and introduce that of a sense-datum statement. This notion is defined, not by reference to a theory of sense-data, but by making it a qualified statement of momentary awareness, such as 'I seem to see . . .'. The phrase 'sense-datum' could now be replaced by 'seems . . .', or 'incorrigible'; its retention is a gesture of piety to the older tradition, a gesture which could possibly mislead. The second step is to introduce the notion of "complete warrant", which is not clearly defined, but is ultimately presented as the sort of warrant that only a "seems-statement" has. A new version of Phenomenalism now follows: any statement which asserts more than can be expressed in "seems-statements" cannot be completely warranted by ordinary sensory observation alone.

Sayre: I had attempted in *Part 3* to show how the notion of sense-datum could be circumvented without removing our formulation of the Phenomenalist's thesis too far from its more traditional statements. To do this, I first provided a fairly explicit definition of what I understand in this context by 'complete warrant', and circumscribed a class of statements the defining characteristic of which is that they are completely warranted on the basis of momentary sensory awareness alone. I suggested then that the phrase 'sense-datum statement' could be used to refer to statements of this sort if convenient, but left this entirely optional. Finally, I offered the illustration of some statements beginning with 'It seems to me!' which I take to be members of this class but not necessarily the only members. I am beginning to fear now that this procedure was not sufficiently clear, since both Mr. Rorty and Fr. McMullin seem to have thought that I *define* 'sense-datum' in terms of the "seems-statement". In fact, I defined the class of statement which might, if one wishes, be taken to answer to the designation 'sense-datum statement' quite independently of any examples at all, and in a way which I had hoped would be beyond controversy. And *then* I suggested that, according to the way my sensory awareness is structured, a good example for me of a sense-datum statement would be a seems-statement. It is entirely an open question whether another person would choose an example like this. I am therefore not quite satisfied with the rendition of my thesis which makes it essential that material object statements be expressible in terms of seems-statements if they are to be warranted in a certain way. I have expounded instead the thesis that any statement which is completely warranted on the basis of sensory observation alone can be expressed in statements which are completely warranted on the basis of momentary sensory awareness.

Mc Mullin: There are two remarks I would like to make about this thesis. First, it is a tautology, since it is assumed to follow from the definition of 'complete warrant'. It would be a sad end, indeed, for Phenomenalism if it were to be withdrawn in this way from philosophical debate and promoted to the unassailable but uninteresting rank of tautology.

Second, the whole notion of *warrant* here needs further investigation. Your sense of 'complete warrant' is so strong that no statement of any philosophic or scientific interest seems to possess such a warrant. In that case, to say of some particular assertion (e.g., about a matter-form distinction) that is lacks complete warrant in this sense has little bite. It is not clear that this thesis has any specific negative implications for a doctrine of matter in the way that traditional Phenomenalism had. Moreover, your thesis seems equally to impugn statements in physics—for example, statements about electrons, which certainly are not given in appearances. If you would want to say that such statements could possibly be completely warranted, you would have to admit that a theoretical warrant might be sufficient. To be more concrete here, statements in electron theory, which we might take to back up statements about electrons, would themselves have to be translatable into statements which can be warranted on the basis of appearances if the theory is to have the type of warrant you are talking about. So you seem to be denying that there is a type of theoretical warrant which is ultimately different from your empirical warrant. If so, then you seem to be undercutting the ground of statements not only in metaphysics but in physics as well. I don't think you intend this, but that is the impression given by the end of the paper.

Sayre: I would like to clear my remarks from your charge of triviality or tautologousness which has been raised also by Mr. Rorty. There are various ways in which a philosophic thesis might be trivial, some but not all of which are probably bad. It would be trivial in a bad way to propound, with an air of being informative, a proposition which no one would contest. It is clear I have not done this. It would be objectionably trivial also to begin an argument with a set of premises and to conclude with a mere reformulation of these premises. I am sure my paper does not do this, although this may be the sense in which it has been accused of triviality. In my definition of what we have been calling the 'sense-datum statement' I use the expression 'momentary sensory awareness'. In my general thesis I use the expression 'sensory observation'. I tend to be convinced by Mr. Rorty that my thesis in some way commits me to an assertion of some sort about the constituency of sensory observation out of momentary sensory awarenesses, or momentary experiences. Such an assertion may be phenomenologically unfounded, but it clearly is not trivial. Neither, consequently, is the thesis which commits me to it. In another sense, a conclusion may be called trivial if it follows deductively from its premises. If my argument were trivial in this sense, I would be rather pleased. At any rate, I do not detect a bad sort of triviality in my thesis.

Your suggestion that I have made my sense of 'complete warrant' too strong to be of interest concerns me a bit more. According to my definition, the true statement A introduces complete warrant for B if there is no other statement C such that if C were true B would be false. If there were other statements of this sort, the truth of B would be contingent upon them since if any one of them is true B would be false. Complete warrant for B, in such a case, would include the falsehood of each of these statements. It follows from this definition that a statement which is not completely warranted may indeed be false. That is, for any statement B which is not completely warranted there is a statement C which may be true, and which is such that if it is true then B will be false. What I have in mind here most definitely is not Descartes' demon with his world of possible delusions, but rather the real world in which we do not always know whether our statements are true or false because we do not always have complete survey of all the conditions upon which their truth is contingent. And what I am saying about this world is that when we do not know whether there are conditions which in fact render one of our statements false, we do not know whether in fact our statement is true or false, and consequently we certainly cannot consider it to be completely warranted. We can consider it completely warranted only when we know that the conditions which could render it false do not obtain and can truly assert that this is so. The statements in which we assert this, then, provide complete warrant for our original statement, which without them is incompletely warranted or not warranted at all.

I do not believe that this definition, as I have used it, justifies our concluding that no philosophic or scientific statements are completely warranted. It would surprise me if this were so, and I would have to be persuaded to believe it. Some statements a scientist might make (reporting his observations) might be completely warranted by sensory observation, and others might be warranted by their theoretical context or by a combination of observation and theory. I am not sure whether there are any scientific statements quite like your example about electrons, which *are* completely warranted. But there very well might be such statements, and if there are I think it is safe to say that their warrant is primarily of a theoretical rather than an observational sort. Nothing in my paper tends to deny this. I have not said much about electrons and their ilk, since electrons necessarily have properties different from those possessed by ordinary material objects which removes them from ordinary observation. About all I have said that bears upon the justification of statements about electrons is that insofar as their warrant cannot be exhibited entirely in sense-datum statements it depends upon their context in a theory, and that this theory itself cannot be *entirely* warranted on the basis of sensory observation. This should not disturb a scientist. The only person who might properly be disturbed by the consequences of my remarks, I believe, is one who advances a statement which he claims is based on ordinary sensory observation, but who cannot exhibit a set of statements, in his own chosen terminology, each of which is completely warranted, and which together express the same thing that his original statement expresses.

Eslick: I would like to suggest that, at least from your own point of view, pure sense-datum statements are not possible. I think that both Plato and Kant show clearly that unless you have access to categories that utterly transcend sense-data—sameness, difference, being, which are not really given in sense-data—you can't judge at all. If you are going to understand sense-data as one kind of flux of Humean impressions, I think you cannot, strictly speaking, understand how it is possible to make judgments.

Sayre: I certainly would not want to consider sense-data to be Humean impressions alone. In fact, I leave the question of the nature of sense-datum statements open. I would like to emphasize also that I think there are many statements which cannot be analyzed in terms of sense-datum statements. And among these surely are statements about the transcendent properties you mention. These statements would not be statements about material objects at all, and my thesis concerns only statements about physical objects.

Eslick: Nevertheless, it is crucial to know what you understand by 'sense-datum', because if you are going to limit it to what Whitehead calls "presentational immediacy", for example, it seems to me you have no real foundation for scientific endeavor, for predictions about the future, no real explanation of memory, of the way in which the past enters into your present experiences.

Sayre: You are asking me to do something I have deliberately avoided doing, to give a prefabricated definition of 'sense-datum'. I leave the question of the phenomenological character of his experiences entirely up to the person who finds it useful to use the phrase 'sense-datum statements' in reference to what I define as statements which can be completely warranted on the basis of momentary sensory awareness. It would lead to philosophic embarrassment if I were to attempt to tell you or anyone else how to describe his experience.

Hanson: I think the question of warrant and the question of just exactly how one is going to characterize a sense-datum are connected. It seems to me that sense-datum experiences have been on occasion treated as events, as things which take place. There are times at which an individual is said to be "having an experience" describable in a certain way. Common language often speaks this way. There is a whole class of literature having to do, in an oblique way, with the phenomenanalist's position—and I think that your paper falls in this class—where the sense-datum is really a kind of shadowy referent. It seems to serve as the limit of a series of decreasing empirical claims, such that each one becomes less and less vulnerable by the systematic removal of those vulnerable empirical contents. The question of how one would treat, for theoretical purposes, this limit of a series of decreasing empirical claims, of how one would relate this to the actual events that take place in people, is, to some extent, connected with the question of warrantability. And I felt that the way in which you resolved this (this might be a little unfair) was, in a sense, not to let it come up in too serious a way, namely, to stick where the paper belongs, with the logical-analytical issue. That's why I think Mr. Eslick's question ought to get in your side somewhere and bother you.

Sayre: It does get in my side to the extent that I try to exclude it. I have avoided using 'sense-datum' as an independent term. And if there is any call to introduce the term, I would want it to be taken as referring to whatever it is that warrants assertions that fall under my class of assertions which can be completely warranted on the basis of momentary sensory awareness.

Hanson: One further point about this. The question of whether or not anyone ever has a sense-datum experience seems to me thoroughly independent of the question of whether or not the claim that an individual did have a sense-datum experience makes sense. By analogy with physics, the question of whether or not anyone has discovered an ideal gas is independent of the question of whether descriptions of an ideal gas are sensible descriptions. It seems to me that you are talking about the second sort of thing and that Mr. Eslick is raising questions about the first. The issue of warrantability can't be resolved unless you are both talking about the same thing.

Sellars: I would like to return to the question raised about warrant, and to concentrate for the moment on the notion of an incomplete warrant. My suspicion is that if sense-datum statements are taken as you define them then *any* warranted statement about material objects would also involve incompletely warranted statements. In other words, you can only go from your completely warranted statements to an assertion that there is a material thing over there by adding *incompletely* warranted assertions about what other physical objects there are—good background perception, what the light is like, and so on. My suspicion is that incomplete warranty is an essential part of the very meaning of physical objects, and I think it is by an illegitimate twist that you have concluded that the meaning of physical objects must in some sense be expressible in terms of completely warrantable statements. This, I think, is the very crux of the matter, and this is where I would disagree with you. I think that any statement which is not about the appearances of physical objects, if it is to be warranted, will be incompletely warranted. Among the premises that we presuppose would also be other incompletely warranted statements, and this is part of the very logic of physical object statements. I think that what you show is that statements such as, 'There appears to be a physical object', can be completely warranted, but I do not think you show that any statement like,

'There is a horse in the corner', can be completely warranted. Now, I would argue that you must give an account of the warranting of the latter, and I would argue that any warrant of this statement is going to presuppose statements, which are not themselves completely warrantable, about the circumstances of perception.

I sympathize with the general drift of the paper, but I think that the Phenomenalism you are offering really consists in the tautology that the only completely warranted statements are those which are so set up that we do not challenge them. Nothing will take you from that to any conceptual analysis of physical objects in terms of such statements.

Sayre: Your comments are close to the mark, but I believe they add up to a clarification of rather than an objection to my thesis about the warrant of material object statements. It seems, however, that you may have overestimated my ambitions. I have not attempted to provide a conceptual analysis suitable for all types of material object statements. Phenomenalists traditionally have attempted something like this. But I have not aspired to join forces with them in this respect, and have in fact been rather critical of their attempts at such a conceptual analysis. I have been concerned only with what I take to be a dominant theme of traditional Phenomenalism, namely the claim that everything we say about material objects with a warrant of a certain type (*sensory* warrant) can be said in a certain type of statement (*completely* warrantable). To escape moot questions about the "nature of sense-data", I suggest a rather neutral definition of 'sense-datum statement' in terms of warrant, and in these terms expound the thesis that all statements which can be completely warranted on the basis of sensory observation can be expressed in sense-datum statements. You suggest that it is a consequence of this way of putting the matter than *any* statement about the existence of material objects involves incompletely warranted statements. You are probably right in this. To point this out is to give a reason for rejecting Phenomenalism as an analysis of material object statements generally. I would agree that it is inadequate generally, but would maintain that it is less obviously inadequate for some material object statements than for others of the sort you stress. Incidentally, to say that it is a consequence of my thesis that some (or all) statements about material objects cannot be completely warranted, assuming this is not a tautologous thing to say, is to give reason also for being very hesitant to consider my thesis a tautology. Tautologies do not have consequences which are not tautologous.

Cohen: I don't see how one can avoid sympathizing with the general drift of the paper. Its point is to try to persuade us not to believe that for which there is no reason to believe. Now, if you follow up the Sellars-McMullin qualification of this, however, I take the conclusion to be: that total phenomenalism is useful only insofar as any position which can be shown to reduce to that loses everything for us. It would be a good way, then, of refuting somebody to show that their position entails the demand for *complete* warrant in observation statements . . .

MATTER AND EVENT

Richard Rorty

§1 *Introduction*

Most fundamental controversies about the nature and status of matter are episodes in a struggle between Aristotelian realism and the tradition of subjectivist reductionism which stretches from Descartes through Berkeley and Hume to Russell and Goodman. This struggle has shifted ground many times, but there is a recognizable persistence in the sort of arguments employed, and the sort of distinctions invoked, on both sides. Realists hold that matter-vs.-form is going to have to be a basic distinction in any adequate cosmology, whereas reductionists hold that this distinction does more harm than good. If they give a place to matter, it is not matter-as-opposed-to-form, but matter in some meaning of the term which has little more than the name in common with what Aristotelians are talking about.

Whitehead viewed the grand opposition between these two schools as a reflection of the opposition between "two cosmologies which at different periods have dominated European thought, Plato's *Timaeus*, and the cosmology of the seventeenth century, whose chief authors were Galileo, Descartes, Newton and Locke".[1] He thought of Aristotle as having filled in and rounded out the *Timaeus* (*CN*, 24), and of the post-Kantian epistemological controversies between positivistic empiricists and idealists as the inevitable outcome of a search for the presuppositions and consequences of the Newtonian cosmology (*PR*, 76 ff., 123 ff.). He thought of "the philosophy of organism" as replacing *both* cosmologies, and as being as different from either as either was from the other.

Nevertheless, from the point of view of realistic philosophers, insisting

---

[1] *Process and Reality*, New York, 1929, p. ix. Future references to this book will be to *PR*, and will usually be inserted in the text. The following abbreviations will be used in occasional references to Whitehead's other books: *Adventures of Ideas*, New York, 1933 as *AI; Science and the Modern World*, New York, 1925 as *SMW; The Concept of Nature*, Cambridge, 1920 as *CN; Essays in Science and Philosophy*, New York, 1947 as *ESP*. William Christian's *An Interpretation of Whitehead's Metaphysics*, New Haven, 1959, which I have used heavily, will be cited as "Christian".

I very much regret that I had not read Ivor Leclerc's "Form and Actuality in Whitehead" (in *The Relevance of Whitehead,* ed. Leclerc, London, 1961) at the time this paper was written. Any future comparison of Aristotle and Whitehead should take its point of departure from Leclerc's essay.

upon the irreducibility of the distinction between substances and qualities, and between substances and relations, Whitehead's philosophy usually looks like one more variant of subjectivist reductionism. His cosmology, with its ingression of "eternal objects" into sub-microscopic "actual entities", seems one more attempt to blend hard-headed atomistic materialism with the elegance of Platonic logicism—a combination whose possibilities have fired the imagination of philosophers ever since the rise of modern mathematical physics (and have become dazzling since the invention of symbolic logic). To reduce substantial forms to "conceptual prehensions", and to reduce the particularity and concreteness of actualities to patterns of relatedness with other entities, as Whitehead seems to do, is apparently to abandon all hope of a *rapprochement* with Aristotelian realism. Whitehead's notion of "subjective aim" and his analysis of "concrescence" are, to be sure, reminiscent of some key Aristotelian terms and themes, but the resemblances do not seem to come to much. Whitehead's polemics against the quest for a "substratum", his assertion that "Creativity" should replace Aristotle's category of "primary substance" (*PR*, 32), as well as many of his historical allusions and judgements, suggest that Whitehead viewed his system as climaxing the revolt against the Aristotelian world-view which Descartes, Newton and Locke had begun. Further, the affinities of Whitehead's atomism of actual entities with the logical atomism of Russell and the nominalistic *Aufbauen* of Carnap and Goodman seem obvious.[2]

If one accepts these affinities at face value, however, one may lose sight of two other sets of affinities. In the first place, Whitehead's cosmology is at least as close to Bergson and James as it is to Russell and Goodman. One cannot ignore his Bergsonian insistence that taking time seriously—the substitution of process for stasis as the inclusive category—permits one to demolish Aristotelian substances in the *right* way, whereas everybody else (Russell and Goodman, as well as Descartes, Newton, Hume *et al.*) has been demolishing them in the *wrong* way. In the second place, one needs to notice that Whitehead's criticisms of these wrong-headed attempts draw on the same sort of arguments as those employed by Aristotle's defenders. Most of the usual points made by Aristotelian realists against nominalists, materialists, sceptics, and like, are strongly echoed by Whitehead. Whitehead stands be-

---

[2] Cf. Christian, 247, who points out that in 1919 Whitehead was still toying with the "class-theory of particulars" (cf. §3 below) characteristic of the Berkeley-Russell-Goodman tradition, but that by 1925 he had realized that he wanted to break with it. The Whitehead whom I shall be discussing in this paper is the Whitehead of *Process and Reality*, and I shall make no attempt to cover the shifts in Whitehead's views. (This, incidentally, is the reason why I refer throughout to "actual entities"—the term used in *PR*—rather than to "events", the term used in the earlier writings.)

tween two reductionistic philosophical movements—Bradley's and Bergson's (*ESP*, 116)—in the same way in which Aristotle stood between Plato and the materialist successors of Heraclitus. Both men find themselves insisting, against the simplistic analyses of such reductionisms, that a "critique of abstractions" is required. (Cf. *PR*, 253)

Contrariwise, Whitehead and Aristotle are attacked by reductionists for the same reasons. Both make heavy use of the distinction between potentiality and actuality to resolve cosmological dilemmas, and "potentiality", as all reductionists know, is merely an anthropomorphic vestige of pre-scientific picture-thinking. (As is teleological explanation, to which Aristotle and Whitehead are equally devoted.) Aristotle, in the distinction between matter and form, and Whitehead, in the distinction between actual entities and eternal objects, make heuristic use of the actuality-potentiality distinction to bifurcate the universe in (so the reductionists say) arbitrary, unempirical, and unnecessary ways. ("Aristotle without specific forms" and "Whitehead without eternal objects" are almost equally popular slogans.) Thus one might expect, on the principle that one's enemy's enemy is, at least temporarily, one's friend, that Aristotle and Whitehead would have interestingly similar aims and strategies.

In this paper, I shall try to spotlight some of the anti-reductionist features of Whitehead's cosmology, in order to show how Whitehead's critique of the Newtonian world-view, and the philosophical systems which presuppose this world-view, resembles the critique offered by Aristotelians. Then, on the basis of these similarities between Aristotle and Whitehead, I shall discuss some differences between them. On the basis of these differences, I hope to exhibit the significance of Whitehead's "taking time seriously" for a discussion of the concept of matter.

§2 *Reductionism and Distinctions of Level*

I have said that both Aristotle and Whitehead are realistic philosophers who build their respective cosmologies around distinctions which, in the eyes of reductionist philosophers, seem arbitrary. I now wish to define (dogmatically and curtly) "reductionism" and "realism" in terms of the presence or absence of a certain sort of distinction. These definitions are no more than rough hints, to be developed and given sense in what follows, but formulating them here will give us some pegs on which to hang the Whiteheadian doctrines which we need to extricate and examine.

Reductionism, as I shall use the term, is the position which adopts what Whitehead calls "the unreformed subjectivist principle"—the principle that "the datum in the act of experience can be analysed purely in terms of universals". (*PR*, 239) Holding to this principle, and defining a "universal" as "that which can enter into the description of many particulars" (*PR*, 76), leads to the dissolution of particularity itself. For any candidate for the status of an ultimate particular is confronted with the alternative of either disclosing itself as a congeries of universals or condemning itself to unexperiencability. Making a virtue of necessity, reductionism then claims thát particularity is either unknowable or unreal. The history of philosophy since Descartes, in Whitehead's eyes, is the history of the failure of reductionism—of the foredoomed attempt to develop an adequate cosmology with only one type of basic entity—*viz.*, *repeatable* entities. The notion of an "unrepeatable entity" has, since Descartes, been taken to be an absurdity, and the admission that such entities exist has been taken as either a proof of scepticism or the mark of an "incomplete" analysis—which is why the Cartesian tradition can end only in Humean scepticism or Bradleyan idealism (cf. *PR*, 85).

Realism, as I shall use the term, is the position which holds that an adequate cosmological account can be achieved in terms of an irreducible distinction between two sorts of entities—entities of radically distinct categorical levels. The cash-value of the phrase 'distinct categorical levels' is that entities of these two sorts are such that any given arrangement of the first sort of entities is logically compatible with any given arrangement of entities of the other sort. That is, realism is the insistence that explanation must always be in terms of a correlation of *independent* arrangements,[3] and cannot consist in a reduction of entities on one level to entities on another.

These abstract and stark definitions may be given some initial relevance to the usual meanings of the terms defined by noting that the leading candidates for the position of irreducibly distinct levels of entities are "things" and "properties", and that the best efforts of three hundred years of reductionist thought have been devoted to breaking down this distinction. The view that the essence of realism lies in the refusal to reduce kinds of things to the sets of qualia which form the criteria for the application of thing-kind-names has become fairly familiar. The independence of the knower from the known, which forms the common-sense kernel of realism, becomes, in the light of philosophical analysis, the independence of the contexts and methods

---

[3] On this notion of independence of categoreal level as a prerequisite for realism, cf. W. Donald Oliver, *Theory of Order*, Yellow Springs, Ohio, 1951, chaps. 1–3, where the notion is developed in great detail and with a precision which I cannot attempt here. I have attempted to apply the arguments which Oliver presents to problems concerning the nature of philosophical controversy in "The Limits of Reductionism" (in *Experience, Existence and the Good: Essays in Honor of Paul Weiss*, ed. Irwin Lieb, Carbondale, Illinois, 1961) and to problems of epistemology in "Pragmatism, Categories, and Language", *Phil. Review, 70*, 197–223, esp. pp. 217 ff.

by which we *specify* things from those by which we *describe* them. Manley Thompson has argued (successfully, I believe) that the methodological analogue of the cosmological doctrine that "things are not collections of properties (nor, *a fortiori,* of sense-data, ideas, or the like)" is that the question "what *kind* of thing is it?" (which we answer by *specifying*) is not reducible to a series of questions of the form "which thing?" (which can be answered by *describing*).[4] Thompson suggests (and Whitehead would agree) that an attempt to perform the latter reduction can end only in a pragmatism which is indistinguishable from idealism.[5]

Thompson has further shown that specification can only be kept distinct from description if one erects an irreducible distinction between words which function as K-terms (members of a classificatory scheme which classifies things into kinds) and words which function as D-terms (names of properties), *even though* the same symbols may be used for both sorts of terms.[6] This distinction between two sets of terms, where the meanings of the members of one set are independent of the meanings of the other set, is the logical analogue of the distinction between thing and property. Whether the distinction is made metaphysically, methodologically, or logically, its import consists in the refusal to adopt, as it were, a "monism of explanation", in which to "explain" something is to reduce it to an instance of a class—a class definable in terms of universals.

The master argument which realists use against reductionists is that the reductionist position cannot be made intelligible without smuggling in a covert realism. If one attempts to take seriously the notion that all data can be analyzed in terms of universals, one finds oneself faced with the question "What *is* it that can be analyzed in terms of universals?" More spe-

---

[4] "On the Distinction Between Thing and Property" in *The Return to Reason: Essays in Realistic Philosophy*, ed. John Wild, Chicago, 1953, pp. 125–51, esp. pp. 129–33.

[5] Cf. Thompson, *loc. cit.,* pp. 148–51. See also, for a critique of the notion of a "method of pure description", Oliver, *op. cit.,* chap. 4. For the resemblance between pragmatism (as usually conceived) and idealism (as usually conceived), cf. J. A. Passmore, "The Meeting of Extremes in Contemporary Philosophy", *Phil. Review, 69*, pp. 363–75.

[6] Cf. Thompson, *loc. cit.,* pp. 130 ff. This last point suggests the way in which the distinction between independent levels of entities which is central to realism must take account of the fact that language communicates just to the degree that it avoids the use of token-reflexive terms. The strong point of reductionism (and the reason why reductionism came into its own only with the "linguistic turn" and the adoption of the "formal mode of speech") is that any singular terms are always replaceable by descriptions. But, as Strawson points out, the employment of a language in which such replacement is consistently carried out will make *all* reference to particulars impossible (cf. "Singular Terms, Ontology, and Identity", *Mind, 65*, p. 449).

cifically, if one adopts the view that all explanation is a matter of discovering which class a thing is a member of, then one sooner or later finds oneself faced with the problem "What is the analysis of the notion *member of a class*?" The "datum" or the "member" can only be a bare particular, a bare substrate. Just insofar as it is something more than this, it requires further analysis in terms of further universals. Just insofar as this analysis is *not* offered, a covert distinction between levels—the level of the repeatable universals and the level of the bare particulars—is being assumed. As soon as the challenge to reduce this distinction is accepted by the reductionist, however, a potentially infinite regress is generated. Each analysis of the level of particulars into a new pattern of universals calls forth the need for a new level of particulars in which the new pattern of universals can be exemplified. Awareness of this regress may lead the reductionist himself to question the notion of bare particularity. But if he does so, he is no longer able to give a clear meaning to the notion of "universal". Universals, as "repeatable", require something to distinguish their various repetitions.[7] The notion of repeatability is equivalent to the notion of "capable of entering into external relations"; if there are no particulars, there is no possibility of entering into such relationships. If objects are mere congeries of universals then, as idealists never tire of pointing out, all statements which attribute characters to objects are necessary truths, signifying internal relations. We thus wind up with a metaphysical monism as a consequence of our insistence upon a "monism of explanation". Given such a monism, the problem of "mere appearance" or "mere error" takes the place, for the reductionist, of the problem of "bare particular", or "mere substrate", and is equally baffling.[8] This is the *reductio ad absurdum* of the reductionist's attempt to simultaneously explain the datum and reduce it away, and the confirmation of the realist's contention that "order", "explanation", "knowledge", and "analysis" are only intelligible as long as we hold the entities *to be* ordered, explained, known, or analyzed apart from the entities *in terms of which* the ordering, explanation, knowing or analysis is to be performed.

In what follows, I shall try to show how Whitehead's awareness of the need to avoid this sequence of reductionist absurdities led him to adopt

[7] Cf. Charles Hartshorne, "The Compound Individual", in *Philosophical Essays for Alfred North Whitehead*, New York, 1936, p. 202: "Universal and individual are ideas that are clear only in relation to each other, and where either conception is neglected the other will suffer also."

[8] Cf. *PR*, 78, on how the misinterpretation of the doctrine of universals paves the way for Kant's "degradation of the world into mere appearance", and *PR*, 85, on Santayanian scepticism or Bradleyan idealism as the only possible outcomes of reductionism. Cf. also *PR*, 349–50.

certain key doctrines, and to show the analogies between these doctrines and certain key Aristotelian doctrines, adopted from the same motives. In the course of expounding these doctrines, I hope to put some flesh on the bare bones of the notion of "distinction of categoreal level" which I have introduced. Having done so, I shall be in a position to compare Aristotle's "form-matter" distinction with Whitehead's "eternal object-actual entity" distinction. Both distinctions are attempts to locate ultimate and irreducible categoreal distinctions whose discovery and exposition will provide a stable foundation for realism. I shall be arguing that Aristotle's hylomorphism was an attempt to establish such a distinction, and that this attempt failed because of Aristotle's identification of "definiteness" with "actuality". I shall try to show that Whitehead's substitution of "decisiveness" for "definiteness" as the criterion of actuality permits him to succeed where Aristotle fails.

Comparisons between Aristotelian and Whiteheadian concepts do not lend themselves to lucid exposition, for Whitehead's critique of alternative cosmologies is so radical as to systematically transform the meaning of almost every traditional philosophical term. This paper is intended as an attempt to plot these transformations.

§3 *Final Causality and Atomism*

Early in *Process and Reality,* Whitehead remarks that: "Final causality and atomism are interconnected philosophical principles." (*PR,* 29) The faintly paradoxical air of this remark is due to our habit of associating atomism (in its logical and psychological, as well as its physical, forms) with a doctrine of external relations. We associate teleology, on the other hand, with a doctrine of internal relations—a thing which is striving to realize its end constitutes itself by that striving, and would not be the same thing were it not so striving. Saying that atomism and final causality are interconnected, then, would seem to be saying that something can only sustain external relations if it also sustains internal relations, or the converse, or both. Now Whitehead does, in fact, want to say the former. He holds that only because actual entities sustain internal relations to goals—their "subjective aims"—are they capable of sustaining external relations to other actual entities.[9] What prevents an actual entity from being "reduced" to the sum

---

[9] As we shall see more clearly below, the external relations sustained by an actual entity A are not A's prehensions of other actual entities which are "objectified" by A; on the contrary, *these* relations are *internal* to A. The only external relations sustained by A are prehensions of A by "later" actual entities, for which A is objectified. The relation, X-prehending-Y, is always internal to X but external to Y.

of its physical prehensions of other actual entities (and thus what separates the "philosophy of organism" from Absolute Idealism) is the individuality and unrepeatability of its subjective aim. Whitehead holds that an actual entity cannot be analysed without remainder either into its physical prehensions of other actual entites (the domain of efficient causality; cf. *PR*, 134) or into its conceptual prehensions of eternal objects (the domain of final causality; cf. *PR*, 159). Nor can these two poles be disjoined in order to be interpreted as two independent actual entities in their own right.[10] An actual entity can retain its integrity only by being interpreted in terms of both levels of entities—the eternal objects and the other actual entities— *at once*. Particularity is safe, and the reductionist implications of the "subjectivist principle" are avoided, only if this distinction of level is maintained (Cf. *PR*, 128, 228).

This Whiteheadian doctrine should be compared with Aristotle's doctrine that a material substance cannot be reduced either to its form or to its matter, that it has matter only because it has form, and that "the actuality of a substance is its goal". (*Meta.* IX, 1050a 9)[11] Because a substance has a goal, a goal which is (in the case of "things which exist by nature") identical with its form, it cannot be analysed as "the sum of its qualities", in the sense of the sum of the properties which form the criteria for specifying it as a substance of such-and-such a species plus the sum of the properties which describe its accidents. To form the sum here involves a breakdown of categoreal level—a "category-confusion" in Ryle's sense of the term. The attempt to form such a sum is based upon the "unreformed subjectivist principle" and the ensuing myth of a "bare substratum"—a substratum which supports "essential" and "accidental" attributes in the same external way. The "formula of the definition" of which Aristotle speaks does not name a complex property; it names the substance, and names it directly.[12] "Each thing

---

[10] For Whitehead's critique of the attempt to thus disjoin them, cf. *PR*, 108.

[11] Compare also Aristotle's identification of *ergon* with both *telos* and *energeia*, Whitehead's identification of "decision" with "actuality" (*PR*, 68), and Leibniz's appeals to teleology against the mechanistic reductionisms of the Cartesians (in order to prevent the monads from dissolving into space-time points). For a comparison of these three philosophers on a related topic, cf. J. H. Randall, Jr., *Aristotle*, New York, 1960, p. 170.

[12] Cf. Thompson, *loc. cit.*, p. 136: "K-terms . . . signify but a single kind of entity and thus have *simple and direct signification* which is neither denotation nor connotation." Wilfrid Sellars arrives at the same conclusion in connection with an analysis of the forms of Aristotelian substances; cf. "Substance and Form in Aristotle", *J. of Phil.*, 54, p. 695: ". . . thing-kind words are . . . common names of individuals, not proper names of universals".

itself and its essence are one and the same in no merely accidental way."
(*Meta.* VII, 1031b 19) The internal relation of the substance to the species
of which it is a member (and thus to its goal) is what permits the substance
to retain its independence (its atomic character)—an independence which
permits it to be externally related *to* its accidental properties, rather than
dissolving *into* them.[13]

Both Aristotle and Whitehead invoke teleology in order to explicate the
distinctions of level which save the primary actualities (the "atoms") of their
respective cosmologies from dissolution. However, this resemblance is ob-
scured by, so to speak, a difference of scale. It is pointless to compare a
Whiteheadian "actual entity" with an "episode of accidental change" occurring in an Aristotelian substance. Although there is a sense in which
these two notions do explicate the same pre-analytic phenomenon, the radical
differences between Aristotle's and Whitehead's categoreal schemes make
such a comparison produce only paralogisms and misunderstandings. The
proper comparison is between an actual entity and an Aristotelian primary
substance—a substance which, however, is distinctive in being a species unto
itself (resembling, in this respect, angels as characterized by St. Thomas).
No two Whiteheadian occasions have the same subjective aim (cf. Christian,
310), and thus there is no distinction in Whitehead between specification
and individuation. There is, however, an analogue of the distinction between specification and description—namely the distinction between an
actual entity's description as "subjectively immediate" and as "objectively
immortal" (*PR*, 34). As we shall see in §§ 6 and 8 below, this analogue
preserves the requisite Aristotelian distinction of level, while dismissing the
Aristotelian problem of the relationship between the secondary substance
"*X*-hood" and the substantial forms of individual *X*'s. Thus, actual entities
are, as it were, miniature Aristotelian substances; they owe their unity and
their irreducibility (their "atomic" character) to their internal relationship
to a goal—a goal which is characterizable only in terms of entities of a different categoreal level than those which characterize their external relations
to other actualities.

---

[13] The problem of how the difference between the relation of a substance to its
form and its relation to its other attributes should be formulated is the major metaphysical problem which Aristotle willed to his heirs. (The trouble with taking seriously
the *identification* of substance and essence is, of course, that such an identification
seems to condemn the accidents, *à la* Plato, to the realm of "mere appearance".)
We have the record of Aristotle's own unsuccessful struggles with this problem in
Metaphysics VI–IX; see especially the discussion of the difference between *ti sēmainei*
and *to ti ēn einai* in VII, c. 4, and compare Thompson, *loc. cit.,* pp. 133 ff.

§4 *Unity as Requiring Categoreal Diversity*

A second crucial Whiteheadian doctrine is, to an adherent of the unreformed subjectivist principle, as paradoxical as the interconnection of teleology and atomism. This is the doctrine that the ultimate unit of actuality
must be internally complex; a unit, in short, must be the *unity of* something.
"Each ultimate unit of fact is a cell-complex, not analysable into components
with equivalent completeness of actuality." (*PR*, 334, italics added) The
implications of this doctrine are spelled out in an explication of the "category
of objective diversity":

> The category of objective diversity expresses the inexorable condition—that
> a unity must provide for each of its components a real diversity of status,
> with a reality which bears the same sense as its own reality and is peculiar
> to itself. In other words, a real unity cannot provide sham diversities of
> status for its diverse components . . . The prohibition of sham diversities of
> status sweeps away the "class theory" of particular substances, which was
> waveringly suggested by Locke . . . was more emphatically endorsed by
> Hume . . . and has been adopted by Hume's followers. For the essence of a
> class is that it assigns no diversity of function to the members of its extension.
> . . . The "class", thus appealed to, is a mere multiplicity. (*PR*, 348; cf. Chris-
> tian, 248)

The fundamental importance of this point for process philosophy is shown
by Whitehead's statement that:

> This doctrine that a multiple contrast cannot be conceived as a mere disjunc-
> tion of dual contrasts is the basis of the doctrine of emergent evolution. It is
> the doctrine of real unities being more than a mere collective disjunction of
> component elements. This doctrine has the same ground as the objection to
> the class-theory of particular substances. (*PR*, 349)

In this doctrine of internal diversity we have perhaps the clearest expression
of Whitehead's rejection of the reductionist notion of explanation as "placing a datum within a class". As Christian says: "This principle of individuality [the subjective aim as defining a mode of togetherness of actual entities and eternal objects], in Whitehead's metaphysics, supersedes in importance the principle of classification" (Christian, 252). The "ontological
principle" that "actual entities are the only reasons", (*PR*, 37) combined
with the definition of "actuality" as "decision amid potentialities", (*PR,* 68)
produces the most fundamental justification of Whitehead's insistence on
the irreducibility of categoreal levels. The attempt to find an ultimate unit
without internal complexity and without the teleology involved in the
"decision" of an actual entity is simply one more form of the search for bare
particulars—for "vacuous actuality". The impulse for such a search can only

be the confusion of ease of classification with cosmological priority,[14] and its outcome can only be a tyranny of internal relations and the loss of real unity.

This protest against the confusion of the distinction between unity and plurality with the distinction between simplicity and complexity is also to be found in Aristotle, in his protests against both materialistic reductions of form to matter[15] and Platonic reductions of matter to form.[16] In both sets of protests, he is led to insist upon the irreducible complexity involved in the hylomorphic analysis of substance, and the loss of unity which occurs when it is proposed to replace this analysis with something simpler. However, in Whitehead's eyes, Aristotle betrayed his own better insight when, in *Metaphysics* XII, he made room for the Unmoved Mover—the perfect case of a "vacuous actuality". Aristotle's break with the realistic requirement of a distinction of level in the case of the Unmoved Mover gave fatal encouragement to the assumption that "satisfactory explanation" demands that "substances with undifferentiated endurance of essential attributes be produced" —an assumption which Whitehead calls "the basis of scientific materialism". (*PR*, 120: cf. *PR*, 241)[17]

However, if we put the Aristotelian notion of "immaterial substance" to one side for the moment, we can see that both Aristotle and Whitehead are, in their theories of the nature of real unity, conforming to the demands which we outlined in §2: the demand that recourse to bare particulars be avoided by establishing two categoreal levels, and that the unitary character of the ultimate cosmological unit should consist in its unification of those two levels—a unification which is possible, and is given meaning, only in virtue of their irreducible difference.

§5 *Potentiality and Actuality*

The paradox involved in the phrase 'unity which is internally complex' fairly cries out for resolution by means of a distinction between potentiality

---

[14] Cf. *PR*, 85, on "the assumption, unconscious and uncriticized, that logical simplicity can be identified with priority in the process of constituting an experient occasion". At *PR*, 202, Whitehead echoes Bergson in remarking that: "We may doubt whether 'simplicity' is ever more than a relative term, having regard to some definite procedure of analysis." For further polemics against reductionist notions of explanation, cf. *PR*, 246, 253, 120.

[15] Cf. *De Gen. et Corr.* I, c. 2, esp. 317a 19 ff.

[16] Cf., e.g., *Meta.* I.

[17] For an account of what Aristotle *should* have said about God in order to avoid abandoning his commitment to categoreal diversity (an account with which Whitehead would be in hearty accord), see Randall, *op. cit.*, p. 143 f.

and actuality, and both Aristotle and Whitehead do invoke this distinction for that purpose. Aristotle holds matter and form together in substantial unity by the formula: "the proximate matter and the form are one and the same thing, the one potentially and the other actually". (*Meta.* VIII, 1045b 18 f.) Whitehead's use of the distinction is summed up in the following passage:

> Just as *potentiality for process* is the meaning of the more general term 'entity' or 'thing', so *decision* is the additional meaning imported by the word 'actual' into the phrase 'actual entity'. "Actuality" is the decision amid "potentiality". (*PR*, 68)

Both Aristotle and Whitehead agree that the actuality-potentiality distinction is unavoidable, but their application of it is so drastically different as to have called forth the suggestion that their respective systems may be transformed into each other simply by following the rule: "What is potential for Aristotle is actual for Whitehead, and conversely".[18] In the previous two sections we have dwelt upon anti-reductionist doctrines on which Aristotle and Whitehead concur. Here, we begin to see how Whitehead separates himself off from both Aristotle and reductionism in the quest for a better reply to reductionism than Aristotle achieved, and thus for a more adequate formulation of realism.

Whitehead claims that

> Some chief notions of European thought were formed under the influence of a misapprehension only partially corrected by the scientific progress of the last century. This mistake consists in the confusion of mere potentiality with actuality. Continuity concerns what is potential; whereas actuality is incurably atomic. (*PR*, 95)

The immediate thrust of this remark is against Newtonian notions of space and time, but the extent of the confusion in question is much wider. The confused notion of the distinction of act and potency against which Whitehead is protesting here is the assumption that there are two equally atomic sorts of things: actual X's and potential X's. This notion, on reflection, drives one to the paradox that (1) "certain possible so-and-sos are not actual so-and-sos", yet (2) "the only possible entities are actual ones".[19] When the absurdity of an attempt to resolve this paradox by finding the "extra some-

---

[18] Cf. R. S. Brumbaugh, "A Preface to Cosmography", *Rev. Meta.*, 7, pp. 53–63. Cf. also Leo A. Foley, *A Critique of the Philosophy of Being of Alfred North Whitehead in the Light of Thomistic Philosophy*, Washington, 1946, p. 120.

[19] I borrow this formulation of the paradox from Nelson Goodman, *Fact, Fiction, and Forecast*, Cambridge, 1955, p. 55. Chapter Two of this book—"The Passing of the Possible"—is a good statement of what becomes of the actuality-potentiality distinction when it is considered under the aegis of the unreformed subjective principle.

thing" which transforms a potential X into an actual one becomes evident, we are tempted to give up the actuality-potentiality distinction altogether. When we do give it up, as reductionists do, we lose the ability to make intelligible the notion of "unity which is internally complex", and thus we have to fall back upon "bare particulars". Whitehead tells us that if we are going to keep and use this distinction, we must start all over again. We must get rid of the notion of potential X's, substitute the notion of "potentialities for X", and abandon the assumption that "the only possible entities are actual ones". A possible entity, for Whitehead, is not a half-baked version of an actual entity; to think of it this way is like thinking of the datum about which one decides as itself a half-baked decision. The actuality-potentiality distinction is in danger, in Whitehead's eyes, whenever one attempts to use the vocabulary appropriate to atomic individuals in describing potentiality. The temptations exerted by language, a language built around pragmatically convenient abstractions,[20] will almost inevitably engender the reduction of the potential to the actual, and thus the loss of the distinction of categoreal levels upon which realism depends. The last person to really struggle with language in order to keep the distinction viable was Aristotle, who, Whitehead thinks, largely failed.[21] The key to his failure, and the reason for his acceptance of "vacuous actuality" in the doctrine of the Unmoved Mover, was the illicit transition from the doctrine of form-as-the-actuality-of-the-matter to the notion of form-as-the-actuality-of-the-composite-substance. This transition evolved into the notion of form-in-isolation contributing something called "actuality" to the composite substance, where-

---

[20] Cf. *PR*, 253. For Whitehead's views on the relation between language and philosophical speculation, cf. *PR*, 16 ff., esp. p. 18: "A precise language must await a completed metaphysical knowledge".

[21] Cf. fn. 12 above. Whitehead's attitude toward Aristotle is full of mixed feelings. Although he regards the subject-predicate model as largely responsible for the popularity of "vacuous actualities", he seems to feel that it is mediaeval philosophers, rather than Aristotle himself, who are chiefly blamable. "The exclusive dominance of the substance-quality metaphysics was enormously promoted by the logical bias of the mediaeval period. It was retarded by the study of Plato and Aristotle. These authors included the strains of thought which issued in this doctrine, but included them inconsistently blended with other notions. The substance-quality metaphysics triumphed with exclusive dominance in Descartes' doctrines." (*PR*, 209) Cf. also *PR*, 45, 81, 85, 122; *AI*, 356; A. H. Johnson, *Whitehead's Theory of Reality*, Boston, 1952, pp. 123–24. Foley (*op. cit.*, p. 109) says that Whitehead is arguing only against the "Lockian or Cartesian" notion of substance as inert substrate, and not the Thomistic-Aristotelian notion of substance as activity. This is largely true. Whitehead knows there is a difference between the two traditions, but he usually seems to think it not worthwhile to distinguish them, since he is convinced that the seeds of decay are already present in Aristotle's confusion of *energeia-as-ergon* with *energeia-as-morphē*.

as matter-in-isolation contributed the element of "potentiality". For White-head, the point of the contrast between actuality and potentiality is lost as soon as one begins to think of "actuality" as something other than "actual-ization *of the potential*".

This analysis permits us to formulate a first rough sketch (to be revised in §7 below) of the central contrast between Aristotle's and Whitehead's approach to the distinction between form and matter. For Aristotle, this distinction is a special case of the distinction between actuality and poten-tiality. For Whitehead, the form-matter distinction, when applied to the process of coming-to-be of an actuality, is the distinction between two sorts of potentiality. The criterion of actuality is found, for Whitehead, neither in pure form (eternal objects in themselves) nor in pure matter (the ex-tensive continuum of real potentiality), but in the unification—under the conditions of the "category of objective diversity" (cf. §4 above)—of both levels.[22] This unification is itself a member of neither level. In other words, Whitehead agrees with Aristotle that an irreducible distinction of level is requisite to realism, but for Whitehead such a distinction, in order to be irreducible, must be strictly correlative. By 'correlative' here I mean that each level must be essentially incomplete, and completable only by interpre-tation in terms of the other level—an interpretation which cannot be achieved wholesale and *a priori* by philosophical inquiry (or by God, for that matter), but only at retail and *pro tempore* by the concrescent activity of actual entities. (This latter point is illustrated by the fact that both 'order' (*PR*, 128) and 'actual world' (*PR*, 102) are, for Whitehead, token-reflexive terms.) Philosophy is thus not the study of "order" nor of "the structure of the actual world"—such a proposal would be analogous to a proposal to study "here" or "the structure of yesterday"—but rather, so to speak, of the grammar of token-reflexive terms and its relation to the grammar of non-token-reflexive terms.[23] "Philosophy is explanatory of abstraction, and not of concreteness". (*PR*, 30)

---

[22] It is perhaps useful to point out the analogy with Kant, who takes "concepts" as "form" and "intuitions" as "matter", and then devotes himself to showing that *neither* can count as "experience", that both are merely analytic components of (po-tentialities for, as it were) experience, and that scepticism (Hume) or idealism (Leibniz) are the result of speaking of either as if it could be, all by itself, a full-fledged experience. Whitehead, I think, conceives himself as having done cosmo-logically and completely what Kant did epistemologically and incompletely: namely, developing the implication of the "reformed" subjectivist principle that "the whole universe consists of elements disclosed in the analysis of the experiences of subjects". (*PR*, 252; on Kant, cf. *PR*, 234–37)

[23] On the importance of token-reflexivity, cf. Hartshorne, "Process as Inclusive Category: A Reply", *J. of Phil. 52*, pp. 95–96.

In place of Aristotle's contrast between form-as-actuality and matter-as-potency, Whitehead contrasts "pure" potentiality (the realm of eternal objects) with "real" potentiality (the extensive continuum: "one relational complex in which all potential objectifications find their niche" (*PR*, 103)). Both levels are required for the explanation of actualities, but neither ·level can do its job if the entities which compose it are thought of as quasi-atoms; as would-be actualities.[24] These two levels will be discussed in the next two sections, but first it will be useful to draw one more comparison between Whitehead and Aristotle. I have cited a suggestion that Whitehead attributes all the characteristics to actuality which Aristotle attributes to potentiality, and conversely. This suggestion cannot be explored in detail here, but there is much truth in it. The most signal instance of this inversion is, of course, that definiteness is for Aristotle a characteristic of form, and form is actuality, whereas for Whitehead definiteness is characteristic of eternal objects, which are pure potentialities. This identification of pure definiteness with pure potentiality seems paradoxical to Aristotelians because for Aristotle (in the *Metaphysics,* and thus in the "orthodox" Aristotelian tradition, though not in the scientific treatises) definiteness is the criterion of actuality. For Whitehead it is not. His criterion of actuality is *decisiveness,* which is not the same thing. This shift of the criterion of actuality is perhaps the most important feature of the difference between a realism built around the notion of *stasis* and one built around the notion of *process. Definiteness,* in Whitehead's eyes is primarily a *logical* notion—although, to be sure, one which needs to be grounded in cosmological notions. The seed of the reductionists' confusion of logical simplicity with "priority in the process of constituting an experient occasion" (cf. fn. 14 above) are already present in Aristotle's inability (despite desperate efforts[25]) to avoid the Platonic mastery of *logos, idea,* and *morphē* over *physis*—of the terms which are necessary to discuss actualities over the actualities themselves. An actuality is a decision about *how to be* definite—a decision which, from the point of view of *other* actualities for which the first is a datum, looks like one more instance of definiteness. But if this exterior point of view is adopted as the point of view of philosophical analysis, then the interior decisiveness of the actuality will be analyzed away into patterns of definiteness. It will begin to seem, as it did to Aristotle, that

---

[24] The level of real potentiality is, of course, constituted by objectified actualities—past, but objectively immortal. But *as* past, and thus as potential, they lose their individuality. Having lost the power of decision, they become matter for decision. Because they *have been* actual, they are not would-be actualities, but simply suggestions to present actual entities about how to be actual.

[25] Cf. Randall, *op. cit.,* p. 116.

superior actuality consists not (as it does for Whitehead; cf. *PR*, 142) in making more and more important and far-reaching decisions, but in being so definite as no longer to have to make decisions at all. From here, it is but a step to the "unreformed subjectivist principle" that "the datum in the act of experience can be analyzed purely in terms of universals".[26] Aristotle shrank from this step, and his attempts to avoid taking it have been the foundation of anti-reductionist thinking ever since, but the "misunderstanding of the true analysis of 'presentational immediacy' " (*PR*. 43) which he shared with Plato left Aristotelianism too weak to withstand the assaults of the reductionist revolt of the seventeenth century.[27]

§6 *Real Potentiality: Matter as Objectified Actuality*

For Whitehead, there are two ways of describing any actual entity—as the culmination of process and as potential for process, as present and as past. The Eighth Category of Explanation tells us that:

> Two descriptions are required for an actual entity: (a) one which is analytical of its potentiality for "objectification" in the becoming of other actual entities, and (b) another which is analytical of the process which constitutes its own becoming. (*PR*, 34)

This distinction is, *prima facie,* the closest Whiteheadian parallel to the Aristotelian distinction between the-matter-in-a-substance and the form of that substance. If one calls the aspect described by (a) the "objective" reality of the actual entity and that described by (b) its "formal" reality (as Whitehead sometimes does[28]), then one can think of its objective reality as the entity considered *qua* matter and its formal reality as the entity considered

---

[26] Cf. Hartshorne, "The Compound Individual", p. 200: "Those who today defend Aristotelianism as the "commonsense philosophy" are simply inviting us to begin the foredoomed process all over again. Every new Aristotle can only usher in a new Berkeley . . ." Cf. also *PR, 79.*

[27] Whitehead's vision of the absurdities of modern reductionism as traceable to an initial misstep taken by Plato and Aristotle is very like Heidegger's. Cf. *Introduction to Metaphysics,* New Haven, 1959, pp. 180 ff., esp. p. 182: "The crux of the matter is not that *physis* should have been characterized as *idea,* but that the *idea* should have become the sole and decisive interpretation of being." Heidegger's analysis, both in *Introduction to Metaphysics* and in *Sein und Zeit,* of Western ontology as dominated by the identification of "being" with "presence" (*parousia, Anwesenheit*)—cf. *Sein und Zeit,* sec. 6—should be compared with Whitehead's suspicion of the ultimacy of "presentational immediacy". Compare also Heidegger on *Zuhandensein-vs.-Vorhandensein (Sein und Zeit,* Secs. 15–17) with Whitehead on causal efficacy-vs.-presentational immediacy.

[28] Cf. *PR,* 118. William Alston ("Internal Relatedness in Whitehead", *Rev. Meta.* 5, pp. 535–58), equates "formal reality" with "existence" and "objective reality" with "actuality". The latter identification, I should want to argue, is very seriously misleading.

qua form. Differences appear, however, when one pursues the analogy. If one takes the general difficulty about "potentiality" to be summed up in the question "How can one ever talk about potentiality, since all we ever find are actualities?", then one is tempted to translate this into Whiteheadian terms as the question: "How can one speak of the objective reality of an actual entity, when all that it *really* is is a process of becoming?". On reflection, however, one should realize that this question is wrong-headed. The *objective* reality of an actual entity is precisely what, ninety-nine times out of a hundred, we *do* speak of. The real problem for Whitehead is how to speak of the *formal* reality of an actual entity. The inversion of modality which, as we have noted, characterizes the contrast between Aristotle and Whitehead, is here marked by replacing the question "How can you talk about potentiality?" with the question "How can you talk about anything else?" In place of the reductionist's paradox that "the only things that can possibly be encountered are actualities, but not all possibilities are actual", Whitehead asks us to reflect on the fact that everything that can be encountered by an actuality is *objectively* real, and thus is merely a potentiality, yet actualities *do* influence other actualities.

This inversion is not a mere verbal twist, although it might seem so. Both Aristotle and Whitehead, after all, seem to identify actuality with immediacy, and it might seem obvious that the immediately encountered—the "given"—is actual, and that its function as a potentiality for further process is derivative, the product of reflection and analysis. But this is not obvious to Whitehead; on the contrary, viewing the situation in this way is as we shall see, a symptom of the failure to "take time seriously". For Whitehead, "immediacy" is of two sorts—the private subjective immediacy of present enjoyment, and the public objective immediacy of the given past. Although the former is a criterion of actuality, the latter is not. The objective reality of an actual entity is "the actual entity as a definite, determinate, settled fact, stubborn and with unavoidable consequences" (*PR*, 336), but stubborn facticity is not the same as actuality, any more than is definiteness. The *decisiveness* which marks the *formal* reality of the actual entity is the *reason* for the objective reality of that entity being as stubborn as it is (cf. *PR*, 68-9), but stubbornness is not the same as decisiveness.[29] In

---

[29] On the derivative status of stubbornness, compare Heidegger, *Sein und Zeit*, s. 210: "Widerstandserfahrung, das heist strebensmässiges Entdecken von Widerständigem, ist ontologisch nur möglich auf dem Grunde der Erschlossenheit von Welt. Widerständigkeit, charakterisiert das Sein des innerweltlich, Seienden. Widerstanderfahrungen bestimmen faktisch nur die Weite und Richtung des Entdeckens des innerweltlich begegnenden Seienden. Ihre Summierung leitet nicht erst die Erschliessung von Welt ein, sondern setzt sich voraus." The attempt to construct *Erschlossenheit* out of *Widerständigkeit* is a product of the attempt to make *Vorhan-*

a system in which "Creativity" is "The Category of the Ultimate", "stubbornness" is a mark of poteniality, not of actuality. The inversion of modalities here, far from being verbal, is part and parcel of Whitehead's campaign against the confusion of logical simplicity with ontological priority, and against the (pragmatically useful, but cosmologically disastrous) confusion of a thing's consequences with its nature.[30]

The importance of this inversion appears when we turn to the traditional dilemmas about the notion of matter. "Matter" has always been asked to play two distinct, and apparently incompatible, roles. On the one hand, matter is supposed to be cuddly, malleable, and receptive—it seeks form as the female seeks the male. On the other hand, it is resistant, obstreperous and stubborn—it needs, Aristotle tells us, to be "mastered".[31] In Aristotle, emphasis teeters back and forth between these two roles. The first appears when "matter" (as "material cause") is being used to make substantial change intelligible, and the second when "matter" is being invoked to explain accidents, individuation, and monstrosities. With the subsumption, in the cosmology of the seventeenth century, of substantial change under changes of quantity and quality, matter adopts the second role almost exclusively. Now when one focuses on this second role, one sees that matter is not resistant because of its indefiniteness, but precisely because of its definiteness. It is not because a lump of marble is "formless" that it resists the sculptor, but because it has the *wrong* form. Again, reflection from the seventeenth-century point of view on the malleability and feminine complaisance which Aristotle attributes to proximate matter makes one realize that these features are due not to the "materiality" of the material cause but simply to its possession of the "right" form. So, following the lead provided by the "unreformed subjectivist principle", it begins to look as if one could analyze both the cuddliness and the stubbornness of "matter" away into congeries of forms, and thus eliminate altogether the need to give indefiniteness an ontological status.

Thus, *if one identifies actuality with definiteness,* one finds it absurd to think of "potentiality" as anything but a name for the confusion among our ideas (which was how the seventeenth century philosophers did think of

---

densein prior to *Zuhandensein* (cf. fn. 27), as is the reductionism ensuing from an acceptance of the "unreformed subjectivist principle". (Compare also Peirce, on the impossibility of constructing "Thirds" out of "Seconds".)

[30] Cf. *PR*, 336: "The "formal" aspect is functional so far as that actual entity is concerned: by this it is meant that the process involved is immanent in it. But the objective consideration is pragmatic. It is the consideration of the actual entity in respect to its consequences."

[31] Cf. *De Gen. Anim.* IV, 769b 11 ff., 788a 5 f.

it). The actuality-potentiality distinction thus evaporates altogether. Without the heuristic aid provided by this distinction, all distinctions of categoreal level tend to evaporate. But if, with Whitehead, one distinguishes between definiteness and decisiveness, then one can unite the two roles played, in Aristotle, by "matter" by saying that matter ("objectified actual entities") is definite, stubborn, and resistant precisely because it *is* potential; and conversely. Only the definite and resistant is malleable—that is, only a perfectly definite feeling (cf. *PR*, 338) can be a datum for further feeling, but once a feeling *is* definite, it is over, past, and thus, though objectively immortal, *no longer* actual (cf. *PR*, 130). Instead of interpreting the distinction between matter and form as one between real indefiniteness and real definiteness, Whitehead interprets it as the distinction between the past and the present.

§7 *Pure Potentiality: Primary Matter as Abstract Multiplicity of Forms*

At this point, however, one may wish to raise questions of the following sort: is there then no place for the vague, the indefinite, and the muddled in Whitehead's system? Is there any point in using the notion of "potentiality" if all one means by it is "past actuality"? Doesn't a philosophy of creativity entail real indefiniteness as an ultimate categoreal level?

The answer to these questions, I shall argue, depends once again upon the contrast between "definiteness" and "decisiveness". There is no such thing in Whitehead's system as "real indefiniteness", but there distinctly *is* real indecisiveness, and this indecisiveness does form a distinct categoreal level. For it is here that the eternal objects come into the act. The Whiteheadian analogue of "primary matter" is the "barren inefficient disjunction of abstract potentialities" (*PR*, 64). Primary matter, in Aristotle, is the ultimate background against which substantial change—generation and destruction—takes place. In Whitehead, " 'Change' is the description of the adventures of eternal objects in the evolving universe of actual things" (*PR*, 92), and it is against this background of the bare unstructured multiplicity of eternal objects that actualities evolve. The concrescent processes which are the formal realities of actual entities order this multiplicity, and change it from a *mere* multiplicity into a pattern of *relevant* potentialities. The measure of an actual entity's actuality is the measure of the extent to which it succeeds in establishing such a pattern.[32] The analogue of the Great Chain

---

[32] Cf. Christian, 267: "This ordering of eternal objects is an actual occasion's contribution to its future," and *PR*, 132: ". . . "order" in the actual world introduces a derivative "order" among eternal objects". (But this does not mean that the eternal objects are "changed" by being thus ordered; cf. fn. 39 below.)

of Being which (for Aristotelians, if not for Aristotle[33]) stretches from Primary Matter to the Unmoved Mover(s) is an hierarchy stretching from "the-eternal-objects-as-mere-multiplicity" (an ultimate abstraction, which is precisely as inconceivable as "Primary Matter"[34]) to God (who *positively* prehends *all* the eternal objects, whereas all other actualities prehend positively only a tiny fraction of them). "Each occasion exhibits its measure of subjective intensity. The absolute standard of such intensity is that of the Primordial Nature of God, which is neither great nor small because it arises out of no actual world." (*PR,* 75) Each temporal actual entity inherits suggestions, as it were, about relevant patterns of form from past actual entities, and it rearranges these patterns in accordance with its own *telos*—its own subjective aim. The more inclusive its subjective aim, the more it will be able to do toward restructuring inherited patterns, and the more individuality, decisiveness, and actuality (three synonymous terms, for Whitehead) it will have. When we think in terms of "change" (that is, of the creative process in the large) rather than of "enjoyment" (the concrescence of an individual eternal object), we see the eternal objects as playing the role of that-which-becomes-arranged and the individual actual entities as doing the arranging. From this point of view, the whole creative process can be seen as the attempt to find the pattern of relevance among eternal objects which will produce the greatest subjective intensity of enjoyment.

Taking this conclusion together with the discussion of real potentiality in §6, we can now revise the description of the contrast between Aristotelian and Whiteheadian treatments of the form-matter distinction which we offered in §5. Our new set of analogies is as follows:

(1) the *specific form* of an individual material substance S has no Whiteheadian analogue, since (cf. §3) there is no distinction in Whitehead between specification and individuation;

(2) the *form* of S is analogous to the formal reality of an actual entity A;

(3) the *proximate matter* of S is analogous to A's past—the actual entities

---

[33] This parenthesis expresses my reservations about how seriously to take the (very rare) references to *prōtē hylē* in the Aristotelian corpus—reservations, however, which need not be discussed here. Cf. Hugh King, "Aristotle Without *Prima Materia*", *J. Hist. Ideas, 17,* pp. 370–89.

[34] Cf. Christian, p. 265 f., the references there cited, and *PR,* 42. As Christian points out, "multiplicity" is used in a highly technical sense by Whitehead, and must be sharply distinguished from "nexus" and "proposition". "Every statement about a multiplicity is a disjunctive statement about its individual members". (*PR,* 45)

(I should like to note that this section could not have been written without the help of Christian's discussion of the doctrine of eternal objects in Part Two of his book. The reader is urged to consult this discussion for a full account of the process of "establishing patterns of relevance"—a subtle topic which can only be sketched in the present space.)

from which A inherits, considered as objective realities (the "extensive continuum");

(4) the *matter-in-S* or *S-qua-matter* is analogous to A's objective reality—that is, A-as-prehended-by-later-actual-entities;

(5) primary matter is analogous to the bare multiplicity of eternal objects, by reference to which A and its ancestors may be described, and which A "orders" by reordering its inheritance from its ancestors.[35]

In §5, we treated the extensive continuum (3) as the analogue of *matter* and the multiplicity of eternal objects (5) as the analogue of *form,* while treating (2)—the formal reality of A—as the analogue of *composite substance.* That treatment expresses the way the situation looks from, as it were, "inside" A's concrescence: the real potentiality (3) is "matter" for the process of being felt under subjective forms dictated by a new "form" (the initial datum of A's subjective aim[36]) plucked out from (5). The table of analogies above, on the other hand, expresses how the situation appears when we step outside of A and look at the whole creative process: from this perspective, we can see both (3) and (5) as "matter" and (2) as "form", thus re-establishing the Aristotelian identification of form with actuality.

This table of analogies, however, should not mislead us into thinking that a concrescent actual entity's contact with primary matter is only by way of the traces of primary matter left in the proximate matter which it is prehending, as the Aristotelian analogues might suggest. A faces both ways—toward the level of real potentiality (3) and the level of pure potentiality (5)—and plays each off against the other in the interest of its own heightened enjoyment. It is in the space between these two levels, so to speak, that creation gets room to occur. Further (to repeat the conclusion of §3) it is by virtue of the independence of these two levels that actual entities can be genuine individuals. The irreducible individuality and novelty of each actual entity is made possible by the inexhaustible array of alternative subjective aims which its pure conceptual prehensions—its prehensions, that is, of eternal objects not yet exemplified among its ancestors—make available to it. Without such pure conceptual prehensions, the new actual entity would merely be the product of efficient causes (cf. *PR,* 75, 134). On the other hand, without the physical prehensions of the realm of real potentiality constituted by the objectified past actualities, it would be merely the product of idiosyncratic final causes, and the solidarity of the universe would be

---

[35] Whitehead himself was inclined to analogize "primary matter" to "Creativity" (*PR,* 46)—which, however, he also analogizes to primary substance (*PR,* 32). The former analogy is probably due to his reading Aristotle as saying that a thing's matter is its internal principle of change—the sort of reading suggested by, e.g., Santayana's "The Secret of Aristotle".

[36] On the notion of the initial datum of a subjective aim (about which Whitehead is rather vague), cf. Christian, 157 f., 215 f., 305 f.

lost:[37] such a cosmology would lead only to Heraclitean resignation. On neither alternative can time be taken seriously: on the first, there is no reason to transcend the past, since it can only be reiterated; on the second, there is no past to transcend.

There is, perhaps, a certain flavor of paradox in our claim that the multiplicity of eternal objects is the analogue of primary matter. For are not the eternal objects consciously modeled on the Platonic Forms? (Cf. *PR*, 69-70) "Forms of definiteness" seem, off-hand, about as far as one can get from primary matter, when the latter is viewed, as it usually is, as a sort of *Urschleim*. Here again, however, the appearance of paradox arises from our Aristotelian habit of identifying both form and actuality with definiteness. But definiteness doesn't decide anything; it is what gets decided *about*. Forms of definiteness are not forms of decisiveness, even though once a decision is taken is can be analyzed into forms of definiteness. We think of definiteness as a mark of actuality because we think that whenever there is something definite, there must have been an indefinite substratum, and a decision which formed that substratum into this definiteness. But this preconception, although it seems like hard-headed realism, in fact leads one down the reductionist path toward the notion of "bare particulars". It is no more arbitrary and irrational for Whitehead to postulate the non-temporal existence of an infinite and unstructured multiplicity of forms of definiteness than it is for Aristotle to presuppose the eternal structure of a finite number of specific forms. After all, *any* cosmology which declines the Hegelian challenge to "deduce", e.g., the colors of the visible spectrum, is going to have to start by letting forms of definiteness in on the ground floor of speculation—as an irreducible categoreal level. The only question is: how, once let in, can they be restrained from swallowing up any other sort of entity to which one wants to assign coeval ground-floor status? That is to say, how can one maintain a categoreal distinction between forms of definiteness and definite actualities?

The most dramatic version of this last question is the Platonic one: if the Forms are as definite as all *that,* what excuse can possibly be found for the sensible world? If actuality is identified with definiteness, then this problem is insoluble. Aristotle, tipped off by Plato to the existence of the problem,

---

[37] Cf. *PR*, 249: "The one eternal object in its two-way function, as a determinant of the datum [i.e., of the objectified past actual entity] and as a determinant of the subjective form [under which the present actual entity prehends the past actual entity], is thus relational. In this sense the solidarity of the universe is based on the relational functioning of eternal objects." (Cf. *AI*, 236) But if there is no datum other than the conceptually prehended eternal object itself, then there is nothing for this eternal object to relate.

but unwilling, in the end, to give up this identification, tried to solve the problem by letting just a *few* forms of definiteness in on the ground-floor —*viz.,* the thing-kind-names of common sense and of primitive science— while consigning all the others to a vague cosmological dustbin called "matter". In doing so, he paved the way for reductions, for when a more advanced science made it thoroughly unclear where substantial change left off and accidental change began, philosophers promptly endowed the forms of accidents with definiteness, and thus with actuality. They thereby (cf. §§5-6) made both the matter-form and the actuality-potentiality distinctions seem pointless.

Whitehead, abandoning the identification of actuality with definiteness, solves the problem by letting in *all* forms of definiteness at once, and by making the criterion of degree of actuality consist in the extent and complexity of the choice among those forms. Using this criterion, the only danger involved in postulating forms of definiteness is that the forms, might, all by themselves, exercise an influence upon the choices which are made concerning them. Insofar as such an influence *is* exercised, the totality of eternal objects would have to be counted as itself an actuality, and a dissolution of the distinction between categoreal levels would ensue. Whitehead was incautious about this danger in *Science and the Modern World*. In that book, he seems to think of each eternal object as being what it is by virtue of its place in a scheme of internal relations with all the other eternal objects. Such a scheme suggests that all the decisions have already been made, and that the concresence of actual occasions must follow rigid guidelines, tiresomely reiterating pre-established patterns. (Cf. *SMW*, 299 ff.) In *Process and Reality,* he takes just the opposite tack and makes the eternal objects, by themselves, utterly unrelated to each other:[38] all "relationships" between eternal objects are now interpreted as derivative from choices ("valuations") made by actualities.[39] Since *every* possibility of order is itself an eternal object (and therefore incompatible possibilities of order are equally pos-

---

[38] As Christian (p. 262) has noted, Whitehead may be construed as speaking, in chap. 10 of *SMW*, of the eternal objects as they are related in the Primordial Nature of God. The latter—a "non-temporal actuality" which provides the primordial valuation of all eternal objects—raises many serious and complex problems for an analysis of Whitehead's cosmology. I shall make no attempt to go into these problems here, although a full defense of my argument in this paper would require that this Whiteheadian analogue of the Unmoved Mover (cf. *PR*, 522–23) be discussed. I am here treating the Primordial Nature as simply one more actuality which transmits suggestions about relevant eternal objects to the future, just as temporal actual entities do.

[39] I put "relationships" in quotes as a reminder that one should not think of eternal objects as in any way "changed" by being thus re-evaluated. Despite Whitehead's use of terms like 'ingression of eternal objects' and 'adventures of eternal objects', it is clear

sible) and since there can be no such eternal object as Order-in-general (corresponding to The Form of the Good),[40] the multiplicity of eternal objects remains as unstructured as ever Primary Matter was. To sum up, Whitehead's insight is that "pure potentiality" is to be found not in the *absence* of definiteness, but in piling definiteness on definiteness until no guidelines for decision between these alternative definitenesses remain.

§8 *Conclusions*

In this concluding section I shall try to tie what I have been saying together by answering the question: what does Whitehead's philosophy of process contribute to a discussion of the topic of matter? For this purpose, I shall return to my original distinction between realism and reductionism in §2, where I said that reductionism tries, and fails, to analyse experience exclusively in terms of *repeatable* entities. The contrast between the repeatable and the unrepeatable—between the universal and the particular— is traditionally explicated by both Aristotelianism and common sense in terms of the distinction between form and matter. When one attends to the macroscopic objects of common sense, it seems obvious that *man* for example, is a repeatable form, but that Socrates is unique and unrepeatable, and so we consider *man* as Socrates' form, and his unrepeatability as due to his matter.

The most obvious and dramatic difference between process philosophy and common sense is that the units of actuality, for process philosophy, are no longer such macroscopic objects. They are replaced by sub-microscopic unrepeatable entities. Taking time seriously means taking the break between past and present as the border between two different actualities.[41] If one does this, it becomes analytic that "only the non-actual is repeatable".

---

that nothing ever "happens" to an eternal object. The order which Whitehead says is introduced among them is not some sort of quasi-spatio-temporal reshuffling, but is simply the "suggestions" about which eternal objects are more relevant than others made to a concrescent actual entity by its predecessor-actualities or by the Primordial Nature of God. To speak of "an order among pure potentialities" is always an elliptical way of referring to real potentiality. To say that an eternal object has been made more relevant to the creative process is not to say that anything has been added to it, nor that its "position" in the multiplicity of eternal objects (in some unintelligible sense of "position") has changed, any more than to say that logarithms or ultrasonic vibrations have become relevant to human technology is to say that they have somehow suffered qualitative change or altered position.

[40] This is because of the paradoxes of self-inclusion. Cf. *PR*, 128, and Christian, pp. 271 ff.

[41] This is why time cannot be taken seriously until one ceases to think of the present as a knife-edge and begins to think of it as an extended duration.

Given this doctrine, it becomes clear that one must either (a) cease to use the matter-form distinction, or (b) deny that form is repeatable, or (c) deny that form is actual.

Aristotelians have fought against transferring the unit of actuality to this sub-microscopic level on the ground that, since such actualities do not change, one cannot preserve the distinction between matter and form. (This attitude has given Aristotelianism a bad name, since it has seemed to involve an unempirical insistence that science should always accommodate itself to philosophy, and never conversely.) Both the cosmology of the seventeenth century and Whitehead's process philosophy, on the other hand, accept this transference. But whereas the thinkers of the seventeenth century thought themselves well rid of the matter-form distinction, Whitehead is intent upon reinterpreting it so as to make it relevant to unchanging entities.

The way in which this reinterpretation proceeds will be clearer if we note the differences, within the cosmological tradition which we inherit from the seventeenth century, between rationalists and empiricists. Empiricists cling to the principle that only form is repeatable, but they deny that form is actual. They think of form as abstract, and of concreteness as inhering in an inexpressible and unintelligible "given". The "given" stands to form in the relation of "exemplification of", even as did Aristotelian matter, but unlike Aristotelian matter, (which was describable, even if not "knowable" in the honorific sense of the *Posterior Analytics*), it is undescribable, and thus its postulation involves the absurdities inherent in the notion of "bare particulars". Rationalists, on the other hand, clinging to the identification of form with both repeatability and actuality, interpret the form-matter distinction by reference to the Platonic distinction between reality and appearance—so that "givenness", like "time" and "individuality", are treated as illusions. They thus can reply that the "given" is not truly concrete, and that only the Concrete Universal is truly actual. Although avoiding the absurdities of the notion of "bare particulars", rationalists encounter analogous absurdities in their attempt to explain how Reality manages to disguise itself as Appearance. Rationalists and empiricists thus either gratefully abandon talk of matter-vs form altogether, or else (following up certain Aristotelian leads, but neglecting others), they speak of matter-as-the-unfortunately-indescribable-actual (in empiricism) or of matter-as-unfortunate-and-non-actual-illusion (in rationalism) .

Now Whitehead wants to say that:

(1) The seventeenth century was right, against Aristotle, in ceasing to regard middle-sized common-sense objects as paradigmatic of actuality.

(2) Rationalists are right, against empiricists, in saying that the "given" is not actual.

(3) Empiricists are right, against rationalists, in saying that what is repeatable is not actual.

(4) Aristotelian realists are right in insisting, against everybody, that the form-matter distinction, or some analogue thereof which will preserve the ability of actualities to sustain both external and internal relations, is needed.

Whitehead's treatment of the relations between actual entities and eternal objects—that is, his reinterpretation of the form-matter distinction—is dictated by the need to reconcile these four positions. His point of departure is (3): the repeatable is not actual. Empiricists, having grasped this, are misled by their acceptance of the unreformed subjectivist principle into inferring that the actual is indescribable. What is required in order to get around this principle, Whitehead says, is the recognition that actualities are describable in terms of other actualities. (Cf. *PR*, 76 ff.) Empiricists, however, are afraid that such description will lead to idealism, and thence to monism. They fear that admitting that actualities are "present in" (cf. *PR,* 79) other actualities will dissolve actualities into congeries of internal relations. Whitehead replies that a congeries of internal relations is not the same as a congeries of universals. Relations are as unrepeatable as anything can be (cf. *AI,* 296; *PR,* 349-50; Christian, 236), and an actuality which consists of internal relations to other entities is unrepeatable precisely by virtue of being a congeries of such relations. The idealist attempt to construe relations as universals is just another product of the unreformed subjectivist principle.

But this still leaves us without an account of external relations (without becoming involved in infinite regresses). Since entities which can sustain external relations must be repeatable, and since the repeatable is non-actual, we have to invoke repeatable potentialities to serve as "forms of definiteness" by which actualities, as terms of relations, may be characterized. But postulating such forms—the eternal objects—raises the traditional dilemma of the Platonic Forms: either (a) the forms postulated to characterize the actualities are so much like actualities as to make them indistinguishable from their actualizations, or (b) they are so different from actualities that they cannot do an adequate job of characterizing them. Now it is in the resolution of this dilemma that the significance of process philosophy emerges most clearly. The dilemma is resolved by distinguishing definiteness from decisiveness, a distinction which is unintelligible if one does not take time seriously. For if one does not, one will be unable to see a difference between the evaluative decision of a presently concrescent actual entity and the inheritance from that actual entity received by its successors between, in other words, formal and objective reality. The latter *is* indistinguishable from a form of definiteness, and if all actual entities were past (or, what comes to the same thing, if they could be viewed *sub specie aeternitatis*) there would be no difference between the actual world and the extremely complex eternal object which described it. If time is taken seriously, however, and it is thus recognized that 'actual world' and 'actuality' are token-reflexive terms, then one can escape the first horn of the above dilemma by distinguishing between the definiteness of an entity's characterization (its objective reality) and the decisiveness of its concrescence (its formal reality). The latter is actual, and therefore non-repeatable. The former is "repeatable", and therefore potential, in the sense that it is related (externally to it, although internally to each entity which prehends it) to a potentially infinite number of subsequent actual entities by being "present in" them. The apparent dilemma is now seen to rest upon a confusion between past and present—between characterization and entity characterized.[42] Its second horn is escaped by replying that the difference between the characterization of the actual and the actual entity is no greater, though no less great, than the difference between past and present—which, if one takes time seriously, is precisely the difference which one would expect.

We saw in §6 how the distinction between formal and objective reality resembles the traditional distinction between the form and the matter of a substance. In §7 we saw how the multiplicity of eternal objects resembles primary matter. We can now see that the crucial *difference* between Aristotle's and Whitehead's doctrines of matter is that Whitehead retains the advantages of the matter-form distinction while avoiding the disadvantages of the species-individual and essence-accident distinctions. Specifically, he (a) retains Aristotle's categoreal distinction of level while avoiding both (b) the "unscientific" character of Aristotle's distinction and (c) the latent tendency toward the subjectivist principle (cf. §5) inherent in Aristotle's use of teleology in regard to enduring and changing objects:

(a) The distinction between formal and objective reality permits one to save realism by (1) enabling the primary realities to bear both external and internal relations, (2) maintaining the independence of the level of pre-analytic (non-objectificatory) reference to an entity (the self-satisfaction of an actual entity in its subjective immediacy) and the level of its analysis (an actual entity as objectified, and thus as analyzable into a pattern of eternal objects); (3) uniting these two levels by the "two-way functioning

---

[42] Whitehead thus can be seen as replacing the distinction between "specification" and "description" (discussed in §2 above) with the distinction between reference-to-an-actual-entity-as-subjective-intensity and reference-to-an-object. The present-past distinction is thus substituted for the thing-property distinction.

of eternal objects" in conformal feeling (cf. *PR,* 249), so that despite the utter privacy of an actual entity's subjective intensity, it is genuinely "present in" later actualities.

(b) The ordinary seventeenth century objection to the Aristotelian form-matter distinction, interpreted as the distinction between the repeatable form of a species and the unrepeatable combination of accidents which individuate the species' members, was that since any given accident or congeries of accidents was as repeatable as a specific form, any distinction between the repeatable and the unrepeatable was arbitrary and unempirical. Whitehead's distinction between the decisiveness of formal reality and the definiteness of objective reality permits one to make sense of the notion of the ultimate unrepeatability of the actual, while his refusal of the gambit: "But how do they differ *sub specie aeternitatis?"* protects him against reductionist efforts to make the repeatable either unintelligible or illusory. He is thus able, as Aristotelians are not, to welcome the seventeenth century's substitution of a law-event framework of scientific explanation for the Aristotelian species-individual framework.

(c) The Aristotelian identification of actuality with definiteness was a result of Aristotle's unwillinginess to take time seriously. This unwillingness led him to postulate "immaterial substances", to identify form with actuality, and to lay the foundation for the tradition that satisfactory philosophical explanation must be in terms of "substances with undifferentiated endurance of essential attributes". (*PR,* 120) Process philosophy, by breaking with this latter tradition, is able to avoid the reductionisms which flow from an acceptance of the unreformed subjectivist principle while preserving the links with modern scientific results which make these reductionisms so plausible. Whitehead's willingness to accept the Heraclitean view that 'actual world' is a token-reflexive term, and his consequent refusal to identify "actuality" with the results of an ideally long or ideally "objective" analysis in terms of universals, radically transforms the Aristotelian notions of form and matter. The history of futile debate between philosophies which presuppose the Aristotelian cosmology and philosophies which presuppose the cosmology of the seventeenth century shows, I think, that they badly needed such transformation.

COMMENT

RORTY HAS GIVEN US A BRILLIANT PAPER THAT IS MUCH MORE THAN A DISCUSSION OF Whitehead. His general thesis, as I understand it, is two-fold: he argues (a) that Whitehead, like Aristotle, is a "realist", i.e. an anti-reductionist with respect to material things; and (b) that Whitehead's realism is stronger and more satisfactory than Aristotle's, for the reason that Whitehead "takes·time seriously". I am not sure whether I think these general claims are true or not, but the questions I want to raise for discussion concern more particular points. There are three such points.

(1) Rorty says that Aristotle is a realist, that he is able to avoid the reduction of things to qualities (or relations or ideas or sense-data) because he regards things as composed of two irreducibly different elements, matter and form. But I do not see *how* the recognition of the matter-form distinction enables Aristotle to protect things from reduction; Rorty has not, I think, made this clear. Furthermore, I do not see that the matter-form distinction is *necessary* for this purpose. Things *are* not reducible to qualities; reductionism *is* a mistaken doctrine. But if a reason for this irreducibility of things is required, then it surely lies—as Rorty himself suggests—in the logical or grammatical difference between "specifying" and describing things, between referring and characterizing. Referring and characterizing are irreducibly different functions of language, irreducibly different things that people use words to do. It is this distinction that lies behind the distinction between things and qualities, and it is the irreducibility of referring to describing that accounts, ultimately, for the irreducibility of things to qualities. I do not claim that this is obvious or even very clear; that it is even *so* needs to be argued and the point itself needs to be explained, and I have no time here to do either. But I think it is true nonetheless, and hence that what Rorty says is not.

(2) Rorty says that Whitehead is a realist in the same sense as that in which Aristotle is, although on different grounds or in a different way. But I question this: I question whether Whitehead is even interested in protecting ordinary things (the individual man or horse) from reduction to qualities or the like. Aristotle's philosophical task is the explication and description of the scheme of concepts that is actually employed in our thinking about the world. Aristotle is trying to tell the truth about something already there, whose nature is fixed independently of his own inquiry. But Whitehead's task is quite different. He is not trying to describe a system of concepts that is already there for examination but rather to invent a new system that will be superior to the old in certain specific ways. Whitehead's philosophical task is that of construction or creation

249

—"imaginative generalization" as he himself calls it at one point in *Process and Reality*. This difference of philosophical procedure between Aristotle and Whitehead, this difference in what the two do *as* philosophy, has two consequences that are relevant here. First, Whitehead is a realist with respect, not to ordinary things (Aristotle's primary substances), but to actual entities, which are like things in being ontologically ultimate in the system in which they figure but in very little else. Hence that on behalf of which reductionism is being resisted, that which is being saved, is different in the two cases. Second, since actual entities are made and not found, Whitehead's realism is not so much a defense of the established order as a free, creative act or stipulation, one which can perhaps be supported or justified by citing reasons of a sort, but not the reason that this is how the world is. Whitehead is a realist because he wants to be (for certain reasons), not because he has to be, on pain of thinking what is false.

(3) Rorty says that Whitehead's realism has a different basis from that of Aristotle, that he is able to resist the reduction of actual entities to qualities or relations by some means different from that chosen by Aristotle, the matter-form distinction. But I do not think that Rorty has made clear in his paper what this means is, largely because he seems to say that it is different things at different times, and I cannot tell which of the apparently suggested alternatives is supposed to be the analogue of Aristotle's matter-form distinction, the device whereby the slough of reductionism is circumvented. I have just argued that Whitehead's task is different from Aristotle's and hence that the question: "What is the basis of philosophical realism?" has not the same sense when asked of Whitehead's actual entities that it has when asked of Aristotle's primary substances. So I would prefer to say that my present difficulty concerns not the basis of Whitehead's realism, as if there were some feature of actual entities that made it a mistake to reduce them to qualities, but rather the stipulated connections between the concept of an "actual entity" in Whitehead's system and other concepts in terms of which certain features of actual entities are explicated. The features in question are the individuality, particularity, and uniqueness or unrepeatability of actual entities. The question is: "What is it about Whitehead's system of concepts that guarantees or explains the presence in actual entities of these features?" Now sometimes Rorty answers that it is the fact that Whitehead distinguishes two sorts of elements or components of things, just as Aristotle distinguishes matter and form. But then to the further question: "What are the two sorts of element in question?" Rorty appears to give different answers. Sometimes he seems to say: actual entities and eternal objects; sometimes physical and conceptual prehensions; and sometimes the formal and the objective reality of an actual entity. At other times, however, Rorty seems to answer the question: "What is it that guarantees the individuality, etc. of actual entities?" by citing, not a distinction or pair of elements, but some single feature of actual entities. But here again he seems to give different answers to the question as to what this feature is. Sometimes he seems to say that it is the fact that each actual entity has a unique subjective aim provided

for it by God (but this is a mistake: a subjective aim is not unique for White-head, except accidentally, since it is composed entirely of eternal objects); some-times the fact that an actual entity is composed of unrepeatable physical pre-hensions (but the unrepeatability of physical prehensions is derivative and depends upon the unrepeatability of the prehending and prehended entities, hence cannot be used to explain the unrepeatability of the latter); and some-times the fact that actual entities enjoy "subjective immediacy" and exemplify "decisiveness" (but these are metaphors, whose sense needs to be made clear, besides which these features again presuppose and do not explain the individual-ity of actual entities). My reaction to this wealth of suggestions is mainly one of confusion (though some of them are untenable, as I have tried briefly to indicate in the foregoing parenthetical comments); I am just not clear as to what Rorty's view about the basis of Whitehead's realism is. Furthermore, I feel bound to add to the confusion by suggesting a further possibility of my own. Is it not the extensive continuum, an admittedly dark notion in White-head's philosophy, in terms of which the individuality of actual entities is to be explained? Is it not, i.e., rather this than any of the things that Rorty has mentioned?

V. C. Chappell

MATTER AS ACT[1]

Robert O. Johann

There is a growing desire among Catholic thinkers to follow White-head's advice and "take time seriously". They want to be able to see the world as temporally developing, with man, whatever his special prerogatives, part and parcel of the temporal process. In implementing this desire, they are somewhat embarrassed by their Aristotelian heritage, whose cosmology was elaborated against a completely different background and world-view, a "steady-state" view of the universe as opposed to an evolutionary one. The purpose of these pages is to suggest a possible alternative.

§1 *The Problem*

The very idea of evolution appears at first sight an affront to reason. It seems to suggest that you can have more at the end of a process than you had at the beginning, or more simply, that you can get something out of nothing. This apparent violation of the principle of causality is the general basis on which scholastic philosophers have normally rejected the idea in the past and refused even to consider the possibility of life arising from non-life, much less spirit from matter. Actually, however, philosophers holding for evolution have shown no less respect for causality. How else explain their tendency either to reduce the higher to the lower, as in the various forms of materialism, or to claim the presence of the higher in the lower from the outset, à la Teilhard de Chardin, Whitehead and panpsychists in general? For if there are no essential differences between the different levels of perfection in the universe, the problem of causality disappears—but so also, it would seem, does evolution. One may hold for variety in the world without progress, but it is at least difficult to speak of genuine progress if there is no genuine variety.

The problem, therefore, to be faced by any serious philosophy of evolution is how to be reasonable and evolutionary at the same time; how satisfy the exigencies of thought about process and still be true to the irreducible novelty of what supposedly has emerged through process. Life is something essentially new and more perfect in relation to brute matter; man is something

---

[1] This paper is a substantially revised version of the one presented at the Conference on the Concept of Matter. In making the revisions, I profited greatly from the discussion and comments on the original version provided by the participants of the Conference, particularly Dr. L. J. Eslick of St. Louis University.

essentially new and more perfect in relation to other worldly forms of life. This much is certain and the Scholastics have been right in holding on to it. If an evolutionary theory denies it, not only does it become untrue to the clearest evidence of experience; it also deprives evolution itself of any real significance.

On the other hand, to maintain these differences in terms of traditional hylomorphic theory is to maintain them in a way that that would seem to rule out the possibility of evolution. For, according to hylomorphism,[2] essential differences between bodies are explained by the union of specifically distinct forms with a purely passive substrate, primary matter. It is this primary matter which provides the link of passage between one kind of body and another. Thus, an essential (substantial) change occurs when primary matter loses the form which it had and acquires a new one. Since, however, it is of itself pure potency, a bare and wholly indeterminate substrate, primary matter cannot "do" anything itself towards this acquisition. As regards any change it is wholly passive. Hence, the appearance of a new form in matter must necessarily be attributed to the influence of an efficient cause which is already in act, formally or eminently, with respect to that form and which is said to "educe" the same from the potency of matter. And here is the difficulty for any hylomorphic theory of an evolving universe. For the causes responsible for the transformation of matter will themselves be a part of this world, and then there can be no question of the emergence of new forms but merely a redistribution of old ones. Or the responsible cause will transcend the world—and this would have to be the case for the appearance of a spiritual soul which transcends the potency of matter—in which case the new perfections do not really arise by a process of development; they are inserted.

The question, therefore, arises: Is there possibly another way of interpreting the structure of bodies which, while allowing for essential differences—a necessary requirement, we have seen, for a genuine theory of evolution—does not at the same time exclude evolution? And since the difficulty with hylomorphism in this respect stems from its explanation of essential change as the transformation of a purely passive substrate, primary matter, we can narrow down the question to this: Does an understanding of the material world really require such a substrate? As an indication that it does not, we shall suggest in the following section a possible alternative.

---

[2] The theory of hylomorphism to which I refer is the one which has become traditional in manuals of Scholastic philosophy and which is centered on the idea of "primary matter" as pure subjective potency, completely lacking of itself in any formal determination. However, if one thing became clear in the discussions that led to this book, it is that not everyone who admits a doctrine of primary matter will subscribe to this interpretation of it.

§2 *The Scheme*

The basic element in our conceptual scheme for "thinking" evolution is the notion of "imperfect act". By "act" we mean here "source of action", and by "action" we mean "self-affirmation", "self-expression". An act is "perfect" insofar as it is in and by itself (i.e. as divided from the "other") an adequate and sufficient source of action, i.e. self-affirmation. And it will be in and by itself an adequate source of action when the other is in no sense an intrinsic component of its self-position. Perfect act is act that is capable of affirming itself precisely as divided from and opposed to the other. Insofar as it is perfect act, the term of its action is simply itself as affirmed. Perfect act is act that is actively one with itself. Concretely, it is an act that can say "I"; it is personal.

"Imperfect act", on the other hand, is act that in varying degrees falls short of the personal, that in varying degrees is divided from itself in its action. This is to say that the other will always be more or less an intrinsic component of its action. It may be a component only terminally, i.e. where the other is required not as co-principle of the action but as that at which it terminates, as in the case of animal act.[3] Or it may be a component initially, i.e. when act is active only as element in a complex, when it is really the complex of which the act in question is a "part" that initiates action, such that the other is co-principle with act of its action. This is the case with brute matter, and it represents the greatest division of act from itself and consequently the lowest level of act.

Now if what we say is true, viz. that the difference between the personal (spirit) and the impersonal (matter) can be explained in terms of the distinction between perfect and imperfect act, then we may have here a way of conceiving the evolutionary process. For if imperfect act is in a state of division from itself as act, then a progressive overcoming of this division would be at the same time a progressive emergence of spirit. In other words, although matter and spirit would still remain essentially distinct, they would nevertheless be in the same line of perfection, that namely of "act", and it might be possible to see the second as arising from the first through a process of development. Since the development would be precisely in terms of act's overcoming its initial division from itself, it would not demand the presence of spirit or personal act at the outset nor require the injection of such act from the outside.

How then would such development be explained? We have a clue to the

---

[3] See §3.

answer already in the notion of action. Traditionally, "action" has been divided into two kinds: transient and immanent, and both have been conceived in terms of motion. Transient action was the moving of another; immanent action was the moving of oneself. Transient action was a process of perfecting the other; immanent action a process of perfecting oneself. Now it has not always been noticed that this notion of action presupposes a more fundamental one, one more closely akin to our own. For the moving or perfecting of another can only be conceived as the self-affirmation of the act of an agent *in* a patient, whereas the perfecting of oneself requires a self-affirmation of oneself, which as a process abides *in oneself*. In other words, both transient and immanent action presuppose action as the self-affirmation of act. And in terms of action as self-affirmation or self-expression, the way is opened for conceiving evolutionary process. Let us examine how this is so.

Suppose at the outset a non-systematic manifold of imperfect acts. As imperfect, none of these acts is capable of action by itself but only in conjunction with the other elements of the manifold. Each, therefore, as act is essentially involved in a complex, in the manifold of the other. Such involvement, which implies a division of act from itself since each is act only in conjunction with what is not itself, is precisely what we mean by materiality.

Now it belongs to the nature of act to be active, to act. At this level, however, action can be conceived only as interaction, or perhaps better, as the union of imperfect acts in action. For since none of these acts is by itself a sufficient principle of action, what is really active is the complex. It is the plurality of elements taken precisely as a plurality, in which each element is as it were a function of the rest, that enjoys a true sufficiency as act. And it is to this sufficiency, the sufficiency of the complex, that the joint action of imperfect acts gives expression. In other words, the dynamic integration of imperfect acts in action terminates in themselves as integrated. The sufficiency in virtue of which they interacted and which initially existed only in a "fragmented" state, distributed as it were over the elements of the manifold, emerges now as that in virtue of which the imperfect acts form a whole or system. The term of interaction, therefore, includes not merely the original elements themselves, but the fruit of their active synthesis, *sc.* a formative act that integrates them into a unity and which, as their unification, enjoys itself the sufficiency of the original complex. Interaction is thus the way to synthesis, to the emergence of systems whose formative acts will be more and more sufficient as acts—or, to put it another way, less and less divided from themselves as acts.

Regarding this emergence of systems, several remarks are immediately in order. First of all, it should be noted, the new system is not strictly speaking the product of efficient causality. An efficient cause is a reality extrinsic to its effect and is therefore something complete in itself independent of the causal relationship. In the scheme we are proposing however, the dynamic source of the system enjoys no such extrinsic or independent status. For, as we have seen, the source of the system is the non-systematic complex of imperfect acts whose conjunction in action terminates in *themselves* as systematically integrated. In other words, if action is self-position, self-affirmation, then act-as-affirming-itself (source) and act-as-affirmed (term) do not constitute two independent realities. They are rather distinct states of a single reality, a reality whose processive character essentially involves both of them. The system which arises, therefore, is not a reality other than and extrinsic to the complex from which it has arisen; it is rather the complex itself *as other, sc.* as systematically integrated. It is not therefore efficiently caused by the complex. It *proceeds* from the complex. And it proceeds from the complex precisely because the complex proceeds to itself (affirms itself) as system.

This may be all very well, the reader will say, but does it really answer the problem? For a system cannot be reduced to the elements that make it up. The very idea of system implies not only a plurality of elements but also a principle of order and unity in which they participate, an act of a higher order which plays a formative role in relation to the elements it organizes and which is "higher" precisely because it is formative. The question is then: where does the formative act of the system come from? You seem to have more at the end of the process (elements plus formative act) than at its beginning (elements alone). Either this will have to be added from the outside by creation (therefore no evolution), or you are getting something from nothing (goodbye to causality and reasonableness).

Since this difficulty touches the heart of the matter, let us proceed slowly. When the complex is said to affirm itself, it does so in virtue of the collective sufficiency (as act) of its elements (imperfect acts). This collective sufficiency is not at the outset something over and above the elements themselves —it is precisely the sufficiency they enjoy, taken together, to initiate action. Now action is self-affirmation, self-expression. The action of the complex, therefore, is the expression of itself as a complex, i.e., the expression of the active sufficiency which the complex as a complex possesses. As expressed, this collective suffciency will now be an act over and above the bare elements: it is indeed the fruit of their action. However, as the terminal expression of their collective sufficiency, it is not something independent of them but is precisely *their* act, a single act in which they all participate. But such an act is what we mean by a formative act, and the participation of a plurality of elements in such an act is what makes them a system.

The formative act, therefore, cannot be conceived simply as added from the outside by creation.[4] Were it so, it would not be the terminal expression of the collective sufficiency of the complex of imperfect acts; it would not be *their* act. Neither, however, is it a question of getting something from nothing. For if it is the nature of act to be active, i.e., to assert and express itself, then act as terminally expressed belongs to the very reality of act. Unless act gives rise through action to itself as expressed, it would not be act. In like manner, apart from the terminal expression of their collective sufficiency, i.e., the formative act to which they give rise through their joint action, the imperfect acts of the complex would not be acts at all, even imperfectly. To speak, therefore, of a complex of imperfect acts is to speak of that whose nature it is to systematize itself through action, i.e., to give rise through action to an act that is formative of itself (the complex). Finally, precisely as formative of the complex itself and not something independent of it, it cannot, as we indicated above, be said to be caused by the complex. How then describe its origin? Its origin is one of procession; it proceeds from the complex of imperfect acts. And it does so because whatever enjoys the sufficiency of act (here, the complex) is by nature processive, i.e., active. Act is self-affirmative, self-expressive.

But here some further precisions are in order. If it is granted that the idea of "procession" involves no violation of the principle of causality, the question can then be asked whether we are not now back on the other horn of the dilemma. Is not our explanation in terms of the scheme: act-action-act, another example of panpsychism, i.e., of putting the "higher" in the "lower" to start with? If action is self-affirmation, self-expression, and if the various stages of evolution are explained in terms of the progressive self-affirmation of act, then the initial stage of evolution has to contain already whatever is going to "evolve" from it. Hence, once again the price of avoiding contradiction in evolutionary thinking is the apparent impossibility also of any real novelty—and the impossibility of any really significant evolution. In what sense, therefore, are the systems which emerge by action, or more precisely their formative acts, already contained in the manifold of imperfect acts? Or conversely, in what sense are they really new?

The formative act of an emergent system is, we have said, the terminal expression of the collective sufficiency (as act) of the complex of imperfect acts which are now the elements of the system. Prior to action, therefore, it can be said to be "contained" in the complex only tendentially, i.e., as the term of the natural tendency of any active sufficiency or act (here, the collec-

---

[4] It may, however, *also* be created (see §4, and the problem of the human soul).

tive sufficiency or, if one prefers, the collective act) to express itself. In other words, it is there only in the sense that what by nature is ordered to give rise to it is there, and therefore as ultimately pertaining to the complete intelligibility of what is there.

So far so good. But if it is really just an unfolding of what is there, in what sense is what unfolds really new? Is it like an acorn evolving into an oak, or an egg into a chicken? First of all, it is clear that the novelty cannot be absolute. It cannot be that what arises is completely new with respect to what was there, i.e.; having no connection by way of origin with what was there—that would be contradictory. The newness can only be understood in terms of difference, "real novelty" being understood in terms of "essential difference". If we are to speak of significant evolution, it can only be because what arises is essentially different from that from which it arises. Is such the case here? And how can it be, if action is "self-expression"?

The answer to the first question is: yes. The answer to the second will require a little more ingenuity.

At the outset, let us remark that the disappearance of essential differences comes about when the various stages in the evolutionary process are interpreted ultimately in terms of one or other of them. Thus spirit will be reduced to a highly complex arrangement of particles of matter, or elementary matter will be seen as already endowed with a rudimentary form of life, consciousness, liberty, etc. The point, however, of our explanation has been to see the essentially different levels of perfection in the universe as successive stages in a single process. In other words, one level is not reduced to another, but all are included as separate and irreducible phases of a single, transcendent dynamism, the dynamism of act. Thus, for example, act as by itself an inadequate source of action, i.e., as divided from itself and a function of the other (*sc.* matter) is essentially distinct from act that is by itself an adequate source of action, i.e., that is divided from and opposed to the other and undivided from (actively one with) itself (*sc.* spirit). But they are likewise essentially inter-connected by their being separate phases in the single process of action by which act-as-divided-from-itself is enabled to overcome this division. And here, I think, we have the answer to our question.

The process of evolution is the process of act's overcoming its initial division from itself. It is therefore a process of progressive synthesis. The novelty then inherent in genuine evolution can be understood as the novelty of genuine synthesis. For a genuine synthesis of disparate elements, which is achieved through the emergence of a formative act that is the terminal expression of their collective sufficiency, will automatically differ essentially from the elements it synthesizes. Each step, therefore, in such a synthetizing process will be an advance, giving rise to an act that is more sufficient and comprehensive in itself as act[5] than anything that has preceded it. On the other hand, precisely because it arises as the term of action, it will look to what has gone before as its origin and the ground of its emergence. Thus, the fact that action is self-expression will not prevent the term of action from differing essentially from its principle so long as the principle is a collectivity of disparate elements and the term is their genuine synthesis. But the fact that action is self-expression will also assure that whatever emerges will belong to the same world as its predecessors and together with them go to make up one reality.

Summing up this section, therefore, we can say that all the real and essential variety in the world is a variety that is interior to an all-inclusive world-process. There are no higher unities that are not dynamically related to lower unities as the terminal expression of their collective sufficiency. Action (the process of act) thus becomes a sort of transcendent theme of which the variety in the world is the articulation. It will be a progressive articulation, one that takes time, since the higher and more complex syntheses will presuppose the existence of simpler syntheses for their own emergence and functioning. Perhaps more importantly, it will be a non-systematic articulation, proceeding not according to the rigid determinism of universal laws, but more or less at random, with false starts and dead-ends. For world-process, the process of act, is not to be understood as the pre-determined unfolding of a determinate nature or essence. Natures, essential structures, systems, all these are the emergent consequents of process (act's progressive overcoming of its initial division from itself), not its antecedents.

At the outset, as we have indicated, the active sufficiency of act is distributed non-systematically over a coincidental manifold (sc. of imperfect acts). Since the imperfect acts are not systematically related to one another, their conjunction in action (the way to synthesis) is likewise non-systematic. Each synthesis or system, therefore, that emerges will always have the character, not of a step in a logical process, but rather of a unique event. It will be a step forward in world history. And it will, moreover, be a tentative step. For the very fact that systems will arise non-systematically means also that they will find themselves non-systematically related to their environment. In other words, their success and survival as systems, as well as their ability to provide the bases for further syntheses, are not automatically guaranteed by their mere emergence but will depend on the host of coincidental factors outside themselves that constitutes their situation. Thus the only thing assured by the "nature" of act in its initial stage is that systems will emerge and that more complex systems (higher syntheses) will be preceded

---

[5] See §3.

by less complex ones. All that is assured is that the original abundance will strive in countless ways and by countless experiments towards its own synthesis. What these ways and experiments have actually been, which have succeeded and which failed, and what the future is likely to be, all these are elements in the unfinished story of the universe.

§3 *Steps in the Process*

So far we have concentrated on the broad outline of the scheme. Our purpose has been to show how it is possible to "think" evolution, to conceive of process in such a way that its giving rise to genuine novelty will not do violence to the principle of causality. For that reason, we have limited our attention to the general idea of a progressive synthesis of imperfect acts without detailing its stages. Now, however, that we have shown the theoretic possibility of such a process, we must try to fill in the main steps. Our aim, therefore, in this section will be to indicate how the actual and essential variety in the universe of our experience can be interpreted as a progressive articulation of the dynamism of act, i.e., as successive stages in the process by which act overcomes its initial division from itself and expresses itself in terms of ever increasing self-sufficiency and comprehensiveness.

The progressive sufficiency of act is measured according as the other is less and less intrinsically required for its being and exercise as act. In proportion as act becomes more and more "collected" in itself (less and less divided from itself), in the same measure it becomes more and more divided from the other in its activity, more and more self-sufficient. Now, the lowest level of sufficiency is that of wholly imperfect act, act in its greatest state of "uncollectedness" or division from itself, where it functions only and wholly as an element in a complex. Not itself a body, it is the root of bodiliness, that sheer multiplicity of which a body is the systematic unification. Next would come an act which while still radically insufficient and imperfect (i.e., while still needing the other precisely as co-principle of its action) would nevertheless have a certain role of its own independently of the other, namely, as formative of a system of wholly imperfect acts, or of a system of systems of such acts. Still unable to initiate an action attributable to itself as source, still brutely material, it is nevertheless a notch above the elements it organizes and of whose collective sufficiency it is the integration. Once, however, the threshold of life is passed, the radical insufficiency (as act) of brute matter is overcome. For it is the characteristic of living things that they act not only in conjunction with the other, but, in some sense, in opposition to the other. They are not merely subordinate elements in a larger complex; they begin to subordinate the other to themselves.

This can be seen, for example, on the level of vegetative life. In the vegetable, the formative act of the system of elements which is the vegetable does not look to what is outside the system for the completion of itself precisely as act, but for new elements to be integrated into the system and ultimately to be formed into new systems similarly organized. In other words, vegetative act looks to the other not as co-authoring its action but precisely as providing material in which to assert itself. Without such material, there can be no self-assertion. On the other hand, the assertion is of itself alone, not of itself plus the other. It must act *in* the other, but it authors the action by itself. Vegetative act, therefore, manifests a greater sufficiency as act than does act on the level of brute matter. But if its action is attributable to itself, as in the case of perfect act, nevertheless, unlike the latter, it still finds the other intrinsic to its action as its necessary matter. It is still, therefore, divided from itself in its activity and, in that sense, still "imperfect".

A higher level of sufficiency is found in the act which is the principle of animal life. The formative act of the system of elements called an animal does not look to the other to complete itself precisely as act; nor does it look to the other simply insofar as it is in the process of being integrated into its own system and supplies the matter for its own self-assertion. Rather it looks to the other precisely as distinct from the system which it itself informs and as that to which that system as a whole is referred in its action. Whereas vegetative act is related to the other only as the potential subject of its own formative influence, animal act is related to the other in its otherness. The other is not merely that in which it asserts itself, but that with which it deals in its action, that at which its action terminates, that which objectively specifies its self-affirmation. Since the animal is thus more clearly divided from the other than is the plant, it is correspondingly, to a greater extent than the plant, something in itself. Since, however, its act does not yet terminate at itself in its action but only at the other, so that the other is still an intrinsic complement of its action, it is still in a measure separated from itself in its activity; it still can affirm itself only outside itself; it is not yet actively one with itself, not yet perfect act.

Perfect act is reached for the first time in the human person. The formative act of the system of elements which is man does not look to the other merely as supplying potential elements for that system, nor again as the intrinsic term of its action. The action of this act is able to terminate at the very act which is its source, at itself. It is interior to itself, actively one with itself, aware of itself. In its ability to utter the electrifying "I am", it at last manifests an act that is no longer divided from itself in its action. And being no longer divided from itself, being at last complete in itself, it also manifests itself as perfectly divided from the other. It is an act for which the other is no longer, as it were, an aspect of itself, but for which the other exists precisely as other. Man, it should be noted, is not disjoined from the other on all the levels of his being. Insofar as he is a system of imperfect acts, of elements, he is wholly involved in the manifold of the other, a part of the world that surrounds him. Again, as sharing in the life of plant and animal, he is still more or less continuous with the other and divided from himself even as they are. It is only as person that he confronts the other as other. And even here, he still needs the other. A being that is in itself only as opposed to and facing the other, still needs the other in order to be itself. The human person defines himself and acquires his historical identity by freely relating himself to what is distinct from himself. His affirmation of himself can take place only in relation to the other. Without the other, in whose presence he is called to take his personal stand, there is no self-affirmation and no human person. In other words, the formative act which makes man a person is not Pure Act to which nothing is opposed and whose self-affirmation is wholly unconditioned. Since, on the other hand, what conditions personal act is the other *as other*, personal act manifests itself as wholly divided from the other and therefore as perfect act, complete and sufficient in itself as act.[6]

The interrelationship of the various levels of being in the universe, which we have described in terms of the progressive "sufficiency" of act as act, can likewise be seen in terms of the progressive "emergence" and "comprehensiveness" of act. The two ideas of *emergence* and *comprehensiveness* are connected. By "emergence" I mean the liberation of act from the state of being simply an element in a larger complex, simply a part included in (or comprehended by) the manifold of the other. Since, however, this liberation from being part of the other cannot be conceived as isolation from the other —we have seen, for example, how personal act, which is in no sense a mere element in a complex, is still not isolated from the manifold of the other— we are forced to say that in the measure that an act ceases to be subordinated and included as an element in the manifold of the other, to that extent it begins to subordinate the manifold to itself, to include it within the range of its activity, to be "comprehensive" of it. "Emergence" and "comprehensiveness", therefore, are two sides of the same coin and each is directly propor-

---

[6] This being the case, any further synthesis in which the person himself would be involved will have to take a different direction. Since personal act is complete in itself as act, it cannot be conceived as being further subsumed under some higher formative act. A further synthesis will have to be a synthesis of absolutes, one that maintains in their integrity and independence the realities it unites. As Teilhard de Chardin points out, there is one way that lies open, the way of love. If human experience is finally to fulfill its promise as being totally comprehensive, an active synthesis in which nothing at all is lost, this it will do only in so far as it becomes more and more an experience of communion.

tionate to the "self-sufficiency" that we considered above. A brief analysis will show how this works out.

Wholly imperfect act, act that is act only in conjunction with what is other than itself, is in no sense comprehensive of the manifold but is wholly comprehended by it. Imperfect act that is nonetheless the formative act of a system—such as you would have, for example, in the case of an atom—although still essentially an element in a larger complex, begins nevertheless to comprehend something of the manifold under itself, to include it within its proper range of influence. It is precisely formative of the elements within the system it unifies. What is thus comprehended, however, is not comprehended as other than itself, but only as its own matter. In the case of vital act, the range of comprehension is larger. Not only are the actual elements of the system it informs included under its scope, but the potential elements as well, i.e. those which, while actually other, are nevertheless in the process of being integrated as parts of the system. The activity proceeding from the formative act of the animal is more comprehensive still. Not only does this act assert itself in relation to what is actually or potentially a member of the system it unifies, but also to what is wholly outside the system. As we have seen, the act of the animal reaches the other in its otherness, although not precisely as other. Since the other is still an intrinsic complement of the animal's action, only that portion of the manifold of the other will be included within the active synthesis of animal life which is directly proportioned to the structure of the animal in question, i.e., only that portion which can be attained by its particular senses.

Coming at last to personal act, however, we have that which is totally comprehensive. In the active life of the person, the whole range of the other is included and included precisely as other. Human experience, as the active synthesis of the self and the other than self, is, as it were, all-inclusive and the whole of reality. For, to grasp the other as other, as distinct and independent of the self, is to grasp it as implicated in, and as itself implying, the self-sufficient totality that confronts the self and in relation to which the self is called upon actively to define itself. This totality, it is true, is only gradually discriminated by the self but right from the outset it is included in the self's experience as the "horizon" of whatever is explicitly attained in action. The range of personal act, therefore, is equal to the whole of the real. Instead of being comprehended as an element in the whole, it is itself comprehensive of the whole.

From what has been said it should be clear that the various levels of perfection in the universe constitute a veritable analogy of act. They are not related to one another simply as so many discrete structures or essences which only have this in common, namely, that each is proportionately related to existence. On the contrary, even in their very differences from one another they are rather just so many steps in the unfolding of the one perfection, the perfection of act. Each level is but a fuller articulation of the self-affirmation of act. Moreover, as we have seen, not only is there a progressive self-sufficiency of act as act; on each successive level of act, there is greater and greater comprehensiveness of the whole in which it is implicated. Looked at another way, since the various levels of perfection all pertain to the one universe, we might say that on each successive level the universe itself exists in a state of greater and greater synthesis, is less and less divided from itself. Instead, therefore, of seeing the various levels as constituted by successive differences added from the outside, we should rather look upon each level as a new step in the progressive possession of itself by the whole. Each level is but the universe itself coming more and more into its own.

§4 Some Objections and Answers

In the scheme we have proposed, evolution is explained in terms of the progressive self-affirmation of act. From this point of view, matter is not ultimately purely passive; matter itself is act, albeit imperfect act, or better, it is a manifold of imperfect acts. The active sufficiency which is initially distributed over this manifold asserts itself through the interaction of the original elements and progressively expresses itself more and more synthetically. Finally in man it completely overcomes its initial division from itself and we have the birth of self-presence and freedom. Henceforward the question of progress is placed in man's hands. He is called upon to master the universe and shape it into an abode of love.

Such is our proposed solution to the problem we started out with. Before we let our case rest, however, it may be well to forestall certain objections that can (and have been) raised against it.

The first has to do with the question of causality and concerns both the relationship of our theory of procession to the traditional doctrine of the creation of the human soul and, more generally, the effect of our position on the whole theory of secondary causality.

Concerning the creation of the soul, let me say at the outset that I do not intend the position I have outlined to be in any way a denial of that fact, nor do I think that there is any inconsistency between them. My whole point has simply been to indicate man's place in the universe as the natural term of a transcendental process, i.e., a process that includes not only man but all the other distinct grades of perfection present in the world and which is precisely the process of "act", the collective sufficiency of the universe, progressively affirming itself. The human soul belongs to the universe as a terminal expression of its collective sufficiency. Now since, as we have seen, its terminal expression belongs to the very intelligibility of act as source, it cannot be said to be produced by act but proceeds from it. For this reason the universe cannot be said to be the efficient cause of man nor to produce man, neither his body nor (much less) his soul. It is rather that whence the human composite arises as the highest individual synthesis of its collective sufficiency. This, however, is not to deny that the soul does have an efficient cause, i.e., that each soul is also the product of a divine creative act. The idea of creation comes into play (and must do so) when the soul is viewed not in terms of its position in world history (i.e., as a formative act that expresses the collective sufficiency of its antecedents and is the term of the process by which this sufficiency finally overcomes its initial division from itself) but simply in terms of its reality as a subsistent spiritual principle.

We might recall here the distinction which St. Thomas makes between the soul as *principium quod* and the soul as *principium quo*. For just as the assertion that the soul is a form (a *quo*) relationally integrated with the body does not preclude the fact that it is also a spiritual subsistent (a *quod*) and that as such it is causally dependent on a creative act, so neither does the assertion that the soul as formative act is the terminal expression of the collective sufficiency of the universe and as such is relationally integrated in the process of world history, preclude its also being created. To put together the idea of creation (which says nothing about how the soul is related to the other entities with which it finds itself in the world nor how, together with them, it goes to make up one world) and the idea of process or evolution, we might express it this way. The original creation of the universe as a manifold of imperfect acts automatically commits God to the eventual creation of perfect act (i.e., when and if that point is ever reached in the non-systematic process of progressive synthesis), since both imperfect act and perfect act are but separate stages in a single process of development, the progressive articulation of the transcendent dynamism of act. Thus our theory in no way precludes the creation of the human soul. It would merely insist that the doctrine of creation be complemented by one of procession if the soul's position in the world is to be understood.

Something similar, I think, must be said concerning the general doctrine of secondary causality. Our theory of the processive character of act does not exclude causality, secondary or otherwise. It merely insists that action is essentially self-affirmation and that consequently the relationship between act as source and act as term is not a causal one. This, however, does not prevent causality from being an aspect of certain instances of procession. On the contrary, I would maintain that the processive character of act is the ground of causality. Thus, for example, when act asserts itself in a pre-existent other (which is what happens on the level of secondary causality), such that the other is transformed into a likeness of the active entity, a causal relationship is set up. But this causal relationship is not between act as source and its self-expression in the other, but between act as source and the transformation of the other. As St. Thomas himself observes, secondary causes do not cause the form which the patient acquires but rather its acquisition by the patient; they do not cause the form as form but that this matter acquire this form. The form itself however, which now is newly present in the patient, is the terminal expression of the self-affirmation of the act of the agent. Thus, although action may in some circumstances involve motion, it is not primarily motion or transformation, but self-affirmation. And the point of our paper has been that only when it is seen in this more fundamental light, i.e., as grounding causality but not reducible to it, can it provide a clue to understanding evolution.

A second line of difficulties can arise from a misunderstanding of the unity of process. We have spoken of a single process of development that embraces all the variety in the world and have referred to it as a sort of transcendent theme that is progressively articulated, the theme of act's self-affirmation. From this it might be concluded that there is somehow "only a single Act of the whole universe, Reality with a capital 'R', of which finite acts are only the incomplete manifestations or appearances".[7] Evolution would then be merely the temporal unfolding on the level of appearance of this one eminent Reality which itself transcends time. Thus instead of taking time seriously, which was our stated purpose in this paper, we wind up relegating it to the order of mere Appearance, and since this order is completely dominated by that transcendent Reality of which it is the appearance, all honest contingency and genuine novelty disappear.

Needless to say, such an objection, for all its seriousness, completely misses the point of our explanation. For us, there is no single Act of the universe. There is indeed a total active sufficiency, but this, as we have insisted, is distributed over a non-systematic manifold. It is not something over and above the imperfect acts themselves, but is precisely their collective sufficiency. And this being the case, there is no such distinction as the one suggested between finite acts as Appearances and eminent Act as Reality, with all its baneful consequences. How contingency and novelty are genuine features of the world, we have already indicated (see §2). The only word to be added now is one about the unity of the process, and its transcendence with respect to the entities to which it gives rise. We speak of the process as one, as indeed a world process, because the original manifold of imperfect acts,

---

[7] From Dr. Eslick's "Comment" below.

due to the inter-dependence of the latter, constitutes a totality (the universe) which, through the union in action of its elements, is progressively systematized. There is nothing in it which is not related to all the rest or has not arisen as a terminal expression of the collective sufficiency of the original manifold. Moreover, since it is the nature of this collective (or total) sufficiency, precisely as active, to overcome its initial dividedness and since in the course of this it gives rise to entities essentially different from their antecedents, we may view the essential variety in the world as intrinsic to the process of self-affirmation of the original sufficiency and this process itself as transcending the different entities to which it gives rise.

There is one final difficulty. It concerns the idea of wholly imperfect act, and the sense that can be attached to a manifold of such acts. That the idea is obscure, I readily admit, for we are here at the lower limit of thought. As to whether or not it makes sense, I can only let the foregoing analyses stand on their own. One caution, however. Since imperfect act is act as divided from itself, it would be a mistake to try to understand it as a substance complete in itself. From the foregoing pages it should be clear that neither is imperfect act a substance, nor is the manifold of imperfect acts a substance. The coincidental complex of imperfect acts is rather a sort of matrix whence the substances we do encounter in our experience arise.

COMMENT[1]

THE THEORY OF MATTER AS IMPERFECT ACT IS PROPOSED BY JOHANN IN THIS PAPER primarily as a philosophical explanation of evolution, if evolution be conceded as a fact. In commenting upon Johann's paper, there are three main lines of questioning I would like to develop. The first concerns problems arising from the distinction between causal production and procession of the higher from the lower. The second is about the unity and plurality of act. The third suggests certain difficulties which I, at least, feel about the doctrine of matter as wholly imperfect act, or acts.

The attempt to account for evolutionary progress by *causal* explanation, in the line of efficient causality, is regarded by Johann as precipitating dilemmas apparently defying solution. Explanation by efficient causes presupposes acceptance of the ancient maxim that out of nothing, nothing comes. Either the higher grades are reduced to the lower, at the cost of an essentially differentiated world, or the higher is regarded as present in the lower from the beginning, at the same cost. Explanation of evolutionary progress by efficient causality seems, according to Johann, possible only in either materialistic or panpsychic terms, and in either case there is nothing more at the end than at the beginning. The result is an un-dappled world, without real progress. A static dappled world is conceivable, but also at the cost of genuine evolutionary development—it would be a Porphyrian "tree", but not a living organism subject to true growth. Specifying perfections on the higher levels would be added from without, but such addition is either an unintelligible case of the universe lifting itself by its own bootstraps, or it is the result of divine intervention, but neither allows for real evolutionary advance.

In order to philosophically accommodate evolution, therefore, Johann moves from the level of efficient causality to that of "procession", and from the plane of *being* to that of *act*. The scholastic analogy of being cannot, according to Johann, provide the key which will allow for a dappled, *evolving* universe. (His conception of the analogy of being is, I think, simply wrong insofar as he regards it as the product of abstraction.) The analogy of being does not, Johann thinks, tell us anything about the dynamic relationship of essential structures to each other. The so-called analogy of "act", however, escapes this limitation. Evolutionary dynamism seems, for Johann, to be restricted to the essential order, and

---

[1] This Comment is directed to the paper in the form in which it was originally given at the Conference. (The paper in this form was subsequently published in *Thought*, 36, 1961, 595–612). In the final version of the paper printed here, the author has made some changes in the light of the Conference discussion. (Editorial Note)

hence to formal causality of some sort, although a "formal causality" which appears to have much more affinity with Hegel and the post-Hegelians (or, perhaps, Plotinus) than with Aristotle. The "analogy of act", and the internal "procession" of act based upon it, is thus non-existential, concerned only with "progressive" specifications of act defined in terms of self assertion. "Act" is a "single formality underlying, running through, and finally embracing the whole hierarchy of forms".

The lower levels of self-affirmation cannot cause or produce the higher, ultimately personal levels. Rather, there is a *single reality*, of which the higher levels constitute the essential completion of the lower. The lower levels are insufficient and essentially deficient of themselves, and their essential or formal completion as acts requires the higher ones, which in this sense are said to "proceed" from the lower. This is all in the dimension of "immanent intelligibility", whereas efficient causality is said to pertain to extrinsic intelligibility. Nevertheless, the question of efficient causality with respect to evolutionary progression cannot, it seems to me, be simply shelved, so that *all* philosophical explanation is in terms of some kind of dialectical exigency or implication which the formally incomplete and partial has for the essentially complete whole, "the total act of the universe". Johann admits this with respect to the creation of soul as perfect act, which nevertheless is thought to "proceed" from the original manifold of imperfect acts in the sense that God, in creating the original manifold, is "automatically" committed to the creation of perfect acts. But must there not be a dimension of efficient causality immanent within the universe, secondary or caused efficient causes? Can a philosophical account of evolution completely ignore this dimension, however formidable the difficulties may be?

Secondly, there are crucial questions raised by Johann's theory bearing upon the ancient mystery of the One and the Many, here in the context of "act". The process of "act" is said to be the total act of the universe, progressively affirming itself. Even souls, or perfect acts, are relationally integrated in the dynamic structure of "total act". There seems clearly to be the assumption that there is an *essential* unity to the universe as a whole, and that it is something more than the unity of order which St. Thomas Aquinas ascribes to it. This is overwhelmingly reminiscent of the "Absolute" of Hegel, or of Bradley, Bosanquet and Royce, and it seems to me that Johann should unflinchingly face the classical difficulties of such a position. There is, it seems, only a single Act of the whole universe, Reality with a capital 'R', of which finite acts are only incomplete manifestations or *appearances*. Can the dialectical "procession" through appearances (imperfect acts) to the Absolute really be identified with evolution? How can there be *real* contingency, *real* novelty in such a world, in which every entity or act, no matter how trivial, *essentially* implies every other, and ultimately, a completed totality which must utterly transcend time and passage? It would seem that a satisfactory evolutionary philosophy must somehow manage to "take time seriously", as A. N. Whitehead was always urging. But on Johann's grounds, it would seem difficult not to relegate time, as in F. H. Bradley, to the side of

Appearance, rather than Reality. Can a merely formal implication, whether Aristotelian, Hegelian, or any other kind, provide the dynamism which evolutionary development seems to demand? Nor is it clear, on the basis of Johann's paper, why *procession* upwards out of the material matrix of wholly imperfect acts should be the preferred direction of dialectical movement rather than *emanation* downwards from an eminent, transcendent reality.

In either case, such an eminent reality, the total act of the whole universe, seems to be given and presupposed, and not really the term of an evolutionary progression. Indeed, in the metaphysical family to which Johann's theory seems to belong, the differentiation and multiplication of finite "acts" should be the function of the *negation* of primal act, of essential selfhood. I can only mention in passing two special difficulties which arise from this in Johann's theory. If, following Spinoza's famous law that all determination is negation, one also asserts a plurality of *perfect*, personal acts, is it possible to significantly differentiate them? And, finally, how can there be a significant *plurality* of wholly imperfect acts, on the level of matter?

The third line of questioning I would propose for Johann concerns the designation of matter as wholly imperfect *act*. Can the extension downward of the so-called analogy of "act" to this ultimate abyss and negation of all essential selfhood be justified? "Being active" is defined "as affirming or asserting oneself", but a wholly imperfect "act" cannot, as such, be self-assertive. It cannot, it would seem, be "act" in any significant sense whatsoever, and to speak of the "interaction" of such non-acts (or even to regard them as somehow a manifold, and even a manifold which can form a unified whole, albeit non-systematic) is surely to beg the question. A hypothetical manifold of such "acts" cannot in any way be differentiated or articulated. Wholly imperfect "act" can "act" only by negating, like Plato's material principle, or Whitehead's Creativity. And even Whitehead's Creativity, regarded in complete abstraction from its conditioning by already constituted actual entities, is barren.

Leonard J. Eslick

EPILOGUE:
EIGHT PROBLEMS

Ernan McMullin

In the Introduction, eight problem-contexts in Greek philosophy were identified as those for which a concept of matter was invoked in some descriptive or explanatory role. We have seen in the course of the essays making up the book that these problems have been so transformed in the course of the last three centuries that one might well be tempted to drop *matter* from the list of concepts of speculative interest to philosopher and scientist today. As an epilogue, it may then be of value to isolate some major problem-contexts where the notion of matter is still invoked. Eight of these will be briefly described; they are for the most part different from the eight that dominated the first development of natural philosophy. In each case, it will appear that a sort of "dematerialization" has been operating; "matter" has in some sense become more problematic. How far has "dematerialization" gone?

§1 *Matter and Objectivity*

As we have seen in the Introduction, matter was often defined in seventeenth-century thought by the properties agreed upon as ontologically primary. Some form of primary-secondary distinction was almost universally admitted; its main purpose was often to define the "essential" properties of matter. Berkeley challenged this assumption, and argued that the alleged O-primary properties of Newtonian mechanics were just as observer-dependent as any other. Most philosophers today would be skeptical of the possibility of drawing a sharp distinction between O-primary and O-secondary properties. The latter would be described as *dispositions;* they are real properties, but they display themselves only under certain circumstances (such as, for example, the presence of light and of an organism with a certain sort of sensitivity). The O-dependence is thus one of *evocation:* a certain context is required in order that the disposition should produce some observable effect. The question of a primary-secondary distinction can now be rephrased by asking: are there non-dispositional prop-

271

erties in which the dispositional ones are rooted and which in some way express the permanent reality of objects?

No one would suppose nowadays that a simple answer in terms of the "quantitative" properties would suffice. The notion of quantity is too vague; all physical properties have measure-aspects, but these are only partially expressible in mathematical terms. Irreducibly physical concepts, like *resistance* or *energy,* will also be needed. Nor can one rely simply upon the classic Newtonian state-variables—mass, length, time. It is now clear that these are theory-dependent in a most complex way, that what one picks out as the "length" of a body will depend on one's theory of length-measurement. More important, length is no longer quite so obviously an intrinsic property, independent of the state of the rest of the universe. Relativity theory has shown that the length of a body depends on its velocity, and also depends on the distribution of gravitating masses in the universe. The same is true of the other apparently simple Newtonian parameters. It is by no means certain, therefore, that there are some O-primary properties which define a body entirely independently of the state of the rest of the universe, i.e. in a non-dispositional way.

Furthermore, the notion of a purely non-dispositional property, one which is constantly actual and does not require any specific context for its evocation, raises some very difficult questions. In particular, since we would get to know such properties only through some dispositional features of the bodies possessing them, and could only reach out to the non-dispositional properties in a tentative way through theoretical reconstruction, there would always be some uncertainty as to just what the O-primary properties were, and to what extent they really were O-primary. It is true that some properties are much more dispositional than others. Solubility is more dispositional than color, for instance, and color more dispositional than mass. But whether a sharp cut can be made between primary and secondary is another matter. The unquestioning assurance of nearly all seventeenth-century philosophers in this regard was due to the assumption (Newton's "analogy of nature") that some simple and intuitively definable quantitative properties (like extension, mass, velocity) could be directly attributed to bodies at all levels of size, in an entirely non-hypothetical way. Now that the construct-aspect of concept formation is better understood, the problem of defining O-primaries has come to seem an almost intractable one.

N. R. Hanson has analyzed one special feature of this problem.[1] He

---

[1] "The dematerialization of matter", in *The Concept of Matter*, ed. E. McMullin, Notre Dame, 1963. This interpretation of the implications of quantum theory has been widely assailed by Soviet physicists as "idealist" (i.e. anti-materialist) in tendency.

argues that quantum theory has shown that the influence of the observer-detector enters, in a complex and indissociable way, into every statement about the microscopic realm. Since it is in this realm (he assumes) that one would ordinarily wish to anchor the predicate 'material', he concludes that the Berkeleyan denial of the validity of the primary-secondary distinction, and the consequent challenge to the existence of an autonomous "matter", can now be supported on scientific grounds. Matter is necessarily detector-dependent; there are no intrinsic absolute properties. Not only is the description of the properties dependent upon the means used to obtain it, but the properties themselves attributed to the object on the basis of the observation have been shown (he claims) to be somehow dependent upon the means of observation. The detector *imposes* these properties, at least partially, upon the object.

This is the so-called "Copenhagen interpretation" of the quantum formalism, often also labeled the "orthodox interpretation" because it has been so widely accepted, in Western theoretical physics, at least. But there are several other interpretations of the formalism from which this implication would not follow. Chief among these is the "statistical interpretation" defended by Landé, Einstein and others, which takes the quantum state-description to represent an ensemble of similarly prepared systems, rather than being a complete description of an individual system.[2] The uncertainty principle is interpreted as a statement of the minimum dispersion in a state-preparation, rather than as having to do with measurement-interference. Most defenders of this view maintain, further, that subatomic entities are to be thought of as particles more or less in the classical sense, possessing perfectly well-defined position and momentum; the wave-particle dualism so heavily stressed on the Copenhagen side is thus rejected as unnecessary. Where the Copenhagen defenders emphasize the two-slit experiment, the other side relies on the Einstein-Podolski-Rosen experiment (which is taken to show that the state-vector cannot constitute a complete description of an individual system).

It is not possible to pursue this unresolved debate. A further, and even less resolved, discussion bearing on Hanson's argument, has to do with the possibility of a new formalism replacing the quantum one, in which complete predictability would be in principle once again attainable. Such a formalism would presumably allow the sharp separation between detector

[2] For a recent summary of the arguments in favor of this interpretation, see L. E. Ballentine, "The statistical interpretation of quantum mechanics", *Reviews of Modern Physics*, 42, 1970, 358–81. See also D. I. Blokhintsev, *The Philosophy of Quantum Mechanics*, Dordrecht, 1968; E. McMullin, *The Quantum Uncertainity Principle*, Louvain, 1954, unpublished dissertation.

and object characteristic of classical physics. A recent theorem due to Bell has shown (as von Neumann had earlier unsuccessfully attempted to prove) that a "hidden-variable" theory would necessarily yield predictions at odds with the experimental results of the present theory.[3] But his theorem cannot be used to exclude a theory which would restore full predictability, not simply by adding "hidden variables" of some kind to substantially the present formalism, but rather by introducing a radically new formalism. Whether the variables in such a case would be anything like the O-primary position and momentum of classical theory it is impossible to say. Perhaps enough has been said, however, to make it clear that the nature of measurement at the quantum level and specifically the nature of the appropriate "cut" between apparatus and object are a matter of continuing intense debate. Hanson's thesis that the basic dependence of quantum properties on the detector has been definitively shown is premature. The contribution of quantum theory to the discussion of the primary-secondary distinction is not yet clear. But we have already reason to question whether O-primary properties can be defined in the unproblematic way that Newton took for granted.

In the first years of the quantum theory of matter, some of the pioneers, notably Bohr, thought that a dependence of measurement upon the *observer* had been shown, and they used this in support of a Kantian type of philosophy. It would be fairly generally agreed today that this was a mistake.[4] The O-dependence of the variables corresponding to the non-commuting operators of quantum theory is at most a dependence on the measure-instrument, not on the observer as such. There is no question, then, of inferring from quantum theory that even the apparently most "material" properties are dependent upon mind or spirit, and that not only the primary-secondary distinction but even the matter-spirit distinction could on this account be challenged. From this point of view, quantum theory is just as objectivist as any physical theory that preceded it. Its significance lies rather in the epistemological domain.

---

[3] Provided that the EPR conclusion also be accepted, i.e. that separated systems, not interacting in the normal way are to be treated as independent. See J. S. Bell, "On the Einstein-Podolski-Rosen paradox", *Physics*, *1*, 1964, 195–200; A. Shimony, "The status of hidden-variable theories", in *Logic, Meth. and Phil. of Science*, ed. P. Suppes *et al.*, Amsterdam, 1973, 593–601.

[4] This would still be contested by some, especially by E. P. Wigner. See, for example, "Two kinds of reality", *Monist*, *48*, 1964, 248–64. For critical evaluations, see H. Margenau and J. L. Park, "Objectivity in quantum mechanics", *Delaware Seminar in the Foundations of Physics*, ed. M. Bunge, Berlin, 1967, 161–87; A. Fine, "The two problems of quantum measurement", in Suppes, *op. cit.*, 567–81.

§2 *Matter and Reduction*

The primary-secondary distinction can, as we have seen in the Introduction, also be drawn in *epistemological* terms. An E-primary property is one which is primary in our *knowledge* of matter; in its strongest form, it is irreducible, incapable of being explained in terms of simpler properties. In the seventeenth century, reductive assumptions were commonly made about the sort of explanation the new mechanics, or the corpuscular philosophy that often accompanied it, could provide. It was assumed that macroscopic properties, like density, temperature, color, could be explained in terms of microscopic entities lacking in these properties, thus eliminating these properties from the list of E-primaries needed by science. Instead of the discrete irreducible forms of the Aristotelian tradition, there are complex physical structures whose behavior can be understood by appealing to postulated simpler elements possessing only a few basic properties from which all others can be derived. This Democritean reduction suggested a Democritean notion of matter: a permanent stuff of which the universe is made, whose properties maintain themselves through all changes. And these properties would be discovered by looking at the most basic science.

There is clearly a tension between this notion of matter, defined in terms of E-primary properties, and the reductionist model of explanation that went with it. There seems to be no inherent reason why any given property or element should be taken to be basic in any other than a temporary sense. The atoms of today are the reducible structures of tomorrow, structures which still more basic elements will serve to explain. On this view, there need be no permanently irreducible properties or elements; if "matter" is to be characterized in E-primary terms (as the seventeenth century generally assumed), it will have to be conceded that such a characterization is at best provisional. This realization was slow in coming. It was helped along by the development of the electromagnetic theory of light in the nineteenth century; it became increasingly clear that the attempt to define a common world-stuff in terms of some irreducible property was not likely to succeed.

The recent history of science supports a sense of the term 'matter' that resembles Aristotle's notion of *material cause* in some respects. One can equate "matter" with the "given" elements in a particular physical theory, the postulated entities in terms of whose combinations and interactions the theory will be constructed, the "materials", in other words, that the theoretician has at his disposal. Since the notion of material cause was utilized by Aristotle in the analysis of different facets of physical explanation, it is appropriate that "matter" here should be the matter of *theory*. It is the theoretical electron, the construct whose properties are fully defined in terms of the theory, that serves as the "material" for explaining light-emission. The physical referent of the construct, the "real" electron which causes cloud tracks and light emission, is only imperfectly grasped by means of the theory. But matter in this sense always escapes full statement in any theory; hence, if we wish to define matter in E-primary terms, it will have to be as a theory-dependent construct and not as the inexhaustible physical object. To suppose that one can define the matter of the physical world in E-primary terms is equivalent to supposing that one has a science whose conceptual structure is definitive and exhaustive of the real, a supposition which Newtonians sometimes were tempted to make, but which no modern scientist would concede.

When, in his analysis of the making of a statue, Aristotle made bronze the material cause, he assumed that the concept *bronze* accurately and exhaustively described the real matter of the statue. Hence, he could take the real physical matter to be without remainder the "matter" of his "explanation" of the change. Today, this would be called not an explanation but a description. In an explanation of the hypothetico-deductive sort that characterizes modern science, the "material" element of the explanation is necessarily a postulated construct, an incomplete rendering of the physical matter which defines the task of the physicist.

Matter in the sense of "material cause" is, therefore, the as-yet unreduced element of theory, specified in terms of E-primaries that are primary relative to *this* theory.[5] If the theory is the broadest possible physical one of its day, like Newtonian mechanics or general relativity theory, the "given" elements, mass-points or energy-momentum tensors, and the concepts in terms of which these are specified, constitute or describe the basic "matter" with which the science of the day has to deal. These matter-elements being fully specified by the theory, further reduction can come only by turning back to the physical world, to the real matter, and finding new depths in it. The fact that further reduction so far, at least, always seems to be possible, opens up a sort of endless horizon with E-primary "matter" always receding before us. The dream of the first reductionists, that there would

---

[5] This is the acceptable sense of M. Jammer's comment: "Matter, as such, if in fact there is any need for such a concept in science, necessarily remains an uncomprehended and incomprehensible residuum of scientific analysis and as much unfathomable", *Concepts of Mass,* Cambridge, Mass., 1962, p. 5. For a fuller treatment of this aspect of the matter-concept, see E. McMullin, "Material causality", *Historical and Philosophical Dimensions of Logic, Methodology and Philosophy of Science*, ed. R. E. Butts and J. Hintikka, Dordrecht, 1977, 209–41.

be a limit to their reduction, an irreducible atom or ether at which science would terminate, has not been realized. It is possible that some limits of this sort still lie ahead for the physicist, limits dictated by the energies available for continuing to penetrate into the interior of the nucleus, perhaps, or by the non-individual character of the subatomic entities that are the "matter" of current quantum theory, or by the dwindling resources of physical imagery, or by the neural structures of the human brain. But these are highly speculative issues.[6]

§3 Matter and Complexity

The Aristotelian tradition made substance a composite of two "principles", primary matter and substantial form. The analysis of unqualified changes (like the death of a living thing or transmutation from one element to another) seemed to show that primary matter had to be regarded as totally indeterminate in itself, for if it possessed any enduring determination in its own right, unqualified change would be illusory, there would be only a single substantial "stuff" which took on different accidental forms, and the unity of the living composite in particular would thus be destroyed. Already in the medieval period, Scotus and others criticized the notion of form on which this argument depended for being too narrow; they argued that there was no reason to reject the idea of a composite form, involving a hierarchy of structures and activities. Substantial change would thus not need to be thought of as a total change of *all* formal determination; one could suppose that various constituents retained their identity, thus accounting for the manifold continuities (of color, size, weight, not to mention more essential characteristics) involved in all alleged substantial changes. The "primary matter" serving as the principle of continuity throughout any change no matter how basic (whether water to air, or living animal to corpse) would not, therefore, have to be regarded as totally indeterminate. In the later medieval period, even the Aristotelians themselves gradually modified the original primary-matter doctrine to allow the presence in primary matter of "virtual forms" and to make it possible to speak of a *quantitas materiae* that remained unaltered throughout change.

But the seventeenth-century challenge went much deeper than this, because the implicit reductionism of the mechanical philosophy excluded any sort of reality to the form of a complex composite such as a living body. To

---

[6] For a discussion of them, see R. Schlegel, *Completeness in Science,* New York, 1967; and E. McMullin, "Limitations of scientific enquiry" in *Science and the Modern World,* ed. J. Steinhardt, New York, 1966, pp. 35–84.

many, such a composite seemed to be analyzable as nothing more than a collection of atoms. The only proper formal reality would thus belong to the atoms or other micro-constituents; the form of a macroscopic complex, like man, is totally reducible to the lower, simpler forms. Descartes conceded this reduction of substantial form for all complexes other than man, and postulated a spirit-matter dualism (or psycho-physical parallelism) for man that served as a working hypothesis for many psychologists up to our own century. Other philosophers in the Aristotelian, the romantic, the idealist, and various other philosophical traditions, attempted to maintain the primacy of the whole over the part, and specifically the irreducibility of living forms, in both the ontological and epistemological senses.

In recent decades, this problem has been extensively discussed, and neither of the earlier extreme solutions, the "totally reducible" "nothing-but" view of Hobbes and the "totally irreducible" notion of spirit found in Descartes or Hegel, find very much favor. The issue is joined at two levels. First, there is the general problem of reduction: what does it *mean* to claim that in principle biology, say, can (or cannot) be "reduced" to physics and chemistry? Obviously, the claim has a wide variety of senses.[7] There are certain features of the living complex that can be discovered, it would seem, only in the context of the study of life; not all features of the behavior of a hydrogen atom in the human body, for instance, can be inferred from a simple study of the hydrogen atom in isolation. These dispositional properties do, of course, *belong* to the isolated atom, but if they are evoked only when it enters into certain sorts of complex, they may well be impossible to ascertain except by observing the behavior of the complex. To speak of reduction in such a case, or to assert that one can specify a set of E-primary properties of the part from which the properties of the whole could be derived, is quite ambiguous. It is true that if one includes dispositions, one could derive an explanation of the whole from an account of the parts. But in one respect this is not a true reduction, because the evidence about the dispositional properties as well as the concepts in which this evidence is expressed are acquired only by experience with the whole; an adequate theory of the whole could not, then, be reached if one were restricted to observation of the parts in isolation. This holds for complexes at all levels, not just organisms but hydrogen atoms as well (could one derive the theory of the hydrogen spectrum if all one knew were free electrons and free protons?). There is a sense, therefore, in which biology is irreducible to physics and chemistry because the range of observations and theoretical

---

[7] See E. McMullin, "The dialectics of reduction", *Idealist Studies*, 2, 1972, 95–115; *id.*, "What difference does mind make?", in *The Brain and Human Behavior*, ed. A. G. Karczmar and J. C. Eccles, Berlin, 1972, pp. 423–47.

concepts that characterize biology will always remain indispensable. But there is another sense in which physics may be enlarged to incorporate the biological domain, as we come, through our study of biology, to know more about the dispositional properties of the physical elements that make up living complexes.

The second problem is more specifically concerned with man; it is usually called the "mind-body problem" in English and American philosophy, where it has been at the center of discussion in recent decades.[8] The word 'body' denotes the material component of man; 'mind' is usually taken to refer to all those capacities of knowledge and willing that distinguish man. The problem is whether it can be shown that these latter are in principle irreducible to the proper activities of the material elements. The view that mind and body are identical (the "identity thesis") runs into formidable difficulties at the level of language, since no one has been able to produce (or come close to producing) translation schemata for even the simplest "mental-state" assertion (e.g. "I am in pain"), sufficient to reduce it without remainder to a collection of "physical-state" statements. Besides, our knowledge of our own mental states seems to differ in kind from the sort of knowledge we have of physical events. To explain color-vision in terms of optics and physiology and atomic energy-levels does not tell someone what it is like to see if he has never seen. There seems to be an irreducible level in situations where subjects are involved, i.e. where one can properly speak of knowledge-acts. Though discussions in recent years have centered upon the *human* subject (partly because of the emphasis among philosophers on language, kinds of statements, modes of translation), it would seem that there may well be a logical absurdity involved in supposing that an exhaustive objective analysis in physical terms (no matter how broadly the term 'physical' be taken) of an animal's brain would tell one what the animal "is thinking" or "is seeing". The "mind-body" problem does not just concern human minds, but it happens to be more easily formulated by us in their regard.

Though identity theories which reduce the material to the mental or the mental to the material are not in much favor among philosophers, it cannot be said that the dualistic or the monistic alternatives are very satisfactory either. Some kind of dualism of subject and object, of mental and material, seems intuitively plausible. But what sort of ontology can one specify in its support? There are two broad possibilities. One is to suppose that the men-

---

[8] See the bibliography given in H. Feigl, "The mental and the physical", *Minnesota Studies in the Philosophy of Science*, 2, 1958, 370–497; *Dimensions of Mind*, ed. S. Hook, New York, 1961; J. Shaffer, "Recent work on the mind-body problem", *Amer. Philos. Quarterly*, 2, 1965, 81–104.

tal and the physical are ontologically quite diverse. But how, then, are they to interact? Are there causal relations between them? Would not this lead to violations of physical law? The uneasy stratagems that causal inter-actionists have been forced to adopt, ranging from Descartes' pineal gland to Compton's partially undetermined electrons, are enough to call this en-tire approach into question. Because of this, some dualists have simply denied that there *is* an interaction, and talk of a parallel series of events in the two orders. This solution, though it has had some value as a method-ological simplification for physiologists and psychologists, has no support today; it is too obvious that the mental and the physical are closely cor-related, that the state of one is directly connected with the other. The older ideas of pre-established harmony between the two (Leibniz) or of God as furnishing the causal link directly (Malebranche) seem to have derived their main support from a quite sharp initial ontological separation of mental and physical; once the grounds for such a separation are chal-lenged, the implausibilities of the relationships postulated to hold between them are accentuated.

The second possibility is a qualified dualism that would maintain on the one hand that mind and body are irreducible to one another, but would also claim that there can be no problem of causal interaction between them because they are no more than diverse aspects of the same reality, not separate substances as the interactionist problematic assumed. They relate, therefore, not as interacting with one another, but as activities at different levels of the same entity. But how is the notion of "level" to be specified? In terms of complexity only? Then we are back to an identity theory. In terms of separable activities? Then we encounter some of the main ob-jections against the strong dualist theories. In terms of the irreducibility of the conceptual systems in terms of which each level is described and ex-plained? But what sort of ontology does this imply? Is it the case that every time I deduce the theorem of Pythagoras or detect the flavor of gasoline, the same series of events occurs in my brain, considered as a physical system capable of being specified in terms of an expanded physics? If so, do we not have a difficulty in maintaining human freedom, since the transition from one brain-state to another could be analyzed in purely physical-physiological terms? It is of no use to invoke quantum "indeterminism" here, since free decision cannot (it would seem) be strictly correlated with neural sequences that are specifiable in either deterministic or indeter-ministic terms. But if strict correlation between the levels does not hold, does it mean that to know what sequence of neural events will occur, we must specify the psychological state of the subject first? These and similar dif-ficulties have led some (like Russell and Ayer) to propose a special form of identity theory in which both mind and matter are reduced to a third element, a neutral "stuff", which exhibits no intrinsic differences of level. This solution, however, encounters all the objections that are raised against the phenomenalism upon which it has usually been based.

It would seem, then, that the concept of the *material* (or the *physical*) needs much closer and more sophisticated analysis if the nature of those material beings who *know* is to be understood. There is a tension between the conditions of materiality and the interiority and unity of the knowing act, a tension that has somehow to be resolved conceptually, if not in the classical dichotomy of *material* and *immaterial*, then in some other way that takes account of the troublesome complexity of a being who is at once a physical object among physical objects, and a subject who also relates to these objects in a quite different way.

§4 *Matter and Substratum*

It is frequently said that the progress of physical science, especially of field theories, makes the notion of an underlying substratum seem unnecessary. The "substratum" here is taken to be a non-describable underlying something. The critic will say that there is no sign of such an entity in his theory. To which the answer is: of course not, since explicit mention in the theory would preclude its being a non-describable factor. The fact that science continues to push back the horizon tells us nothing about what lies beyond the horizon. More significant is the suggestion that the notion of an underlying substantial substratum in which properties inhere, has been eliminated. One does not ask: in what does electromagnetic-field strength inhere? what is the carrier of light waves in "empty space"? Russell in this context speaks of "the disappearance of matter", suggesting that it has been "replaced by emanations from a locality", while "tables and chairs . . . have become pale abstractions, mere laws exhibited in the succession of events which radiate from certain regions".[9]

There is a misunderstanding here. It is true that the apparently continuous objects of sense-experience have been shown to be made up of a myriad of relatively discrete entities with large tracts of relatively empty space between them. But this does not mean that predicates like 'solid' and 'impenetrable' cease to be applicable in their normal senses at the macroscopic level. Again, atoms are seen to consist of further separated entities, themselves no longer individuals. But if any matter is "vanishing" here at all, it is the massy particles of seventeenth-century physics. Finally, the

---

[9] *An Outline of Philosophy*, London, 1927, p. 106.

aether, understood as a mechanical substratum for the transmission of various types of energy, seems to have been eliminated by the Special Theory of Relativity.

One overall comment must be made on these developments. The constructs of the average physical theory today cannot be classified in Aristotelian terms of substance and property. These latter categories are drawn from, and are applicable to, the objects of sense-experience. But a physical theory is an extremely complex indirect way of conceptualizing reality, and we must not look for one-to-one correspondences between it and the structures and categories of the world the theory represents. If it is wrong to substantialize energy and mass, it is equally wrong to make the constructs E and H of Maxwell's theory correspond to properties about which we can ask: in what do they inhere? To say that they do not "inhere in" anything, or that an aether-construct need not be invoked, does not commit us to the view that there are real properties "floating" without a concrete substratum. This kind of illegitimate categorial transfer—one might call it the "Pythagorean" fallacy—rests on a faulty grasp of the relation between the constructs and models of physics and the experienced objects whose behavior the former are intended to "explain", in a quite precise, but not simple, sense of the term 'explain'.

A quite different source for this sort of rejection of an "underlying matter" is to be seen in the philosophical tradition of phenomenalism whose origins we have already discussed in the Introduction. The phenomenalist defines his philosophical position as a denial of the existence of an "incomprehensible somewhat", as Berkeley called it, an inert substratum that itself neither perceives nor is perceived. The enemy is thus any dualistic view which supposes in perceptual objects an unperceivable reality behind the appearances, a substratum to which one must infer, or one which is in itself totally passive so that its activity has to be wholly imposed from the outside.[10] Kant argued that some sort of "matter" was necessary as a guarantor of empiricism against the rationalism of Descartes and Leibniz. In his complex metaphysics, matter appeared in a great many different roles, not easily reconciled with one another.[11] But the basic one was that of the non-constructed, encountered element in sensation, the correlate of the *a priori* forms of sensibility and the source of singularity and contingency. In this respect, Kant's system can be regarded as an attempt to unite the main insights of both empiricism and phenomenalism with respect to matter.

---

[10] See the essay by K. Sayre elsewhere in this volume.
[11] See J. Smith's essay elsewhere in this volume.

Since the work of Russell in our own century, there has been a considerably revival of interest in the phenomenalist case. Is there any argument in favor of it sufficient to outweigh the almost overwhelming counter-argument beginning from the evident unlikelihood of the claim that what we describe as tables and chairs exist only in our perception of them? Phenomenalists have tried to meet this difficulty by allowing unsensed *sensa,* "permanent possibilities of sensation" (Mill), or "objects which have the same status as sense-data without necessarily being data to any mind" (the *"sensibilia"* of Russell). But these ran into all sorts of difficulties: can there be a "possibility" which is not grounded in some actuality, and in the case of sense-data, what is this actuality? if sense-data are dependent upon consciousness, how can the so-called *sensibilia* be autonomous? do not attempts like those of Mill and Russell to retain a modified version of phenomenalism introduce a kind of "quasi-matter" to ground the observed continuity and scientific intelligibility of perceived objects, a "quasi-matter" which is far more mysterious than the "matter" to which the phenomenalists originally objected? The concept of matter in the context of this continuing debate serves to draw attention to certain aspects of our experience: the continuity of the objects of this experience, our ability to understand them in causal terms that have to be laboriously discovered and are not the mere product of association or postulation.

The most recent forms of phenomenalism (Lewis, Ayer, Firth . . .) have been predominantly linguistic in emphasis and aim, reflecting one contemporary style of empiricist philosophy. They attempt to show that all statements about material objects can be translated into equivalent sets of statements about sense-data. In this view, the point of conventional material-object language is to single out for attention the patterns that actually do exist in the sense-data. This diluted form of phenomenalism avoids explicit assertions about the constituents of the world, but there is an implication that if everything that is said with the aid of the concept of material object can be said without it, there is no warrant for supposing that "material objects" exist, i.e. that there are referents for our statements about the physical world which are independent of our perception of them. It is not immediately evident how this follows, or why one should not work in the other direction and dispense with sense-data. And the attempts to find the requisite linguistic equivalences have not, it is generally agreed, been successful.[12]

The main arguments against phenomenalism can be summarized under

---

[12] See R. J. Hirst's essay "Phenomenalism", in *The Encyclopedia of Philosophy*, ed. P. Edwards, New York, 1967.

two heads: first, it assumes an incoherence in the notion of material object which it never really proves; secondly, it does not in fact provide an alternative framework in terms of which those features of our experience, ordinarily understood in terms of enduring material objects, could be handled in terms of discrete sense-data alone. Against the linguistic versions of phenomenalism, there is in addition the standard argument against "translatability" theses of this sort. Either the point being made is a purely linguistic one, in which case it is irrelevant to the traditional ontological claim of phenomenalists about the non-existence of "matter". Or else an illegitimate inference is being made from the equivalence of two languages to a statement about what is. If the two languages are equivalent, then sense-data and material objects have the same status: either both "exist" or neither do. If to be is to be the value of a variable in some acceptable linguistic system, we are forced to admit the referents of *all* such systems, or else argue that one of the linguistic systems is in some way privileged. What the phenomenalist would need to show is not only that sense-datum language can say all that other languages can, but that it can say some indispensable things which other languages cannot.

§5 *Matter and Conservation*

In pre-Galilean physics change was understood to be bounded by certain limits. No matter how radical the change, something was conserved; there was a substratum which carried over at least some restraints from the old into the new. The concept of matter served here as a barrier against the idea that something without antecedents could arise; creation and annihilation had to be excluded as possibilities if scientific explanation were to be at all possible.[13] There were limits to what could happen, limits to the intrinsic changeability of material things. But how were these limits to be specified? We have seen that the efforts to do so in terms of a totally unspecifiable primary matter were not successful; such a substratum, because it left *all* possibilities open, did not seem to do the job for which it was postulated, since all possibilities are obviously *not* open in any change, no matter how basic. In every change, it is clear that there will be *some* continuities. But is there any single feature which is conserved throughout changes of all sorts? Measure-properties are the most obvious candidates, since conservation is most easily specified in measure-terms.

Quantity of matter (or mass) was, as we have seen, the first successful

---

[13] See H. Nielsen's contribution to *The Concept of Matter in Greek and Medieval Philosophy*, ed. E. McMullin, Notre Dame, 1965.

index of conservation. To say in Newtonian terms that "quantity of matter" remains unaltered is to say that no matter how radically a thing changes, if there be no intrusion or loss of matter, then the response to impressed motion and the ability to generate gravitational motion will be the same before and after. In tracing the history of a mechanical system, conserved parameters like mass were so obviously invaluable that from the beginning others were sought. Descartes argued that the quantity of motion in the world is constant, just as is the quantity of matter. But how was this "quantity of motion" to be measured? Newton proposed to do so in terms of the product of mass and velocity. Leibniz, on the other hand, preferred a measure of the *vis viva,* of what ultimately came to be called "energy". After decades of debate, it transpired that both momentum *and* energy could be defined in such a way that they would be conserved throughout "enclosed" changes, changes where no dynamic interaction occurred at the boundaries of the system. Momentum had to be made a vector rather than a scalar quantity, and a distinction had to be drawn between linear and angular momentum, for each of which separate conservation laws hold good.

It was much more difficult to find a definition of "energy" such that the sum of the "energies" of a closed system would remain the same. Since energy was regarded as the ability to do work, not only would one have to include the (kinetic) energies of motion but also the (potential) energies of position. And since work could be done in many different sorts of change, it was necessary to find a definition of energy broad enough to include not only mechanical changes of the familiar Newtonian sort but also thermal and chemical changes, changes induced by electrical or magnetic forces, even changes in living systems. Not until the mid-nineteenth century was there a sufficient theoretical unification of these various processes to allow the postulation of a conservation of "energy" with some degree of assurance.[14] Indeed, to many philosophers of science of that time, especially those in the Kantian tradition, it seemed that the notion of *energy* testified to nothing more than the penchant of the human mind to invent permanences in the flux of the world. In support of this interpretation, they could cite the fact that when apparent failures of conservation were discovered, some way of redefining the concept of energy was always found, in order that the conservation postulate could be retained.

Nevertheless, a closer look at the evolution of the concept of energy shows

---

[14] See E. N. Hiebert, *The Historical Roots of the Principle of Conservation of Energy,* Madison, Wisc., 1962; Y. Elkana, *The Discovery of the Conservation of Energy,* London, 1974.

that modifications in it were not just *ad hoc* attempts to retain a conservation law but were always guided by theoretical analyses of how "work" is performed in different sorts of changes. The fact that "energy" continued to be conserved despite the diversification in the types of change considered indicates with some degree of probability that there is an ontological factor whose permanence is a significant feature of all physical changes. This factor is, of course, only approximately conveyed by the classical ideas of energy and momentum, and by the more recent combined momentum-energy tensor of relativity theory. But the reality of such a factor seems to be required by the long history of fruitful generalization of the dynamic concepts of energy and momentum, a generalization which has brought about an extraordinary theoretical unification of the immense field of physics and chemistry. It is true, of course, that conserved quantities are sought in mechanics because they correspond to mathematical symmetries and thus lend themselves to elegant analysis; it is true too that the mind instinctively seeks permanent factors in its attempt to understand change. Yet the *history* of the concept of energy cannot be adequately accounted for by supposing that it was simply a construct imposed for heuristic reasons upon the data. Were this to have been the case, its history would not have the internal consistency it has.

Does this mean that energy replaces matter as the "stuff" of the new physics? There were those, like Ostwald, in the nineteenth century who thought so. And the mass-energy equivalence of restricted relativity theory reinforced this belief for many, since it was assumed to imply that matter and energy are somehow identical, or perhaps two different states of the same underlying stuff, with matter as a sort of "frozen" energy ready to be released in the form of "immaterial" radiation at any time. But this substantializing of energy (or dematerialization of matter) cannot properly be inferred from the $E = mc^2$ of the theory. The E here is generalized energy which includes not only classical kinetic and potential energies, but also a potential "energy" deriving from the rest-mass of the particle. It is this generalized energy which is conserved. The relativistic variation of mass with velocity means that mass itself can no longer be treated as an exactly conserved quantity; the Newtonian mass conservation-principle turns out to be an approximation holding only at low velocities.[15] But the new combined mass-energy concept of Einstein's theory once again pro-

---

[15] Jammer shows very elegantly that the relativistic "variability of mass" is due to the redefinition of space-time relations in the Lorentz-Minkowski system, and to the consequent shift in the factor expressing the relation of momentum and velocity (i.e. the mass). He concludes that this shift ought *not* be represented (as it often has been) as the discovery of a new property of matter (*op. cit.*, p. 165).

vided a conservation factor, replacing the older conservation laws of mass, energy, and momentum. E and m are of course, operationally different measures, and their dimensionality is not the same. To say that "mass can be transformed into energy" means that the rest-mass of a particle can be decreased and a corresponding amount of radiation-energy given off. The rest-mass of the latter is zero, and so the combined (classical) mass of the system decreases while the (classical) energy of the system increases. But by expanding the notion of mass to allow for an increment of inertial mass with motion, and by expanding the notion of energy to include an energy corresponding to rest-mass, the E factor is still conserved (though both E and m have to be redefined to give the $E=mc^2$ equation).

It was commonly supposed that once energy was found to be capable of transmission through space, it could be regarded as a sort of entity. "Energy, thus disjoined from matter, raised its ontological status from a mere accident of a mechanical or physical system to the autonomous rank of independent existence".[16] But what was discovered, strictly speaking, was not that an entity called "energy" was transmitted, but rather that a certain measure in one place was related to one in another place at a later time. This could conveniently be represented as though a construct-entity whose measure was energy passed from one point of space to another. And since mass originated as a measure of matter, the fact that mass (matter?) could be "transformed into energy" could be cited as further evidence for asserting that "mass and energy are identical, they are synonyms for the same physical substratum".[17] But as we have seen, mass and energy are *not* identical in any sense whatever. And when a nucleus loses part of its mass and radiation is emitted, one must beware of supposing that Einstein's E specifies a physical substratum that remains the same throughout this change.

The extremes to be avoided here are quite familiar in philosophy of science: uncritical reification on the one hand, and an overly critical positivistic refusal of all ontological implications on the other. The "temptation of the noun" is ever-present, the temptation, that is, to take every noun to be the name of some entity, to suppose that 'energy' names a stuff in the same sense in which 'water' does. The temptation is greatly strengthened when the energy-measure proves to yield a conservation law; the form of words 'energy is conserved' almost inevitably leads one to reify. Only a *stuff* of some sort can be the ground for a conservation principle of such generality, it would seem. Yet the constructed character of the concept,

[16] *Ibid.*, p. 173.

[17] *Ibid.*, p. 188. For another example of this sort of reification, see Viscount Samuel's contribution to the chapter, "Energy and ether", in H. Samuel and H. Dingle, *A Threefold Cord,* London, 1961.

its constant modification in the light of new data, its dependence upon a complex network of theoretical concepts, and the implausibility of taking either kinetic or potential energy as a "stuff" of any sort, ought to warn us that the E of relativity theory cannot be taken to name the substratum of physical change any more than did the m of Newtonian physics.

What one *can* say is that the success of conservation principles in the history of mechanics suggests that there is an ontological ground for the matter-motion conservation, one that is imperfectly and provisionally symbolized by the succession of conservation laws of mass, momentum and energy, now unified in a single energy-momentum tensor expression. This ground is not *named* in any proper sense by the term 'energy', so that one cannot think of the relation between terms and ground as a "stuff" one. The ground of dynamic conservation is reminiscent in one way of the primary matter of Aristotle in that no descriptive predicates are certainly applicable to it, nor can it be singled out for reference. It lies permanently beyond the horizon both of naming and of describing. Yet it is very different from primary matter in that it supports a more and more precise set of theoretical constructs as a progressively wider set of dynamic contexts is incorporated into mechanics, and the question of how best to characterize "that which is conserved" is posed in each of them successively.

What status does the conservation principle have? Is it possible that physical processes may be discovered in which it breaks down, i.e. in which no plausible redefinition of the dynamic parameters can be found which will provide a conservation principle? This is hard to answer with any assurance, although the history both of philosophy and of mechanics would suggest that this is unlikely. It is of some interest in this connection to examine the only major physical theory of modern times which explicitly repudiated the conservation of mass-energy. In 1948, Hoyle, Bondi and Gold put forward their "steady-state" alternative to the accepted expanding model of the universe. In order to maintain the uniform distribution of galaxies, they postulated the continuous coming-to-be of hydrogen throughout space at a rate so slow (one atom per cubic meter per five hundred million years) as to be incapable of any sort of direct experimental confirmation, but sufficient to replenish the matter of the universe as the galaxies rushed away from one another. The theory, though ingenious, met with little favor among astronomers and has recently been abandoned even by its original protagonists. The reasons for its failure were quite complex, but one of them was undoubtedly the distaste felt by scientists generally for a "creation" hypothesis, no matter how small the intrusion of matter involved or how regular its rate. The feeling seemed to be that this ran so directly contrary to the entire analogy of physics, to the assumptions about matter that have in the past proved most fruitful, that it would need some overwhelmingly strong reason to warrant its adoption.[18] And this, as it happened, its proponents could not provide.

§6 *Matter and Space*

From Greek times onward, as we have seen, the relationship between matter and space posed some very difficult problems. What reality is to be assigned to the void? Ought matter and void be contrasted? Seventeenth-century rationalism took the risky step of identifying matter with space; the resultant lack of an autonomous matter-factor led its mechanics into inconsistency and ultimate sterility. In recent times, there has been a major effort to formulate a new "geometrodynamics" with the benefit of the much more powerful methods of modern geometry. Its exponents speak of matter as "evaporating into nothingness", of "building mass out of pure geometry", of "a universe of pure geometry", of seeing "the physical world in which we live" as "a purely mathematical construct".[19] These are strong words to the philosopher. The substance behind them is that attempts are being made to formulate a "geometry" of space-time in which particles would appear as geometrical singularities, and electric charges as "wormholes". They would be specified by the metric itself, and would thus not have to be added as "foreign entities". The equations of motion would not have to be imposed *ab extra*, but would follow from the geometrical structure of the space. These attempts have not as yet been successful (though Rainich and Misner have succeeded in finding a purely geometrical representation of electromagnetic fields).[20] The advantage of this approach (if successful) would presumably lie in a simpler and more homogeneous mathematical treatment.

To say that such a model only "simulates" the presence of matter, that it is really "empty space", and that thus we have succeeded in doing away with matter, is another form of the "Pythagorean" fallacy, already mentioned. If the geometrical model gives an adequate theoretical representation of the motion of perceived physical objects, then it is acceptable as a *physical theory;* there will be no need to postulate a model in which the

---

[18] See J. D. North, *The Measure of the Universe,* Oxford, 1965, chap. 10.

[19] J. A. Wheeler, "Curved empty space-time as the building-material of the physical world", *Logic, Methodology and Phil. of Science,* ed. E. Nagel, *et al.,* Stanford, 1962, pp. 361–73; see also C. Misner's essay in *The Concept of Matter.*

[20] C. Y. Rainich, "Electrodynamics in the General Relativity Theory", *Proc. Nat. Acad. Sc.* U.S.A., 27, 1925, 106–36; C. Misner and J. A. Wheeler, "Classical physics as geometry", *Ann. Phys.,* 2, 1957, 525–603.

masses and the geometry have to be separately specified. There is no question of one model "simulating" matter in a way that the other does not. The geometrical model, by hypothesis, does not omit any feature of matter that the other model includes, for if it did, it would fail as a model. So nothing is "left out", there is no "abolition" of matter.

The mistake here arises from several sources. First, there is the classical ambiguity about whether to call space "material" or not. If it be taken as "non-material", then to formulate a physical theory in terms of a "pure geometry" might seem to make its referent "non-material", or to imply that "there is no matter there". There is a confusion here between *physical* space, constituted by the relationship between the concrete objects of perception, and *mathematical* space, which is a mental construct. The latter is called a "space" only because of certain definite *mathematical* properties it possesses. But the mathematical space-concept can be applied to many things besides physical space. To say, then, that physical space is "non-material", tells us nothing whatever about the "materiality" or otherwise of physical systems to which a mathematical space can be applied via a physical theory.

The second source of the difficulty is that "geometry" has been broadened here to include dynamic as well as spatial parameters. Thus the claim that the behavior of matter can now be described in terms of "geometry" alone is ambiguous. Were this to mean Euclidean geometry it *would* imply the Cartesian denial of an intrinsic Euclidean non-geometric mass-factor to physical bodies. The term is so broad here, however, that to say that something is exhaustively describable in terms of a "geometry" tells us nothing whatever about its "materiality" or "non-materiality".

Lastly, the notion of *matter* implicit in the claims made for geometrodynamics is not coherent. Obviously the pure geometric model makes no difference to *physical* "matter", understood as perceptible objects; it would make no sense to say that physical matter has been shown not to exist in consequence of a geometrical discovery. The "matter" which "vanishes" in consequence of the new hypothesis must be a construct-matter within the model itself, or rather one that occurs in the classical theoretical model but not in the new purely geometrical one. But what could this construct-matter be? There are no theoretical predicates that could label it as "matter". To talk of it as a sort of jelly or an inert dot is simply metaphorical. The only difference between the two models lies, not at the level of their referents ("physical matter"), not in a "construct-matter" which occurs in one and not in the other (since there is no way of specifying "matter" that does not apply to both—or neither—equally), but in the methods of mathematical construction used. In one, the mass (not matter) is introduced as a separate non-geometrical factor; in the other, it is deduced from the geometry. Since the mass-measure will be the same in both cases, if matter is "that which mass measures", both models will bear exactly the same relation to it.

§7 *Matter and Mass*

Newtonian physics, as we have seen, encouraged the transfer, both at the commonsense and at the philosophic levels, of the attributes of matter to mass. Just as the term 'matter' had come, after a long evolution, to name the substratum of things, as well as to be a general name for physical objects, so 'mass', which originally denoted a single measurable characteristic of things, was "substantialized" as a sort of scientific synonym for the older term. The world was taken to consist of "point masses", since it sufficed to characterize bodies in terms of their mass and motion in order to make all possible deductions about their behavior (i.e. their movements in space). Mass was no longer regarded merely as a property; it was the substance, conserved through all changes, underlying all other variable (and thus "accidental") properties. *Mass* could thus be regarded as a more scientifically adequate version of the older concept of matter; Newton's idea of mass as a numerical measure for matter (where 'matter' was regarded as the more basic term) gradually lost ground as new ways of defining 'mass', within the broad limits of the Newtonian system, came to be elaborated. Euler transformed the Second Law into an operational definition of inertial mass (IM). Saint Venant showed that it could be measured directly by impact experiments. Maxwell measured it in terms of a constant force (as of a "standard" coiled spring). Mach, who disliked the "metaphysical" concept of force, preferred to define gravitational mass (GM) in terms of mutually induced gravitational accelerations. Besides these different ways of making the Newtonian system operational, many attempts were made to *explain* mass. Max Abraham, for example, tried to prove that IM is entirely of electromagnetic origin. The implication of these various alternatives to the older notion of mass as "quantity of matter" was that the concept of matter could be dispensed with in quantitative operational science.

But the concept of mass itself has more recently encountered some serious challenges to its consistency and operational character. Some of these difficulties were already latent in the theory of measurement contained in the *Principia*. The measurement of acceleration, for instance, depends upon the choice of reference-frame; since the measure of mass is the acceleration it produces, it would seem that the measured value of mass will depend on the acceleration of the observer measuring it. But if this is the case, how is one to define the privileged observer whose measurement gives the "real" value of the mass? Newton was forced by this sort of consideration to introduce all sorts of postulated "absolutes", and to suppose that an inertial system (in practice, a system relative to which his laws of motion exactly held) could somehow be identified without falling into logical circularity. A further challenge to the consistency of the mass-concept was Mach's claim that IM depends upon the distribution of bodies in the universe as a whole. This claim appears to be empirically testable, but so far, the required level of experimental accuracy has not been reached.

But far more serious are the conceptual difficulties raised by the General Theory of Relativity.[21] To jump at once to the most serious of these, if one considers bodies of finite volume (and not just point-idealizations), it becomes impossible even to define mass exactly, let alone measure it. Take active gravitational mass (AGM), for instance, which is the most favorable case. Suppose a very small test-body, B, be introduced, one which is too small to alter the field appreciably. The gravitational effect at B will be calculated by an integral over the effects caused by different regions of the body, A, whose AGM is being measured. But how is this sum to be made? and will it not vary with the direction of B and A? We do not have a unique space-time frame, and the space-time itself (other than in idealized cases) may have non-uniform curvature. So that AGM will depend on the choice of reference-frame, and even then will be variable, depending on the direction and distance of the test-body used. If A is in motion, there will be no satisfactory way of defining "rigidity" for it, and there will also be other problems about the effect of A upon the metric. With PGM, things are even worse, because here we cannot make the initial ideal assumption of an indefinitely small test-body. The two-body problem has not even been solved relativistically for point-bodies; the mutual accelerations and consequent variations of the metric make the way towards a solution very difficult. IM involves forces other than gravitational ones, and so far no satisfactory relativistic treatment of these is in sight; it would quite certainly be even more difficult here than in the case of GM to define a unique value for the mass.

There seems to be good reason to suppose, then, that the notion of mass itself (i.e. an intrinsic measure of response or resistance to motion) is a "classical" one, useful in the Newtonian approximation, but no longer even definable in a full relativity theory. The classical view depended on instantaneous propagation of light, on the making of unique time-cuts, on

---

[21] I am indebted to discussions with Cecil Mast on these issues. See his "Three concepts of mass" in *The Concept of Matter*.

the insertion of observers and test-bodies without disturbing the metric, on the reduction of bodies to point-centers of gravity. Where none of these can be assumed, it seems dubious whether the quest for a unique invariant intrinsic motion-parameter has any hope of succeeding. This ought not surprise us; it was optimistic to suppose that a body could be isolated from the rest of the universe and its mass determined in absolute fashion. If mass is to be the measure of the response of the body to outside influences, it is scarcely to be wondered at that it should depend on the rest of the universe, on its space-time relations with other bodies here and now. And if these bodies are in constant motion and if the space and time factors cannot be neatly separated in Newtonian fashion, the "mass" is not likely to be either invariant or unique.

If one were to be imaginative at this point, one might see in this the interaction between two different traditions of matter: the substratum (with its implicit attributes of conservation; intrinsic absolute measure; close relation with motion) and the Receptacle (involving spatial relations with all other sensible beings; in its singularity incapable of being made intelligible). It is fortunate for the history of physics that at the level of Newtonian approximation, the former prevailed. But it may be that there is a "Platonic" period ahead for science in this regard. The scientists will assuredly seek out new invariances to guide his grasp of beings in motion; but that these invariances will be such that they can be "substantialized" as mass was, is most unlikely.

§8 *Matter and Individuation*

For most contemporary thinkers, the basic reality is the concrete individual; the old Platonic question of what it is that *makes* something an individual, that makes it an incommunicable singular, simply does not arise. Or if it does arise, it will be assumed that individuality can be accounted for in terms of space-time relationship. But there is one domain in which the old problem has come up for intense discussion in the present century, the domain of logic, where the problems of reference and naming force one to adopt a very clear position on what it is that makes something a nameable individual. Russell and Whitehead in their *Principia Mathematica* tried to eliminate names entirely; they assumed that a string of predicates could perform the function of a name, so that "John is ill" became "There is a man such that he is *a* and he is *b* and he is *c* . . . and he is ill", where the predicated *a*, *b*, *c* . . . were supposed to mark off John from all other men. The advantage of this technique was that it allowed all categorical propositions to be reduced to a single form: a quantifier followed by a propositional function of the form: Fx, where F is a specific predicate and x is a dummy variable ranging over all objects.

However, this reduction ran into all sorts of difficulties. It was clear that a variety of complex referring expressions (such as places and dates) had to be smuggled into the so-called predicates in order to make the reduction work. And pronouns like 'this' or 'he' could not be reduced at all. The implicit context-dependence of such terms could not be fully conveyed by any string of predicates, however long. To determine the truth-conditions of statements made by means of sentences like 'This is a table' and 'He is my friend' would require one to know in advance where and who the speaker is. Can this be given by means of global co-ordinates and Greenwich time? Here we are back to Leibniz' problem once again. We somehow need to be *in* the system, to be directly acquainted with various objects in it, in order that this mode of description should work. These seems to be an irreducible referring element in all singular propositions.[22]

To the extent that one can speak of some ontological ground in the singular itself for this curiously irreducible feature of language, one is tempted to recall the primary matter of Aristotle, to which also no predicate properly applied. But the analogy is in some respects a risky one; among other things, it involves the whole question of ontological "constituents", already discussed fully in the earlier volume. The analysis of reference and naming may entitle one to single out that aspect of physical objects which makes it impossible to give an exhaustive propositional account of them in terms of a string of predicates only. But to call this aspect a constitutive co-principle, or to attach to it the label 'matter', could easily mislead. For one thing, there is no assurance that what makes naming irreducible to describing is a "constituent" of the singular in any simple sense. And for another, the notion of matter in its post-Newtonian development does not at all lend itself to the role of individuating element. Nevertheless, having uttered this caution, there is still a sufficient continuity between the old and the new problems of individuation to warrant a further search for resonances between Plato's individuating "matter" and whatever it is that makes proper names semantically irreducible. The controversy over nam-

---

[22] Quine, among others, still argues for the eliminability of singular terms (*Word and Object*, New York, 1960). In recent debates in the logic of demonstratives, the position David Kaplan has dubbed "haecceitist" (and which he defends) finds an irreducible aspect in the reference of indexical terms. Saul Kripke's provocative thesis that proper names serve to designate the same objects in all "possible worlds" (or counter-factual situations) has given new directions to the old debate. See D. Kaplan, "How to Russell a Frege-Church", *Jour. Phil.*, 72, 1975, 716–29; S. Kripke, "Naming and necessity", in *Semantics of Natural Language*, ed. D. Davidson and G. Harman, Dordrecht, 1972.

ing is undoubtedly related to one major part of the matter-problematic we have been pursuing in this book.

§9 *The Concept of Matter and the Future of Science*

We have seen that the concept of matter played a central part in the early development of natural philosophy. In the seventeenth century, the term 'matter' lost much of its significance for the working scientist because (1) from Descartes' time onwards, it was used more and more as an empty but convenient general term to denote the objects of physical science, without any sort of commitment as to the nature of these objects; (2) Newtonian mechanics substituted mass for matter in its analysis of the causes of motion. The term 'matter' does not occur, consequently, in any physical theory today; one finds no symbol corresponding to it in the equations of twentieth-century physics. Does this mean that the concept of matter no longer has any relevance for the work of the physical scientist? By no means. But it is a task of some complexity to show how this is so.[23]

Let us first draw a rough distinction between two levels of language in science. There is first of all the scientific object-language (SOL) as we may call it, comprising those terms which play an *explicit* part in the formulation of theory or the presentation of evidence. Specific concepts (like *energy, gene, valency, acid*) allow specific situations to be described or explained. They are usually operationally definable, or linked in a specified way, at least, with terms that are operationally definable. There are quite precise criteria that enable us to know whether they are applicable to a particular situation or not. As a second level, there is the meta-language (SML) of the scientist, containing such terms as 'theory', 'evidence', 'proof', 'deterministic', which are used to describe not the physical objects on which the scientist works but aspects of the science itself. But this distinction still leaves one broad class of terms unaccounted for, those rather general terms (like 'cause', 'effect', 'progress') which allow a broad characterization of a situation, but which have been replaced in the actual practice of empirical-scientific work by more precisely defined or more operationally manageable terms. They are still used by scientists when discussing their researches in a general way. But one does not find them down at the first level in the constantly modified technical language by means of which experimental

---

23 In the Introduction to *The Concept of Matter* (pp. 38–41), this question was treated in an oversimplified way, leading one reviewer (Mary Hesse, *History of Science*, 3, 1964, 79–84) to criticize the solution given as positivistic in its implications. I am grateful to Dr. Hesse for the impetus her perceptive review gave me to take a closer look at the issues involved.

data or theoretical models are expressed. It is to this pre-scientific language (PSL) that 'matter' belongs: no longer part of the active SOL of science (because replaced by 'mass'), yet not precisely part of the SML either because 'matter' is still an object-term rather than a second-order term like 'explanation'.[24]

But there is a further restriction on the utility to the scientist of the term 'matter' that detaches it from other terms like 'cause' in the PSL. The PSL still serves to provide a sort of general framework of intelligibility within which the scientist speculates, explains his results to non-scientists and relates them with results in other fields. If the term 'matter' retained the sense it had for Newton (denoting the factor which determined an object's responses, passive and active, to motion), then it could still be used (just as 'cause' is used, even though 'force' replaced 'cause' in the *Principia*) to discuss in a very general way the mechanical behavior of bodies. But unfortunately, the Cartesian sense of 'matter' has tended to replace the Newtonian one, with the result that in most modern languages 'matter' is a vacuous term denoting simply that which is accessible to empirical investigation. To say that something is "matter" or "material" in this broad usage, therefore, is to say no more than that it is part of the physical world; nothing whatever is conveyed about the *properties* of the entity so labeled. For Newton, it was still a meaningful question, as we have seen, to ask whether the aether was material or immaterial, or whether the void could be said to be in some sense material. The term 'matter' could thus play an active role in scientific inquiry albeit as part of the PSL rather the SOL. But if it becomes a "transcendental" term, equally applicable to *all* objects of legitimate empirical inquiry, then its utility even in the PSL is of a very limited kind. It is like 'being' or 'thing' or 'object', indispensable but unhelpful.

Of course, scientists sometimes may prescribe a more limited sense of 'matter' for their own purposes. They may take it to denote that which is impenetrable or which has rest-mass, for example, in which case it becomes empirically significant once again whether something be called "material"

---

[24] Except where it stands for material cause (see section 2 above). Since the term 'matter' in this case calls attention to the role that the postulated elements (atoms, charges . . .) play in explanation (i.e. they are the "materials" the theoretician has at his disposal) rather than designating any specific properties that the elements are supposed to possess, it belongs to the meta-language rather than the object-language. But this seems to be the only one of the traditional senses of the term 'matter' where this is true. All of the others specified some kind of ontological principles presumed to be constitutive of the object, and thus pertained in one way or another to the object-language.

or not. But questions about the "materiality" of the electromagnetic field or of the void are rarely asked today. They do not seem to be fruitful because they do not ordinarily suggest any lines of theoretical advance.

Even though the *term* 'matter' seems to be of little technical use to the scientist today, this does not, however, mean that the *concept* of matter no longer influences progress in science. The distinction between term and concept is an all-important one here, if we are to end our inquiry satisfactorily. At the beginning of this inquiry, we saw that although there was no term for matter among the pre-Socratics, it still makes sense to ask about pre-Socratic concepts of matter. The point here was that the concept of matter answers to a certain problematic, or series of problematics, with which the *term* 'matter' ultimately became identified. But the problematics themselves existed and were discussed long before Aristotle proposed a single technical term for the ontological principle that he was postulating as part of the answer to the problems of change, predication, individuation, defect and the rest.

Likewise, the fact that the term 'matter' is no longer part of the working vocabulary of the scientist today does *not* necessarily imply that the *concept* of matter plays no part in his deliberations. But does not this seem an overly sophisticated distinction? How is one to know that a concept of matter is playing a part in a piece of scientific research, if the *term* 'matter' (or some near synonym of it) is never used? It is easiest perhaps to respond by an example. We have seen that conservation principles still play a crucial role in dynamics, though the quantities conserved become more and more theoretically complex. Considerations of mathematical convenience and aesthetic simplicity undoubtedly have something to do with this continuing effort to find quantities that are conserved through change. But there is also a historical continuity between this effort and the discussions in pre-Newtonian natural philosophy of a substratum of change, itself not subject to change. Again, when an astrophysicist finds the Hoyle hypothesis of "continuing creation" repugnant to his intuitive ideas of how Nature operates, he is implicitly falling back on a concept of matter, the guarantor of continuity through change. Or when a psychologist postulates the existence of different levels of being, in some sense irreducible to one another, the relationship between the upper and the lower level may well be analogous to the form-matter relationship of an earlier tradition. When some physicists regard the virtual transitions and time-reversals of recent quantum field-theory as no more than convenient mathematical devices of no permanent theoretical value, once again there is an implicit notion of matter at work.

Behind this assertion on our part is the assumption that a natural philosophy still guides scientists in their speculations, even though it does not show up explicitly in their finished products, or in the language (the SOL or even the PSL) in terms of which these products are formulated or discussed. This natural philosophy unites our pre-scientific experience of the world with the most sophisticated results of recent science; it thus provides an "analogy of Nature" far more complex than Newton dreamt of. It is not a matter of regulating scientific speculation in some absolute way by means of a prior philosophy of nature, of flatly outlawing, for example, the idea that electrons can pass from one point of space to another without occupying the intervening spaces, on the grounds that this violates an intuitive "philosophic" principle of the continuity of motion.

Rather, it is a tentative affair of establishing analogies from one part of experience to another on the assumption that Nature is consistent and intelligible.[25] This means that prior "philosophic" positions rooted in ordinary experience or ordinary language may have to be modified in the light of developments in science (this has happened, for example, to the notion of simultaneity at a distance). On the other hand, it means that the scientist is not absolutely free in the models he brings forward, or at least he must remember that there are degrees of intrinsic plausibility governing such structures and that mere predictive success is not sufficient to guarantee their ultimate fruitfulness. Experience is all of a single piece, and one way of structuring it cannot be wholly isolated from all others. The history of science amply demonstrates that "good" theories and models are governed by criteria much more stringent than mere inductive fit with the data on hand, and that discovery in science, especially discovery at the most basic level (where it is sometimes called "revolution"), is led by an intuitive feel for the analogy of Nature that marks off the great scientist from the merely competent one.

From this point of view, then, the concept of matter still plays a role in the substructure of scientific inquiry, a much more tentative one, to be sure, than it did long ago in the genesis of Newton's *Principia*, but one perhaps not less significant in the long run.

---

[25] It is thus a "mixed" philosophy of nature in the sense defined in E. McMullin, "Philosophies of nature", *New Scholasticism*, 43, 1969, 29–74.

BIOGRAPHICAL NOTES
ON THE AUTHORS

Richard Blackwell, Professor of Philosophy, St. Louis University. Author of *Discovery in the Physical Sciences*, 1969; translator of Aquinas' *Commentary on Aristotle's Physics*, 1963, and C. Wolff's *Preliminary Discourse on Philosophy in General*. Some relevant articles: "Descartes' laws of motion", *Isis*, 1966; "C. Huyghens' *The Motion of Colliding Bodies*", *Isis*, 1978.

Vere C. Chappell, Professor of Philosophy, University of Massachusetts. Editor of *The Philosophy of Mind*, 1962; *The Philosophy of David Hume*, 1963; *Philosophy and Ordinary Language*, 1964; *Hume*, 1966. Some relevant articles: "Stuff and things", *Proc. Aris. Society*, 1971; "Matter", *Journal of Philosophy*, 1973.

Leonard J. Eslick, Professor of Philosophy, St. Louis University. Some relevant articles: "The Platonic dialectic of Non-Being", *New Scholasticism*, 1955; "Substance, change and causality in Whitehead", *Philosophy and Phenomenological Research*, 1958; "God in the metaphysics of Whitehead", in *New Themes in Christian Philosophy*, ed. R. McInerny, 1968; "*Republic* revisited", *Philosophy Forum*, 1970.

Marie Boas Hall, Senior Lecturer in History of Science and Technology, Imperial College, London. Author of *Robert Boyle and Seventeenth Century Chemistry*, 1958; *The Scientific Renaissance 1450–1630*, 1962; with A.R. Hall, *A Brief History of Science*, 1964. Editor (with A.R. Hall) of *Unpublished Scientific Papers of Isaac Newton*, 1962; *Correspondence of Henry Oldenburg*, 1965–, 10 vols; currently completing the edition of the Newton correspondence.

Peter M. Heimann, Lecturer, Department of History, Lancaster University (England). Some recent articles: "Helmholz and Kant", *Studies in the History and Philosophy of Science*, 1974; "Geometry and nature: Leibniz and Bernoulli", *Centaurus*, 1977. Currently working on natural philosophy from Newton to Maxwell.

Mary B. Hesse, Professor of History and Philosophy of Science, Cambridge University. Author of *Forces and Fields*, 1961; *Models and Analogies in Science*, 1966; *The Structure of Scientific Inference*, 1974. Some relevant articles: "Hermeticism and historiography", *Minnesota Studies*, 1970; "In defence of objectivity", *Proc. British Academy*, 1973.

Robert O. Johann, Professor of Philosophy, Fordham University. Author of *The Meaning of Love*, 1955; *The Pragmatic Meaning of God*, 1965; *Building* *the Human*, 1968. Some relevant articles: "Subjectivity", *Review of Metaphysics*, 1958; "The logic of evolution", *Thought*, 1961.

Nicholas Lobkowicz, Professor of Political Science, University of Munich. Author of *Theory and Practice: From Aristotle to Marx*, 1967; editor of *Marx and the Western World*, 1967; *Ideologie und Philosophie*, 1973; *Die politische Herausforderung der Wissenschaft*, 1976. Some relevant articles: "Abstraction and dialectics", *Review of Metaphysics*, 1968; "Historical laws", *Studies in Soviet Thought*, 1971.

James Edward McGuire, Professor of History and Philosophy of Science, University of Pittsburgh. Recent articles: "Neo-Platonism and active principles: Newton and the *Corpus Hermeticum*", in *Hermeticism and the Scientific Revolution*, 1977; "Existence, actuality and necessity: Newton on space and time", *Annals of Science*, 1978. Currently working on a book on Newton's theories of space and time.

Ernan McMullin, Professor of Philosophy, University of Notre Dame. Author of *Newton on Matter and Activity*, 1978; editor of *Galileo, Man of Science*, 1967. Some recent articles: "The fertility of theory and the unit for appraisal in science", *Boston Studies*, 1976; "Vico's theory of science", *Social Research*, 1976; "The conception of science in Galileo's work", *New Perspectives on Galileo*, ed. R. Butts and J. Pitt, 1978.

Richard Rorty, Professor of Philosophy, Princeton University. Editor of *The Linguistic Turn*, 1967. Some recent articles: "Genus as matter", in *Exegesis and Argument*, ed. E. N. Lee *et al.*, 1973; "Overcoming the tradition: Heidegger and Dewey", *Review of Metaphysics*, 1976; "Wittgensteinian philosophy and empirical psychology", *Philosophical Studies*, 1977.

Kenneth M. Sayre, Professor of Philosophy, University of Notre Dame. Author of *Recognition: A Study in the Philosophy of Artificial Intelligence*, 1965; *Plato's Analytic Method*, 1969; *Consciousness: A Philosophic Study of Minds and Machines*, 1969; *Cybernetics and the Philosophy of Mind*, 1976. Editor of *The Modeling of Mind: Computers and Intelligence*, 1963; *Philosophy and Cybernetics*, 1967; *Values in The Electric Power Industry*, 1977.

John E. Smith, Professor of Philosophy, Yale University. Author of *Reason and God*, 1961; *The Spirit of American Philosophy*, 1963; *Experience and God*, 1968; *Themes in American Philosophy*, 1970; *The Analogy of Experience*, 1973. Some recent articles: "The value of community: Dewey and Royce", *Southern Journal of Philosophy*, 1974; "Jonathan Edwards as philosophical theologian", *Review of Metaphysics*, 1976.

James A. Weisheipl, O.P., Professor of Philosophy, University of Toronto. Author of *Nature and Gravitation*, 1955; *Development of Physical Theory in the Middle Ages*, 1959; *Friar Thomas d'Aquino: His Life, Thought, and Works*, 1974. Some recent articles: "Ockham and some Mertonians", *Mediaeval Studies*, 1968; "Motion in a void: Aquinas and Averroes", *St. Thomas Aquinas, 1274–1974;* "The Parisian Faculty of Arts in mid-thirteenth century: 1240–1270", *American Benedictine Review*, 1976.

INDEX OF NAMES